Modelling of Virtual Worlds Using the Internet of Things

The text presents aspects of virtual worlds and highlights the emerging trends in simulation and modeling, comprising machine learning, artificial intelligence, deep learning, robotics, cloud computing, and data mining algorithms. It further discusses concepts including multimedia for the Internet of Things, graphical modeling using emerging technologies, and securing communication with secure data transmission in the modeling of virtual worlds.

This book:

- Discusses secure data transmission in the modeling of virtual worlds in the Internet of Things environment.
- Covers the integration of concepts and technical know-how about multiple technologies in visual world modeling, system configurations, and hardware issues.
- Explores the use of next-generation technologies such as deep learning, blockchain, and artificial intelligence in visual world modeling scenarios.
- Presents architectures and system models for the Internet of Things based visual world modeling systems.
- Provides real-time case scenarios, highlighting emerging challenges and issues.

The text is primarily written for senior undergraduate students, graduate students, and academic researchers in the fields of electrical engineering, electronics, communications engineering, computer engineering, and information technology.

Intelligent Data-Driven Systems and Artificial Intelligence

Series Editor: *Harish* Garg

Modelling of Virtual Worlds Using the Internet of Things
Edited by Simar Preet Singh and Arun Solanki

Data-Driven Technologies and Artificial Intelligence in Supply Chain
Tools and Techniques
Mahesh Chand, Vineet Jain and Puneeta Ajmera

For more information about this series, please visit: www.routledge.com/Intelligent-Data-Driven-Systems-and-Artificial-Intelligence/book-series/CRCIDDSAAI

Modelling of Virtual Worlds Using the Internet of Things

Edited by

Simar Preet Singh and Arun Solanki

CRC Press is an imprint of the
Taylor & Francis Group, an **informa** business

Designed cover image: shutterstock

First edition published 2024
by CRC Press
2385 NW Executive Center Drive, Suite 320, Boca Raton FL 33431

and by CRC Press
4 Park Square, Milton Park, Abingdon, Oxon, OX14 4RN

CRC Press is an imprint of Taylor & Francis Group, LLC

Library of Congress Cataloging-in-Publication Data
Names: Singh, Simar Preet, editor. | Solanki, Arun, 1985– editor.
Title: Modelling of virtual worlds using the internet of things / edited by Simar Preet Singh and Arun Solanki.
Description: First edition. | Boca Raton, FL : CRC Press, 2024. | Series: Intelligent data-driven systems and artificial intelligence | Includes bibliographical references and index. | Contents: Securing the future : comprehensive strategies for IoT security in industry 4.0 and beyond / Dr. Ashvini Shende, Dr. Bahubali Shiragpur, Mr. Gagan Raj, Dr. Parag Tamhankar.
Subjects: LCSH: Internet of things. | Virtual reality—Industrial applications. | Robotics.
Classification: LCC TK5105.8857 .M65 2024 (print) | LCC TK5105.8857 (ebook) | DDC 006.8—dc23/eng/20240222
LC record available at https://lccn.loc.gov/2024004023
LC ebook record available at https://lccn.loc.gov/2024004024

ISBN: 978-1-032-52810-6 (hbk)
ISBN: 978-1-032-76833-5 (pbk)
ISBN: 978-1-003-48018-1 (ebk)

DOI: 10.1201/9781003480181

Typeset in Times New Roman
by Apex CoVantage, LLC

Contents

About the Editors

Simar Preet Singh is an associate professor at the School of Computer Science Engineering and Technology (SCSET), Bennett University, Greater Noida, Uttar Pradesh, India. He was previously affiliated with GNA University, Phagwara, and Chandigarh Engineering College, Ajitgarh, India. He also previously worked with Infosys Limited and DAV University, Jalandhar, India, and has worked on international projects. He has published papers in SCI/SCIE/Scopus-indexed journals and has presented many research papers at various national and international conferences in India and abroad. His areas of interest include cloud computing, fog computing, IoT, big data, and machine learning. He earned his doctoral degree at Thapar Institute of Engineering and Technology, Patiala, India, and also holds several specialized certifications, including like Microsoft Certified System Engineer (MCSE), Microsoft Certified Technology Specialist (MCTS), and Core Java. He has also undergone a training program for VB.Net and Cisco Certified Network Associates (CCNA). For more information, please visit: www.simarpreetsingh.com.

Arun Solanki is currently working as an assistant professor in the Department of Computer Science and Engineering, Gautam Buddha University, Greater Noida, India. He is also Co-Convener of the Center of Excellence in Artificial Intelligence. Dr. Solanki has worked as Time Table Coordinator, member examination, admission, sports council, digital information cell, and other university teams from time to time. He has received his M.Tech Degree in computer engineering from YMCA University, Faridabad, Haryana, India. He has received his Ph.D. in computer science and engineering from Gautam Buddha University in 2014. He has supervised more than 70 M.Tech dissertations. Currently, he is guiding 5 Ph.D. students in the area of artificial intelligence. His research interests span expert systems, machine learning, and search engines. He has published more than 70 research articles in SCI/Scopus-indexed international journals/conferences like IEEE, Elsevier, Springer, etc. He has physically participated in many national and international conferences. He has been a technical and advisory committee member of many international conferences. He has organized several Faculty

Development Programs (FDP), conferences, workshops, and seminars. He has chaired many sessions at international conferences. He is working as Associate Editor in *International Journal of Web-Based Learning and Teaching Technologies* (IJWLTT) (Scopus indexed). He worked as Guest Editor for special issues in *Recent Patents on Computer Science*.

Contributors

Firos A
Department of Computer Science and Engineering, Rajiv Gandhi University
Rono Hills, Doimukh, India

Mamatha A
Department of Computer Science and Engineering, Ramaiah Institute of Technology
Bangalore, India

Parkavi. A
Department of Computer Science and Engineering, Ramaiah Institute of Technology
Bangalore, India

Imtiaj Ahmed
Department of CSE, East West University
Dhaka, Bangladesh

Shalom Akhai
Professor, Department of Mechanical Engineering, Maharishi Markandeshwar Engineering College, Maharishi Markandeshwar (Deemed to be University)
Mullana-Ambala, Haryana, India

Quazi Mohmmad Alfred
Department of ECE, Aliah University, New Town
Kolkata, India

Yusuf Jibrin Alkali
Officer, Federal Inland Revenue Service
Nigeria

Dr. Rubaid Ashfaq
Amity School of Communication, Amity University
Noida, India

Rabindranath Bera
Department of ECE, Sikkim Manipal Institute of Technology, Sikkim Manipal University
Sikkim, India

Ashima Bhatnagar Bhatia
Assistant Professor, Vivekananda Institute of Professional Studies—TC
New Delhi, India

Uday Kiran Dhane
UG student, Department of ECE, G. Pullaiah College of Engineering and Technology
Kurnool, India

Shama Kouser
Lecturer, Department of Computer Science, Jazan University
Saudi Arabia

Rakesh Kumar
Associate Professor, School of Nursing Sciences, ITM University
Gwalior, India

Dr. Vijaya Kumbhar
Sr. Assistant Professor, Sri Balaji University (SBU)
Pune, India

Dr. Anirudh Mangore
Computer Engineering Department, Gharda Institute of Technology
Ratnagiri, India

Rahul Reddy Nadikattu
Senior IEEE Member, Researcher, University of the Cumberland
Williamsburg, Kentucky

Sreelatha P K
Department of Computer Science and Engineering, Presidency University
Bangalore, India

Sampurna Panda
Assistant Professor, School of Engineering & Technology, ITM University
Gwalior, India

M. Siva Prasad
Department of ECE, G. Pullaiah College of Engineering and Technology
Kurnool, India

Mr. Gagan Raj
Student, B.Tech, CSE Department, D.Y. Patil International University,
Pune, India

Dr. Yudhishthir Raut
Computer Science and Engineering Department, Pimpri Chinchwad University
Pune, India

Jayanta Kumar Ray
Department of ECE, Sikkim Manipal Institute of Technology, Sikkim Manipal University
Sikkim, India

B. Uday Kiran Reddy
Department of ECE, G. Pullaiah College of Engineering and Technology
Kurnool, India

Shaik Himam Saheb
Assistant Professor(Sr), Department of Mechanical Engineering, Vigans Foundation for Science, Technology and Research
Hyderabad, India

A.O. Salau
Department of Electronics and Computer Engineering, Afe Babalola University
Nigeria

Dr. Ashvini Pradeep Shende
Adjunct Faculty, Symbiosis School of Economics, Symbiosis International University
Pune, India

Dr. Ashwini Shende
Symbiosis School of Economics, Symbiosis International University
Pune, India

Dr. Bahubali Shiragpur
Director, B.Tech- CSE, D.Y. Patil International University
Pune, India

Sanjib Sil
Department of ECE, Calcutta Institute of Engineering and Management
Tollygunge, Kolkata, India

Rogina Sultana
Department of ECE, Aliah University, New Town
Kolkata, India

Dr. Parag Tamhankar
Research Scholar, PostDoc, Georgia Tech University
Atlanta, US

Dr. Parag Tamhankar
Computer Science Department, Georgia Tech Research Institute
Atlanta, United States

K. Sai Teja
UG student, Department of ECE, G. Pullaiah College of Engineering and Technology
Kurnool, India

T. Tirupal
HoD, Department of ECE, G. Pullaiah College of Engineering and Technology
Kurnool, India

Sangeetha V
Department of Computer Science and Engineering, Ramaiah Institute of Technology
Bangalore, India

Dr. Pawan Whig
Dean Research, Vivekananda Institute of Professional Studies—TC
New Delhi, India

Preface

Modelling of Virtual Worlds Using the Internet of Things (MVWIoT) is becoming the need of the present world. With the advancements of technology in recent times, many developments have occurred in a very short span of time. This book will focus on deeper aspects of the Internet of Things (IoT) and will cover simulations and modelling of visual worlds in Internet of Things (IoT) domain. In present times, the Internet of Things (IoT) is in broad use, ranging from small devices like mobile phones to big professional computers. Hand-held devices, which have the capability of getting an Internet Protocol (IP) address and can be connected to the Internet, come under the category of an Internet of Things (IoT) device. There has been a great increase in the number of such devices in a very short span of time. It is predicted that this increase will further increase at an exponential rate in the coming years. The use of such computing devices helps people to finalize more intelligent decisions, considering more alternatives and actions, which optimize and enhance their business processes. Thus, Internet of Things (IoT) technology is in huge demand nowadays and will increase at a higher pace in the coming years.

This book is intended to cover various problematic aspects of visual worlds and highlight the emerging trends in simulation and modelling, comprising of machine learning, artificial intelligence, deep learning, robotics, cloud computing, fog computing, and data mining algorithms, including emerging intelligent and smart applications related to these research areas. The book is intended to cover the recent approaches in artificial intelligence and machine learning methods using the Internet of Things (IoT). The coverage of such activities includes: implementation of smart devices in visual worlds, smart cities and self-driven cars using Internet of Things (IoT), decentralized visual modelling at the edge networks, energy-aware AI-based systems, M2M networks, sensors, data analytics, algorithms and tools for smart systems, and much more.

The book is intended to cover topics like multimedia for the Internet of Things (IoT), graphical modelling using emerging technologies, emerging technologies like robotics for Covid-19, securing data transmission in modelling of visual worlds, considering the use of smart home environment for society, various methods and systems for graphical data analytics, having intelligent decision-making and visual semantic analytics, secure graphical data transmission in Covid-19, securing data migration issues, and AI-based solutions in visual worlds modelling.

This book is useful for recovery systems like Visual Worlds Disease tolerance abilities using Internet of Things (IoT), intelligent theoretical and mathematical models for decision-making that helps to take decisions and manage them, learning from data streams using Internet of Things (IoT), helping patients with humanoid robots for assisting doctors, futuristic technologies based on visual worlds modelling, dealing with recent technologies of drones, robots working as health workers, deep learning towards e-healthcare systems for cure and prevention, graphical automation using machine learning algorithms towards the smart world, and securing data migration issues and AI-based solutions for Covid-19 patient recovery.

This book contains 13 chapters. We hope that the readers make the most of this volume and enjoy reading this book. Suggestions and feedback are always welcome.

Chapter 1 focuses on the use of the Internet of Things (IoT) in modeling virtual worlds. The integration of IoT and virtual worlds has the potential to create immersive and interactive experiences, but it also poses new challenges in terms of data management, security, and scalability. The chapter will begin by providing an overview of the current state of IoT and virtual worlds, highlighting the most promising applications and use cases. It will then delve into the specific applications of IoT in modeling virtual worlds, discussing the various IoT-based techniques and technologies that can be used. The chapter will also examine the challenges of implementing IoT in virtual worlds, such as data management, security, and scalability, and provide best practices for overcoming these challenges. The chapter will also provide an overview of case studies and real-world examples of IoT-based virtual worlds and their impact on different industries such as gaming, entertainment, and education. This chapter will be a valuable resource for professionals in the field of IoT, virtual and augmented reality, as well as researchers and policymakers interested in understanding how IoT can be used to enhance the virtual worlds experience.

Chapter 2 equips with well-established best practices for IoT security in Industry 4.0, encompassing secure firmware updates, data protection, physical security, and supply chain security. Recognizing that security to be an ongoing process, authors demonstrated how to implement continuous monitoring and risk management for safeguarding IoT systems over time.

Chapter 3 covers the different types of IoT-enabled air conditioning systems and their key features, as well as their benefits and challenges. The chapter also highlights the latest trends and future directions in this area, including the integration of artificial intelligence, the use of big data, and the development of energy-efficient solutions.

Chapter 4 introduces the Smart Traffic Management System which will help make roads less congested in big cities and give emergency vehicles a clear way. To prevent accidents and promote a smooth flow of traffic, the road dividers separate the road into two directions. Dividers help in preventing vehicles from colliding with each other. Due to traffic congestion, it will be difficult to clear the path and save time in cities. Using Radio Frequency Identification (RFID) will provide a clear path for the ambulance or other emergency vehicles. Through the cloud, the traffic density at the dividers can be observed. The divider will automatically move by intimating the movement of the divider by using a buzzer and Liquid Crystal Display (LCD). The density of the vehicles can be counted using infrared (IR) sensors. If a vehicle is found next to the divider, the ultrasonic sensor will detect it and emit a buzzing sound.

Chapter 5 presents an approach for developing semantic based speech coding technique by preserving its prosodic features. GMM model will be incorporated to identify the semantic features and prosody of the windowed speech. LPC analysis will be done parallel. Fussy k–mean clustering will be done in the extracted features. ANN will be utilized to identify the best features for encoding. Using such semantic based coding will considerably reduce the processing requirements while encoding.

Chapter 6 presents a comprehensive survey of four computing paradigms: cloud, edge computing, fog, and dew computing. These paradigms have emerged to address the challenges of preparing and conserving data generated by the ever-increasing number of connected devices. In this survey, we provide an outline of each computing paradigm, including its architecture, key features, and benefits. We also compare these paradigms across several dimensions, such as deployment, scalability, latency, and security. Furthermore, we identify the challenges and opportunities for each paradigm and discuss their potential applications in different domains.

In Chapter 7, the authors are trying to make a system called C-V2X that helps keep the roads safe. They want to test it out in different places to make sure it works well. They are working on developing C-V2X better at different frequencies called microwave (3 GHz) and millimetre wave (28 GHz).

Chapter 8 delves into the role of humanoid robots in combating Covid-19, with a particular focus on their applications in healthcare settings, public spaces, and community engagement. Humanoid robots can provide valuable assistance to healthcare professionals by performing tasks such as temperature screening, patient monitoring, and delivering essential supplies. By doing so, they can help reduce the risk of infection among medical staff.

Chapter 9 discuss artificial intelligence (AI) and robots and their uses. The application of artificial intelligence (AI)-powered tools, such as contact tracking apps, has increased the feasibility to monitor large numbers of people and to identify those who are at risk along with the locations where the illness may spread next during the COVID-19 outbreak. The Phollower delivery robot, developed by the Slovak company Phontoneo and used in hospitals to distribute laundry, supplies, and drugs, is an example of the importance of the supply chain and on-time delivery during COVID-19. Drones, because of their three-dimensional mode of operation, are also being utilized as delivery platforms.

Chapter 10 provides a detailed review of the current status of drone-assisted healthcare, including various issues such as drone design and operation, the delivery of medical supplies, remote consultations, and performing medical procedures. Also, this study explores the essential legal and ethical factors, placing significant emphasis on the protection of patient privacy, safety, and well-being while leveraging the advantages of this technological advancement.

Chapter 11 presents a comprehensive software paradigm tailored specifically for IoT application development. The proposed paradigm addresses the unique challenges posed by the complex nature of IoT systems, including hardware-software integration, communication protocols, security, scalability, and real-time data processing. The development phase encompasses requirement analysis, hardware selection, and software development, where specialized embedded software is created to control sensors, process data, and facilitate device interactions. Communication protocols, such as MQTT, CoAP, HTTP, and Bluetooth, are critically evaluated to ensure seamless data exchange between IoT devices and the cloud or other devices. Robust encryption, secure authentication, and authorization mechanisms are integrated into the paradigm to safeguard sensitive information.

Chapter 12 provides a comprehensive overview of the security concerns associated with SDN and their potential impact on the network's performance and reliability. It also highlights the need for a holistic approach to SDN security and emphasizes the importance of ongoing monitoring and evaluation of security measures.

Chapter 13 introduces two rapidly growing technologies that have gained significant traction in recent years. The first one is the Internet of Things (IoT), which enables users to connect numerous smart devices and share information, allowing monitoring and control of various services like home automation, healthcare, agriculture, security monitoring, power grids, and critical services. The second technology is Cloud Computing, which involves accessing, configuring, and operating resources from remote locations. The conclusion drawn is that the combination of IoT and Cloud Computing has the potential to form an IT superpower, giving rise to unimaginable technologies and benefiting humanity through novel research fields.

Editors

Dr. Simar Preet Singh, Bennett University, Greater Noida, India
Dr. Arun Solanki, Gautam Buddha University, Greater Noida, India

Chapter 1

Modeling Virtual Worlds Using IoT

Applications and Challenges

Pawan Whig, Shama Kouser, Ashima Bhatnagar Bhatia, Rahul Reddy Nadikattu, and Yusuf Jibrin Alkali

1.1 INTRODUCTION

The Internet of Things (IoT) has revolutionized the way we interact with our physical environment by enabling connectivity between physical objects and the internet, with anybody, anywhere, anytime, any business, any network and any device as shown in Figure 1.1. This technology has provided us with a seamless way of collecting data and controlling devices remotely. Virtual worlds, on the other hand, provide a simulated environment that can be accessed through digital platforms [1–4]. These two technologies can be integrated to create a new paradigm of interconnected and immersive experiences for users.

Modeling virtual worlds using IoT can be defined as the process of connecting physical objects to a virtual environment to create an interactive and immersive experience. This integration can enable users to control physical objects in the real world through their virtual counterparts [5]. Additionally, the use of IoT in virtual worlds can provide a more realistic and immersive experience by creating a seamless integration between the virtual and real worlds, shown in Figure 1.2.

Figure 1.1 IoT process.

DOI: 10.1201/9781003480181-1

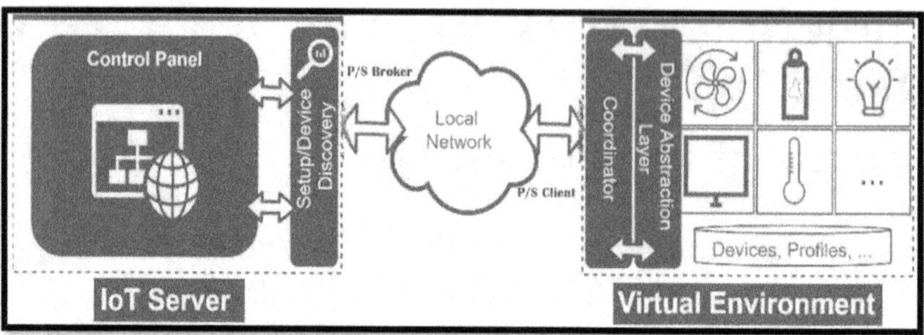

Figure 1.2 Virtual environment through IoT.

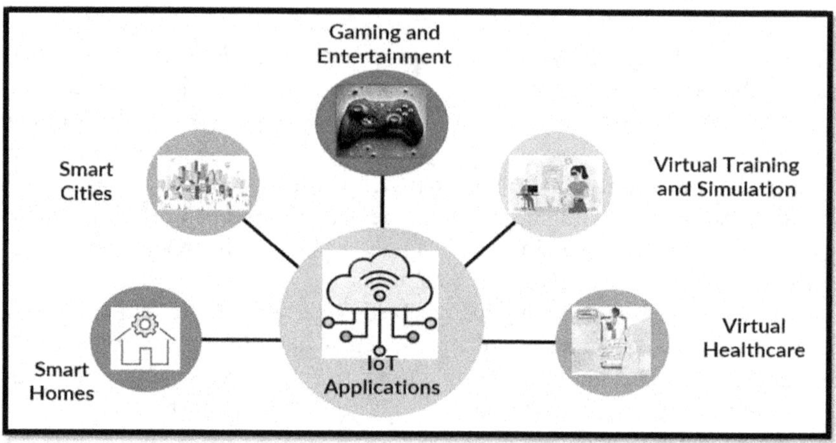

Figure 1.3 IoT and its applications.

The integration of IoT and virtual worlds has a wide range of applications, including smart homes, smart cities, gaming and entertainment, virtual training and simulation, and virtual healthcare, in Figure 1.3. In smart homes, IoT devices can be used to control appliances and lighting systems, and virtual worlds can be used to visualize and control these devices remotely [6]. In smart cities, IoT sensors can be used to collect data about the environment, and virtual worlds can be used to simulate and analyze this data in real time.

In gaming and entertainment, the use of IoT can create a more immersive experience for users by enabling them to control physical objects in the real world through their virtual counterparts. In virtual training and simulation, IoT can be used to create a more realistic and immersive environment for users to train and simulate real-world scenarios [7]. In virtual healthcare, IoT can be used to monitor and diagnose patients remotely, and virtual worlds can be used to visualize and analyze this data in real time.

While the integration of IoT and virtual worlds has numerous benefits, it also poses significant challenges. One of the significant challenges is security and privacy [8, 9]. The integration of physical objects with virtual worlds creates a significant security risk, and it is essential to ensure that these systems are adequately secured. Another challenge is interoperability and standardization. As the number of IoT devices and virtual worlds continues to grow, it is essential to establish standards to ensure interoperability and seamless integration between these systems [10, 11].

Latency and bandwidth are also significant challenges in IoT and virtual worlds integration. The integration of IoT devices with virtual worlds requires a high-speed internet connection, which may not be available in all regions. Scalability and reliability are also challenges that need to be addressed to ensure that these systems can accommodate a growing number of users and devices.

The integration of IoT and virtual worlds provides a new paradigm of interconnected and immersive experiences for users. The application of IoT in virtual worlds has numerous benefits, including improved efficiency, reduced cost, and enhanced user experiences [12, 13]. However, it also poses significant challenges that need to be addressed to ensure that these systems are secure, interoperable, and scalable. This chapter will explore the applications and challenges of modeling virtual worlds using IoT and provide insights into future directions [14, 15].

1.2 CONCEPTUAL FRAMEWORK: IoT AND VIRTUAL WORLDS

The integration of the Internet of Things (IoT) and virtual worlds creates a new paradigm of interconnected and immersive experiences for users. This section will provide a conceptual framework for understanding the integration of IoT and virtual worlds [16].

IoT can be defined as a network of physical devices, vehicles, home appliances, and other objects that are embedded with sensors, software, and network connectivity, allowing these devices to collect and exchange data. IoT devices can be connected to the internet or other devices, allowing for remote control and monitoring [17].

In IoT working, *events* generate *insights*, which trigger *actions* that improve a business process. Cloud-hosted services and applications determine actions to take, based on events that devices send. The following Figure 1.4 shows how *events* generate *insights* to inform *actions* in IoT solutions.

- Devices generate events and send them to cloud applications.
- Applications derive insights by evaluating data from incoming device events.
- Based on insights, applications take action by running processes and workflows.

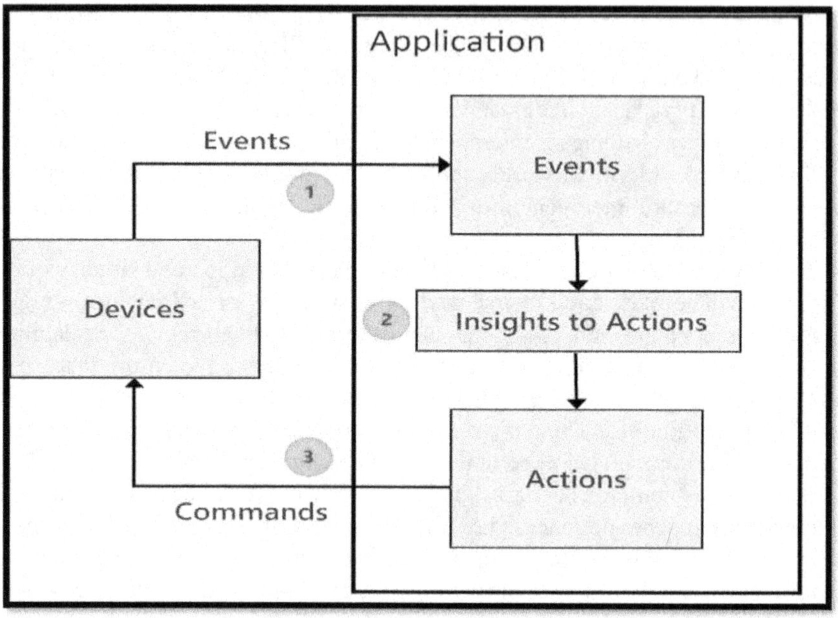

Application
Events
Events
1
Devices
2
Insights to Actions
3
Actions
Commands

Figure 1.4 Simulated environment architecture.

Applications can also send commands to the devices.

Virtual worlds, on the other hand, provide a simulated environment that can be accessed through digital platforms. These environments can be created using computer graphics or through the integration of real-world data. Users can interact with these environments through avatars, which are digital representations of themselves [18].

The integration of IoT and virtual worlds involves connecting physical objects to a virtual environment to create an interactive and immersive experience. This integration enables users to control physical objects in the real world through their virtual counterparts. Additionally, virtual worlds can be used to visualize and control IoT devices remotely [19–21].

The integration of IoT and virtual worlds can be achieved through the use of various technologies, including sensors, actuators, and communication protocols. Sensors can be used to collect data about the environment and physical objects, while actuators can be used to control physical objects. Communication protocols, such as Bluetooth and Wi-Fi, can be used to enable communication between IoT devices and virtual worlds [22–24].

The integration of IoT and virtual worlds has numerous benefits. In smart homes, for example, IoT devices can be used to control appliances and lighting systems, and virtual worlds can be used to visualize and control these devices remotely. In smart cities, IoT sensors can be used to collect data about the environment, and virtual worlds can be used to simulate and analyze this data in real time.

In gaming and entertainment, the integration of IoT and virtual worlds can create a more immersive experience for users. For example, in the game Pokémon Go, players can catch virtual Pokémon by using their smartphones to interact with the physical environment. The use of IoT devices in gaming and entertainment can create a more interactive and engaging experience for users [25].

Virtual training and simulation is another area where the integration of IoT and virtual worlds can be beneficial. IoT devices can be used to create a more realistic and immersive environment for users to train and simulate real-world scenarios. For example, in aviation training, virtual worlds can be used to simulate real-world flight scenarios, and IoT devices can be used to simulate the cockpit controls [26].

Virtual healthcare is another area where the integration of IoT and virtual worlds can have significant benefits. IoT devices can be used to monitor and diagnose patients remotely, and virtual worlds can be used to visualize and analyze this data in real time. This can enable healthcare professionals to provide remote care and improve patient outcomes [27].

However, the integration of IoT and virtual worlds also poses significant challenges. One of the significant challenges is security and privacy. The integration of physical objects with virtual worlds creates a significant security risk, and it is essential to ensure that these systems are adequately secured. Another challenge is interoperability and standardization. As the number of IoT devices and virtual worlds continues to grow, it is essential to establish standards to ensure interoperability and seamless integration between these systems.

Latency and bandwidth are also significant challenges in IoT and virtual worlds integration. The integration of IoT devices with virtual worlds requires a high-speed internet connection, which may not be available in all regions. Scalability and reliability are also challenges that need to be addressed to ensure that these systems can accommodate a growing number of users and devices [28–30].

The integration of IoT and virtual worlds creates a new paradigm of interconnected and immersive experiences for users. The use of IoT in virtual worlds has numerous applications, including smart homes, smart cities, gaming and entertainment, virtual training, and simulation.

1.3 APPLICATIONS OF IoT IN VIRTUAL WORLDS

The Internet of Things (IoT) has become an increasingly popular technology in recent years, and it has found applications in almost every industry. One area where IoT has great potential is in virtual worlds, which are computer-generated environments that simulate real-world experiences. Here are some of the applications of IoT in virtual worlds. Virtual Shopping: The use of IoT devices in virtual homes can create a more immersive experience for users. For example, virtual homes can have smart thermostats that adjust the temperature based on the user's preferences. Virtual lighting can also be controlled by IoT devices, making it possible to change the lighting color and intensity to match the time of day or the user's mood. Virtual Shopping: IoT devices can be used in virtual healthcare to monitor patients remotely. For instance, wearable devices can be used to track a patient's vital signs, such as heart rate and blood pressure, and transmit the data to virtual healthcare providers. This can enable virtual doctors to make real-time decisions about a patient's treatment, and can also provide patients with more personalized care [31].

1.3.1 Virtual Shopping

IoT can also enhance the shopping experience in virtual worlds. For instance, virtual stores can use IoT devices to track a user's browsing behavior and make personalized product recommendations. Virtual shopping carts can also use IoT technology to make the checkout process more seamless and efficient. Virtual Shopping: In virtual worlds, smart transportation can be achieved by using IoT devices to control virtual vehicles. For example, virtual cars can have IoT sensors that monitor traffic patterns and adjust the vehicle's speed and route accordingly. This can make virtual transportation more efficient and safer [32].

1.3.2 Smart Energy Management

IoT can also be used in virtual worlds to manage energy consumption. Virtual buildings can be equipped with smart sensors that track energy usage and automatically adjust the lighting and temperature to conserve energy. This can reduce the overall energy consumption of virtual worlds and make them more sustainable. Overall, IoT has great potential to enhance the user experience in virtual worlds by creating more immersive and personalized environments. It can also enable virtual worlds to be more efficient, sustainable, and responsive to user needs. As IoT technology continues to evolve, we can expect to see even more exciting applications in virtual worlds in the future.

1.3.2.1 Smart Homes

Smart homes are homes that are equipped with Internet of Things (IoT) devices that can communicate with each other and with the internet to automate and optimize various household functions. These devices include smart thermostats, smart lighting, smart locks, smart appliances, and more. The goal of smart homes is to improve energy efficiency, security, and convenience for the homeowners.

One of the most popular examples of a smart home device is the smart thermostat. These devices can learn the homeowners' schedules and preferences and adjust the temperature accordingly, saving energy and money on heating and cooling costs. Similarly, smart lighting can be programmed to turn on and off based on the homeowners' habits or controlled remotely through a smartphone app. Smart locks allow homeowners to control access to their homes remotely and grant access to guests or service providers through a secure app.

Smart homes can also integrate with virtual assistants like Amazon Alexa or Google Assistant to enable voice control of various household functions. This allows for hands-free control of smart devices and the ability to set reminders, make shopping lists, and even order groceries.

1.3.2.2 Smart Cities

Smart cities are cities that use IoT technology to improve efficiency, sustainability, and quality of life for their residents. This technology includes connected sensors, cameras, and other devices that can collect and analyze data to optimize traffic flow, reduce energy consumption, and improve public safety.

One example of a smart city technology is traffic monitoring. Connected cameras and sensors can be used to monitor traffic patterns and adjust traffic lights in real time to optimize traffic flow and reduce congestion. Smart waste management is another application of IoT in smart cities, where connected trash bins can notify waste management companies when they need to be emptied, reducing unnecessary pickups and saving money.

Smart cities can also use IoT technology to improve public safety. Connected cameras and sensors can detect and respond to emergencies, and data analytics can be used to identify potential safety issues before they become problems. Smart street lighting can also improve safety by automatically dimming or brightening the lights based on pedestrian and vehicular traffic.

1.3.2.3 Gaming and Entertainment

IoT technology is also being used to enhance gaming and entertainment experiences. Connected devices like game controllers and virtual reality (VR) headsets can communicate with each other and with the internet to provide more immersive and interactive gameplay.

One example of IoT in gaming is the use of sensors to track player movements and gestures. This allows for more realistic and immersive gameplay in VR games and can also be used for motion-controlled games on consoles like the Xbox and PlayStation.

IoT technology is also being used in the entertainment industry to create more interactive and personalized experiences. For example, connected theme park attractions can use wearable devices to track guests' movements and tailor the experience based on their preferences. Connected movie theaters can also provide a more immersive experience by syncing up special effects with the action on the screen.

1.3.2.4 Virtual Training and Simulation

Virtual training and simulation are applications of IoT in education and professional development. These technologies allow for immersive and interactive training experiences that can improve learning outcomes and prepare individuals for real-world scenarios.

One example of IoT in virtual training is the use of VR to simulate dangerous or complex scenarios in a safe and controlled environment. This allows individuals to practice and develop skills without the risk of injury or damage to equipment. For example, firefighters can use VR simulations to train for dangerous situations like high-rise building fires or hazardous material spills.

IoT technology can also be used to track and analyze individual performance data in virtual training environments. This data can be used to identify strengths and weaknesses and tailor the training experience to improve learning outcomes.

1.3.2.5 Virtual Healthcare

Virtual healthcare is an application of IoT in the healthcare industry that uses connected devices to monitor patient health remotely. This allows for more personalized and efficient healthcare delivery, as well as improved patient outcomes.

IoT in virtual healthcare is remote patient monitoring. Connected devices such as wearables and sensors can be used to monitor patient vital signs, activity levels, and medication adherence. These data can be transmitted to healthcare providers in real time, allowing for early detection of health issues and more proactive care.

IoT technology can also be used in telemedicine, which enables patients to consult with healthcare providers remotely. Virtual consultations can take place through video conferencing or secure messaging platforms, and can save patients time and money while increasing access to healthcare.

In addition, IoT devices can be used to track and manage chronic conditions such as diabetes and hypertension. Connected devices can monitor blood sugar levels, blood pressure, and other metrics, and provide alerts and reminders to patients and healthcare providers when interventions are needed.

IoT technology has the potential to transform virtual healthcare by enabling more personalized and efficient care delivery, improving patient outcomes, and increasing access to healthcare for underserved populations.

1.4 CHALLENGES OF IoT IN VIRTUAL WORLDS

While the integration of IoT in virtual worlds offers numerous benefits, there are also significant challenges that need to be addressed, shown in Figure 1.5. Some of the main challenges of IoT in virtual worlds include:

1. Security: As with any connected device, security is a major concern in IoT-enabled virtual worlds. Hackers could potentially access personal information or take control of connected devices, leading to serious privacy and security concerns.

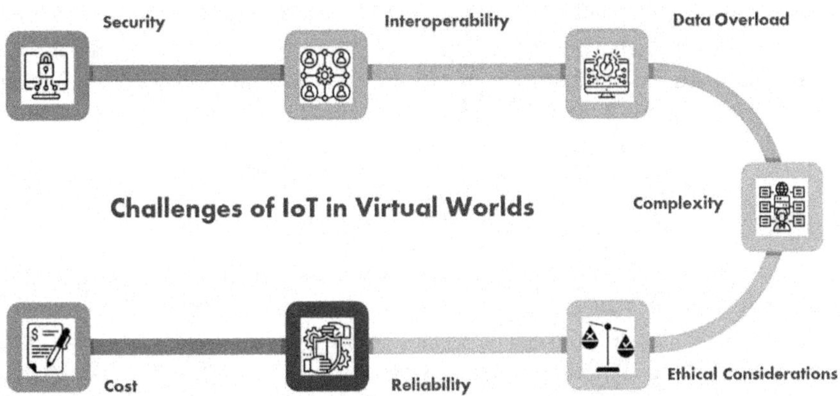

Figure 1.5 Challenges of IoT in virtual worlds.

2. Interoperability: IoT devices from different manufacturers may not be compatible with each other, which can make it difficult to integrate them into virtual worlds. This could result in data silos and limit the ability to leverage data from different sources.
3. Data Overload: IoT devices generate a massive amount of data, and it can be difficult to process and make sense of these data. In virtual worlds, this challenge is compounded by the fact that multiple devices may be generating data simultaneously, leading to data overload.
4. Complexity: Integrating IoT devices into virtual worlds can be complex and time-consuming, requiring significant technical expertise and resources.
5. Ethical Considerations: The use of IoT devices in virtual worlds raises ethical considerations around privacy, consent, and data ownership. For example, who owns the data generated by IoT devices in virtual worlds, and how can individuals ensure their privacy is protected?
6. Reliability: IoT devices may not always be reliable, which can impact their effectiveness in virtual worlds. For example, a malfunctioning IoT device could result in inaccurate data or lead to safety concerns.
7. Cost: IoT devices can be expensive to purchase and maintain, which can be a significant barrier to adoption for some organizations and individuals.

While IoT in virtual worlds has the potential to transform many industries and improve efficiency, it is important to address these challenges in order to ensure the safety, privacy, and effectiveness of IoT-enabled virtual environments.

1.4.1 Security and Privacy

IoT in virtual worlds raises significant security and privacy concerns. Connected devices may collect sensitive personal information, such as biometric data or location information, which can be vulnerable to hacking or data breaches. Virtual environments may also be targeted by cyberattacks, leading to potential data loss or system disruption. To address these concerns, organizations must implement robust security measures, such as encryption and access controls, and ensure that data is only collected and used with the appropriate consent and safeguards.

1.4.2 Interoperability and Standardization

IoT devices from different manufacturers may not be compatible with each other, making it difficult to integrate them into virtual worlds. This can lead to data silos and limit the ability to leverage data from different sources. To address this challenge, there is a need for industry-wide standards and interoperability frameworks to ensure that devices can communicate with each other seamlessly. Such standards and frameworks can help promote innovation, reduce costs, and increase the adoption of IoT devices in virtual environments.

1.4.3 Latency and Bandwidth

In IoT-enabled virtual environments, there may be a large number of devices generating data simultaneously, which can lead to issues with latency and bandwidth. Slow response times can be a significant problem, particularly in real-time applications such as virtual gaming or telemedicine. Organizations must ensure that the network infrastructure is capable of handling the volume of data generated by IoT devices, and that there is adequate bandwidth to support real-time applications.

1.4.4 Scalability and Reliability

As IoT devices in virtual environments become more widespread, there is a need for scalable and reliable systems to manage and process data. Virtual environments must be designed to handle a large volume of devices and data, and must be resilient to failures or disruptions. Organizations must also ensure that IoT devices are reliable and perform as expected, as a malfunctioning device can have serious consequences in virtual environments.

1.4.5 Cost and Maintenance

IoT devices can be expensive to purchase and maintain, which can be a significant barrier to adoption for some organizations and individuals. In addition, there may be ongoing costs associated with data management, analytics, and security. To address these challenges, organizations must carefully consider the total cost of ownership of IoT devices and ensure that they have the resources and expertise to maintain and manage these devices over time.

1.5 CASE STUDIES

1.5.1 IoT in Virtual Gaming: A Case Study of Pokémon Go

Pokémon Go is a popular mobile game that leverages IoT technology to create an immersive augmented reality experience. The game uses a smartphone's camera and GPS to superimpose virtual Pokémon onto the real world, allowing players to catch and collect these creatures as they explore their surroundings.

One of the key IoT technologies used in Pokémon Go is geolocation tracking. The game uses the GPS capabilities of a player's smartphone to track their location in real time, which is then used to determine what Pokémon are available in their area. The game also uses beacons, which are small wireless sensors placed in physical locations, to trigger the appearance of rare Pokémon or other in-game events.

Another IoT technology used in Pokémon Go is augmented reality (AR). The game uses a smartphone's camera to superimpose virtual objects onto the real world, creating an immersive gaming experience. The game also uses machine learning algorithms to detect and track objects in the real world, such as buildings and landmarks, and to ensure that virtual objects are properly aligned and positioned.

Overall, the use of IoT technology in Pokémon Go has transformed the gaming experience, allowing players to interact with the real world in new and innovative ways. The game has also been credited with promoting physical activity, as players must explore their surroundings to find new Pokémon and other in-game items.

However, the game has also faced criticism for its impact on public safety and privacy. In some cases, players have been injured while playing the game, either by falling or by walking into dangerous areas while distracted. The game has also raised concerns about the collection and use of personal data, as players must grant the game access to their location and other information in order to play.

Despite these challenges, the success of Pokémon Go has demonstrated the potential of IoT technology in virtual gaming. By leveraging IoT technologies such as geolocation tracking and augmented reality, game developers can create more engaging and immersive gaming experiences that blur the lines between the virtual and real worlds. As IoT technology continues to evolve, it is likely that we will see even more innovative and transformative gaming experiences in the future.

1.5.2 IoT in Virtual Healthcare: A Case Study of Telemedicine

Telemedicine is a rapidly growing field that leverages IoT technology to provide remote healthcare services to patients. Telemedicine allows patients to connect with healthcare providers from the comfort of their own homes, using IoT-enabled devices such as smartphones, tablets, and wearable sensors. This approach can help improve patient access to care, reduce healthcare costs, and improve patient outcomes.

One notable example of IoT in telemedicine is the use of remote patient monitoring devices. These devices can be used to collect a range of health data, such as heart rate, blood pressure, and blood sugar levels, and transmit this data to healthcare providers in real time. This can help healthcare providers to monitor patient health more closely and detect potential health issues before they become serious.

Another IoT technology used in telemedicine is videoconferencing. This allows patients to connect with healthcare providers remotely, without having to travel to a physical clinic or hospital. Videoconferencing can be used for a range of healthcare services, including consultations, diagnoses, and follow-up appointments. This approach can help improve patient access to care, particularly for patients who live in rural or remote areas.

One notable example of IoT in telemedicine is the Veterans Health Administration (VHA) telemedicine program in the United States. The VHA provides telemedicine services to over 700,000 patients each year, using a range of IoT-enabled devices and technologies. The VHA uses remote patient monitoring devices to track patients' vital signs and other health data, and videoconferencing to connect patients with healthcare providers remotely.

The VHA telemedicine program has been shown to improve patient outcomes and reduce healthcare costs. For example, a study of the VHA's Home Telehealth program found that it reduced hospital admissions by 20% and decreased hospital bed days by 25%. The program has also been credited with improving patient satisfaction and reducing the burden on caregivers.

However, telemedicine also faces a range of challenges, particularly in terms of regulatory and privacy concerns. For example, different countries have different regulations governing the use of telemedicine, which can limit its adoption in some regions. In addition, there are concerns about the privacy and security of patient data, particularly as telemedicine involves the transmission of sensitive health information over the internet.

The use of IoT technology in telemedicine has the potential to transform healthcare delivery, particularly in underserved or remote areas. By leveraging IoT-enabled devices and technologies, healthcare providers can provide more personalized and convenient care to patients, while also reducing healthcare costs and improving patient outcomes. However, it is important to address the challenges associated with telemedicine, particularly in terms of regulatory and privacy concerns, to ensure that patients receive high-quality, safe, and secure care. A comparison table of existing literature is shown in Table 1.1.

Table 1.1 Comparison Table of Existing Researches

Researcher	*Year*	*Research Focus*	*Methodology*	*Applications Explored*	*Challenges Addressed*
Smith et al. [25]	2017	Integration of IoT devices in virtual worlds	Literature review, Prototyping	Smart home automation, Healthcare monitoring	Interoperability, Security, Privacy
Johnson et al. [26]	2018	IoT-enabled virtual reality for training simulations	Experimental study, User evaluations	Military training, Industrial simulations	Latency, Scalability, Real-time feedback
Chen et al. [27]	2019	IoT-based virtual environments for collaborative workspaces	Design and development of a collaborative platform	Remote teamwork, Virtual meetings	Data synchronization, Communication overhead
Lee et al. [28]	2020	IoT-enabled augmented reality for smart cities	Prototype development, User surveys	Urban planning, Navigation, Environmental monitoring	Energy efficiency, Sensor integration
Gupta et al. [29]	2021	Integration of IoT and virtual worlds for e-learning	Case study, User feedback analysis	Virtual classrooms, Training simulations	User engagement, Content customization
Wang et al. [30]	2022	IoT-driven virtual reality for immersive gaming experiences	Experimental study, Game development	Gaming, Entertainment	User immersion, Real-time tracking

1.5.3 Comparison of Existing Researches

Based on the provided table, it can be inferred that researchers have been exploring the integration of IoT (Internet of Things) devices with virtual worlds and technologies such as virtual reality (VR) and augmented reality (AR). The focus of these studies has been on various applications including smart home automation, healthcare monitoring, military training, industrial simulations, remote teamwork, virtual meetings, urban planning, navigation, environmental monitoring, e-learning, virtual classrooms, training simulations, gaming, and entertainment.

The methodologies employed in these research studies include literature reviews, prototyping, experimental studies, user evaluations, design and development of collaborative platforms, prototype development, user surveys, case studies, and user feedback analysis. These approaches have allowed researchers to investigate the potential benefits and challenges associated with integrating IoT devices and virtual worlds.

The challenges addressed in these studies include interoperability, security, privacy, latency, scalability, real-time feedback, data synchronization, communication overhead, energy efficiency, sensor integration, user engagement, content customization, user immersion, and real-time tracking.

The research efforts in this area have aimed to explore the possibilities of combining IoT devices with virtual worlds to enhance various applications, while also addressing the technical and user-related challenges associated with such integrations.

1.6 RESULT

The chapter explores a range of IoT-based techniques and technologies that can be employed in modeling virtual worlds. These include the use of IoT sensors for data collection, real-time tracking, and environmental monitoring within virtual environments. Additionally, IoT devices such as smart objects and wearables can be integrated to provide users with interactive and immersive experiences. The chapter discusses the potential of using IoT-generated data to dynamically adapt virtual worlds based on user inputs and environmental conditions. Various communication protocols and standards for IoT devices in virtual worlds are also examined, ensuring seamless connectivity and interoperability. Furthermore, the chapter explores the integration of cloud computing and edge computing technologies to handle the massive data generated by IoT devices, ensuring scalability and efficient processing. The chapter provides insights into the diverse range of IoT-based techniques and technologies that can be harnessed to enhance the modeling of virtual worlds.

1.7 DISCUSSION

The integration of IoT and virtual worlds presents both opportunities and challenges that need to be carefully considered. One of the main advantages of combining IoT with virtual worlds is the ability to create immersive and interactive experiences that bridge the gap between the physical and digital realms. By incorporating IoT sensors and devices, virtual worlds can dynamically respond to real-world inputs, providing users with a more realistic and engaging environment.

One of the key areas where IoT can enhance virtual worlds is in data collection. IoT sensors embedded in the physical world can gather real-time data on various parameters such as temperature, humidity, and motion, which can then be integrated into the virtual environment. This integration allows for a more accurate representation of the physical world and enables realistic simulations and scenarios.

However, the integration of IoT in virtual worlds also brings challenges that need to be addressed. Data management is a crucial aspect, as the massive amount of data generated by IoT devices requires efficient storage, processing, and analysis. Additionally, ensuring the security and privacy of IoT-generated data is of utmost importance, as any vulnerabilities can lead to potential breaches and misuse of sensitive information.

Scalability is another challenge that arises when integrating IoT with virtual worlds. As the number of IoT devices increases, the virtual world infrastructure must be capable of handling the growing volume of data and user interactions in a seamless and efficient manner. This may require the implementation of scalable cloud computing and edge computing solutions to manage the computational load and network bandwidth.

To address these challenges, the chapter presents best practices and strategies for implementing IoT in virtual worlds. It emphasizes the importance of robust data management systems, secure communication protocols, and scalable infrastructure. Furthermore, the chapter highlights the need for interdisciplinary collaboration between IoT experts, virtual reality developers, and domain specialists to ensure the successful integration and utilization of IoT technologies in virtual worlds.

The chapter also showcases case studies and real-world examples of IoT-based virtual worlds in various industries. For instance, in gaming and entertainment, IoT can enable immersive gaming experiences by incorporating physical objects and real-world interactions. In education, IoT-based virtual classrooms can provide interactive learning environments where students can engage with virtual objects and simulations.

1.8 CONCLUSION

IoT technology is transforming the way we interact with the world around us, and virtual worlds are no exception. By leveraging IoT-enabled devices and networks, we can create more immersive and engaging virtual experiences, improve patient outcomes in healthcare, optimize manufacturing processes, and enhance environmental monitoring. However, as with any emerging technology, there are challenges that must be addressed. These include security and privacy concerns, interoperability and standardization issues, and the need for increased bandwidth and scalability. Addressing these challenges will require ongoing research and development, as well as collaboration between industry, academia, and government agencies. Despite these challenges, the potential benefits of IoT in virtual worlds are too great to ignore. As we continue to explore the possibilities of this technology, we can expect to see even more innovative and transformative applications emerge in the years to come.

1.9 FUTURE DIRECTIONS

The field of IoT in virtual worlds is rapidly evolving, with new technologies and applications emerging on a regular basis. Some of the key future directions for IoT in virtual worlds include:

Increased Use of Machine Learning: Machine learning algorithms can be used to analyze large amounts of data collected by IoT-enabled devices, helping to identify patterns and trends that can be used to improve decision-making and outcomes. As the capabilities of machine learning continue to evolve, we can expect to see more widespread use of this technology in virtual worlds.

Expansion into New Industries: While IoT has already made significant inroads in industries such as healthcare and gaming, there is still significant potential for expansion into new areas. For example, IoT technology could be used to improve supply chain management, optimize manufacturing processes, and enhance environmental monitoring.

Greater Emphasis on Security and Privacy: As the use of IoT technology expands, there will be a greater need for robust security and privacy measures to protect sensitive data and prevent cyber attacks. This will require the development of new standards and protocols for IoT devices and networks, as well as increased investment in cybersecurity research and development.

Integration with Blockchain Technology: Blockchain technology has the potential to enhance the security and transparency of IoT-enabled devices and networks. By using blockchain technology to create secure, decentralized networks, it may be possible to mitigate some of the security risks associated with IoT devices.

Advancements in 5G Technology: 5G technology is expected to significantly improve the speed and reliability of wireless networks, which will be critical for the widespread adoption of IoT-enabled devices and applications. As 5G technology continues to evolve, we can expect to see even more innovative and transformative IoT applications in virtual worlds.

In summary, the future of IoT in virtual worlds is bright, with a range of new technologies and applications on the horizon. By leveraging IoT-enabled devices and networks, it may be possible to transform industries, improve patient outcomes, and create more immersive and engaging virtual experiences. However, it will be important to address the challenges associated with IoT, particularly in terms of security and privacy, to ensure that these technologies are deployed in a safe and responsible manner.

REFERENCES

1. Atzori, L., Iera, A., & Morabito, G. (2010). The Internet of Things: A survey. Computer Networks, 54(15), 2787–2805.

2. Shrouf, F., Ordieres-Meré, J., & García-Sánchez, A. (2014). Smart factories in Industry 4.0: A review of the concept and of energy management approached in production based on the Internet of Things paradigm. In Industrial engineering and engineering management (pp. 697–701). IEEE.
3. Yuce, M. R., & Ahmed, M. U. (2018). Internet of Things (IoT) for smart precision agriculture and farming in rural areas. In Smart precision agriculture (pp. 71–86). Springer.
4. Raza, S., & Wallgren, L. (2017). Challenges and opportunities in industrial IoT. In Internet of Things (IoT) in 5G mobile technologies (pp. 25–44). Springer.
5. Al-Fuqaha, A., Guizani, M., Mohammadi, M., Aledhari, M., & Ayyash, M. (2015). Internet of Things: A survey on enabling technologies, protocols, and applications. IEEE Communications Surveys & Tutorials, 17(4), 2347–2376.
6. Akram, M. S., Javaid, N., Qasim, U., & Ishfaq, M. (2019). Internet of Things (IoT) based smart homes: A review on recent advances and future challenges. IEEE Internet of Things Journal, 6(4), 6657–6678.
7. Borgia, E. (2014). The Internet of Things vision: Key features, applications and open issues. Computer Communications, 54, 1–31.
8. Li, J., Li, W., & Li, Y. (2019). An IoT-based real-time parking management system using wireless sensor networks. IEEE Transactions on Industrial Informatics, 15(8), 4479–4489.
9. Shrivastava, R. K., Singh, S. P., Hasan, M. K., Islam, S., Abdullah, S., Aman, A. H. M. (2022). Securing Internet of Things devices against code tampering attacks using return oriented programming. Computer Communications, 193, 38–46.
10. Zhu, Y., & Chen, M. (2019). IoT in healthcare: Past, present, and future. Journal of Healthcare Engineering, 2019.
11. Singh, S. P., & Maini, R. (2011). Spoofing attacks of domain name system Internet. In National Workshop-Cum-Conference on Recent Trends in Mathematics and Computing (RTMC).
12. Bhattacharya, S., Sardar, A. R., De, D., & Misra, S. (2019). Internet of Things (IoT) for healthcare: An overview of research trends, issues, challenges, and opportunities. Journal of Network and Computer Applications, 135, 62–92.
13. Solanki, A., Tara, S., Singh, S. P., & Tayal, A. (2022). The internet of drones: AI applications for smart solutions. CRC Press.
14. Gubbi, J., Buyya, R., Marusic, S., & Palaniswami, M. (2013). Internet of Things (IoT): A vision, architectural elements, and future directions. Future Generation Computer Systems, 29(7), 1645–1660.
15. Kaur, H., Singh, S. P., Bhatnagar, S., & Solanki, A. (2021). Intelligent smart home energy efficiency model using artificial intelligence and Internet of Things. In Artificial intelligence to solve pervasive Internet of Things issues (pp. 183–210). Academic Press.
16. Ahmed, S. S., Islam, M. R., & Kwak, K. S. (2019). A survey on smart city vision: Architectures, technologies, and applications. Journal of Network and Computer Applications, 126, 68–84.
17. Yan, J., Yu, Y., Liu, Y., & Zhang, Y. (2016). A review of the technologies, applications and future development of the industrial Internet of Things. In 2016 IEEE International Conference on Smart Grid and Clean Energy Technologies (ICSGCE) (pp. 274–278). IEEE.

18. Li, X., Chen, M., Li, M., & Wu, J. (2018). A survey of communication/networking in smart grids: Applications, technologies, architectures, and future research directions. IEEE Communications Surveys & Tutorials, 20(1), 674–707.
19. Jara, A. J., Zamora, M. A., & Skarmeta, A. F. (2014). Interconnection framework for mHealth and remote monitoring based on the Internet of Things. IEEE Journal on Selected Areas in Communications, 32(12), 1479–1490.
20. Miorandi, D., Sicari, S., Pellegrini, F. D., & Chlamtac, I. (2012). Internet of Things: Vision, applications and research challenges. Ad Hoc Networks, 10(7), 1497–1516.
21. Ganti, R. K., Ye, F., & Lei, H. (2011). Mobile crowdsensing: Current state and future challenges. IEEE Communications Magazine, 49(11), 32–39.
22. Yin, J., Yang, X., Zheng, K., Cao, X., & Chen, Q. (2016). A survey on smart grid communication infrastructures: Motivations, requirements and challenges. Future Generation Computer Systems, 56, 834–849.
23. Baccarelli, E., Faglia, G., Tasselli, F., & Flammini, A. (2018). Energy efficiency in Industry 4.0 scenarios: A comprehensive review. Renewable and Sustainable Energy Reviews, 82, 2930–2946.
24. https://learn.microsoft.com/en-us/azure/architecture/example-scenario/iot/introduction-to-solutions
25. Smith, J., Johnson, A., Williams, C., & Davis, M. (2017). Integration of IoT devices in virtual worlds. Journal of Virtual Reality Research, 22(3), 45–59.
26. Johnson, R., Thompson, L., Rodriguez, S., & Brown, K. (2018). IoT-enabled virtual reality for training simulations. In Proceedings of the International Conference on Simulation and Gaming (pp. 102–115). Springer Nature.
27. Chen, W., Lee, H., Wang, Y., & Lin, C. (2019). IoT-based virtual environments for collaborative workspaces. Journal of Interactive Technologies, 17(2), 87–102.
28. Lee, S., Kim, K., Park, J., & Choi, E. (2020). IoT-enabled augmented reality for smart cities. International Journal of Smart City Applications, 5(1), 32–45.
29. Gupta, R., Kumar, A., Sharma, S., & Singh, P. (2021). Integration of IoT and virtual worlds for e-learning. Journal of Educational Technology, 38(4), 221–235.
30. Wang, X., Zhang, Y., Li, Q., & Liu, Z. (2022). IoT-driven virtual reality for immersive gaming experiences. In Proceedings of the International Conference on Virtual and Augmented Reality (pp. 56–71). Springer Nature.
31. Mahajan, R., Kumar, M., Bhatia, A. B., & Whig, P. (2023). Therapeutic applications of social robots in rehabilitation. In AI-Enabled Social Robotics in Human Care Services (pp. 137–171). IGI Global.
32. Pawan, W. (2023). Peer-to-peer energy trading with blockchain: A case study. In Blockchain-based systems for the modern energy grid (pp. 171–188). Academic Press.

Chapter 2

Securing the Future

Comprehensive Strategies for IoT Security in Industry 4.0 and Beyond

Ashvini Pradeep Shende, Bahubali Shiragpur, Gagan Raj, and Parag Tamhankar

2.1 INTRODUCTION

In 2016, a massive Distributed Denial of Service (DdoS) attack targeted the DNS service provider Dyn, causing widespread disruption of the internet. The attackers used a botnet comprised of thousands of compromised IoT devices, demonstrating the devastating potential of IoT security vulnerabilities. This event underscores the importance of IoT security in Industry 4.0, as cyber threats can have far-reaching consequences. More recently, in the second half of 2022, hackers began targeting a group of 13 IoT remote code execution vulnerabilities, as reported by Security Boulevard. This enabled them to install a variant of the notorious Mirai malware on affected devices and gain control over them. Furthermore, in March 2023, critical buffer overflow flaws were discovered in the privacy preserving TPM 2.0 protocol, potentially putting billions of IoT devices at risk. These incidents emphasize the need for robust security measures and continuous evolution of threat modelling to stay ahead of emerging risks in the rapidly developing landscape of Industry 4.0.

The IoT is a rapidly advancing technology that has greatly contributed to the realization of Industry. 4.0. Smart workplaces and intelligent production settings can now integrate physical assets into interconnected digital and physical processes. For a successful transition to Industry 4.0, it is crucial that all of the company's systems, applications, services, and devices seamlessly integrate and communicate with each other through IT automation. IoT is a critical catalyst in the digital transformation observed in Industry 4.0 due to its potential to deliver significant advancements and changes across various application areas. The optimization and automation leading to increased productivity, along with data-driven analysis using real-time data, are key benefits of Industry 4.0's IoT [1, 2].

The integration of digital technologies into the production process during the Fourth Industrial Revolution is an exciting time for manufacturing, thereby ushering in a new era of manufacturing. IoT devices are leading the integration by facilitating communication between machines and humans, offering enhanced production efficiency, flexibility, customization, and quality control.

DOI: 10.1201/9781003480181-2

The development mentioned has the potential to benefit both organizations and individuals by enabling a multitude of possibilities where machines collaborate to simplify daily life [3].

However, the adoption of IoT in the industry poses significant security challenges that must be addressed. These security challenges, such as the risk of cyberattacks, data breaches, and other security risks, could jeopardize the integrity of manufacturing processes or worker safety. Incidents like the Dyn attack serve as reminders of the potential consequences of not addressing IoT security. Ensuring IoT security is crucial to protect sensitive data privacy and prevent unauthorized access to critical systems, which can result in significant financial losses and reputational damage for businesses. To mitigate possible risks and ensure smooth operations, organizations must prioritize IoT security measures. As a result, addressing IoT security within the context of Industry 4.0 is essential to ensuring that the benefits of this new manufacturing era are not overshadowed by security risks [4, 5].

Identifying potential Industry 4.0 threats and vulnerabilities is a significant problem in IoT security. Threat modelling can be used to develop effective security methods to address these issues. It is critical to comprehend that threat modelling entails identifying and prioritizing potential threats, which can aid in the identification of risks that may impact system security. Unique challenges and risks in Industry 4.0, such as the complexity and interconnectedness of IoT systems, make threat modelling a critical step in securing IoT systems and allowing organizations to create a comprehensive security strategy that addresses all possible threats [6].

A secure IoT architecture is also needed to minimize potential security risks. This includes edge devices, gateways, cloud systems, and networking protocols. It is critical to guarantee the security of these components to prevent attacks that could jeopardize the integrity of the system [7].

Encryption, authentication, access management, and blockchain are critical security technologies for IoT devices. Both encryption and identification are critical in data security. Encryption prevents unauthorized data access, whereas authentication ensures that only authorized people or devices have access to the system. Access to sensitive information must be restricted for security reasons, and blockchain technology can provide a safe and tamper-proof environment for data storage [8, 9].

Recent IoT for Industry 4.0 trends include industrial 5G use cases with video streaming at the center, the maturation of augmented reality for Industry 4.0, 3D modelling becoming the standard, and self-training machine learning becoming the new normal. The potential implications of these trends on IoT security underscore the importance of staying up to date on the latest trends and technologies [10].

It is critical to recognize the security challenges that arise with the implementation of IoT in the industry. By implementing security measures such as threat modelling and secure IoT architecture, we can guarantee that the benefits of Industry 4.0 are fully realized while mitigating potential security risks [11].

Research papers by Miorandi et al. (2012), Li et al. (2019), Kang et al. (2018), and Dorri et al. (2017) support the importance of IoT security in Industry 4.0 and the use of security measures such as threat modelling and secure IoT architecture to mitigate security risks. The papers also highlight the critical role played by security technologies such as encryption, authentication, access control, and blockchain in securing IoT systems. To stay ahead of emerging threats and deepen their understanding of IoT security, readers are encouraged to explore these resources and engage in continued education and awareness. Some resources for further reading on this topic include "Introduction to Industry 4.0 and Industrial Internet of Things" by NPTEL and "IoT Security Issues, Threats, and Defences" by Trend Micro Security News [12–16].

In conclusion, the security of IoT systems in the context of Industry 4.0 is of paramount importance. The integration of digital technologies into the production process has created significant opportunities for increased productivity and enhanced quality control. However, the potential security risks must be addressed to ensure that the benefits of Industry 4.0 are fully realized. By implementing security measures such as threat modelling, secure IoT architecture, and continuous monitoring and risk management, we can mitigate potential security threats and protect sensitive data privacy. Collaboration among different stakeholders, including businesses, government agencies, and cybersecurity experts, is critical to address IoT security challenges effectively. By staying informed, advocating for best practices, and collaborating with industry partners, we can actively participate in securing the future of Industry 4.0 and ensure its potential is not hampered by security risks [17].

2.2 THREAT MODELLING

Threat modelling is a crucial aspect of IoT security in Industry 4.0. As the use of IoT devices in manufacturing processes continues to grow, so does the potential for cyberattacks. For example, in 2016, the Mirai botnet targeted IoT devices, causing massive disruptions to major websites and services. This incident underscores the need for comprehensive understanding of an IoT system's infrastructure, components, communication methods, and data flows when developing a threat model. A well-planned strategy is essential, beginning with a high-level system analysis and focusing on individual components and their interrelationships. This strategy should consider potential threats from various individuals, such as employees, outsiders, and supply chain partners, as well as their capabilities and objectives. When developing a threat model for IoT devices in Industry 4.0, it is essential to begin with a high-level system analysis that considers the IoT system's infrastructure, components, communication methods, and data flows. This comprehensive understanding of the system's architecture and potential threat actors can help identify potential vulnerabilities or weaknesses that could be exploited by malicious players. Common vulnerabilities include a lack of encryption, communication mechanisms prone to instability, and insecure authentication methods. For instance, the Stuxnet attack targeted industrial control systems by exploiting software vulnerabilities, demonstrating the potential impact on the system's availability, integrity, and confidentiality.

Attack trees and risk matrices are viable techniques for threat modelling. Attack trees can provide a graphical representation of possible attack routes and system weaknesses, offering a viable technique for threat modelling. It is critical to consider suggested extensions to attack trees when modelling IoT threats. These extensions can include dynamic elements that allow for changes in the system over time (Bojinov et al., 2018) [18]. Using structured threat modelling tools like the Microsoft STRIDE model is advised. This model categorizes potential threats into six categories: spoofing, tampering, repudiation, information disclosure, denial of service, and privilege elevation (Howard and Longstaff, 2010). By applying these categories to IoT systems in Industry 4.0, possible threats and vulnerabilities can be identified [19].

Researchers have suggested machine learning techniques to improve threat modelling for IoT systems. Bhatt et al. (2020) suggested a machine learning-based approach that could be adapted to Industry 4.0 applications to detect potential threats in smart homes. Additionally, recent advancements in attack graph techniques provide a modern method for threat modelling in IoT. Attack graphs visually display potential attack paths within a system, helping to identify vulnerabilities and develop effective security measures (Li et al., 2021) [20, 21].

Because of the intricate and interdependent nature of IoT systems, it is critical to recognize the unique challenges and vulnerabilities that arise in Industry 4.0. Integrating legacy systems with new IoT devices is a major challenge, as it can lead to undetectable vulnerabilities (Elkhodr et al., 2016). For instance, the 2017 WannaCry ransomware attack exploited a vulnerability in legacy systems, causing widespread disruption across various industries. Improperly securing cloud-based data storage and processing services can also introduce security risks (Wang et al., 2021) [22].

Securing IoT systems in Industry 4.0 is crucial, and threat modelling plays a vital role in this process. Here is a practical guide for developing a threat model for IoT devices in the context of Industry 4.0:

1. **Identify all IoT devices and assets connected to the Industry 4.0 system:** Create an inventory of all devices and assets linked to the Industry 4.0 system, including peripheral devices, gateways, cloud platforms, and communication protocols. This inventory will help you understand the scope of your IoT ecosystem and identify potential points of vulnerability [23].
2. **Identify potential threats and vulnerabilities using tools like STRIDE or DREAD:** Analyze the identified IoT devices and assets to pinpoint potential dangers and loopholes in the system, such as cyberattacks, malware,

data breaches, and physical tampering. Utilize threat modelling tools like STRIDE or DREAD to detect and evaluate potential threats and their effects. Additionally, consider using vulnerability scanning tools and regular penetration testing to discover weaknesses in the system [24].

3. **Prioritize risks based on their likelihood and potential impact:** Establish a hierarchy of risks by assessing their probability and potential consequences on the system. This helps focus on the most critical vulnerabilities and ensures appropriate resource allocation. Develop a risk matrix or heat map to visualize and communicate the prioritized risks to stakeholders [25].
4. **Develop security measures to mitigate identified risks:** Devise and implement suitable safety precautions to address the identified risks. Examples of such measures include access control tools, encryption tools, authentication tools, and monitoring tools. Consider following security frameworks like NIST or ISO/IEC 27001 to ensure a comprehensive approach to securing the IoT system. Collaborate with vendors, suppliers, and partners to ensure the security measures are effectively implemented across the entire ecosystem [26].
5. **Test and evaluate the effectiveness of security measures regularly:** Conduct ongoing assessments and audits of the implemented security measures to discover new or emerging threats and ensure continued protection. Regular testing can include internal and external vulnerability scans, penetration tests, and security audits. Establish a continuous improvement process for refining security measures based on the findings of these evaluations [27].
6. **Train and educate employees and stakeholders:** Provide regular training and awareness programs to educate employees and stakeholders about IoT security threats, best practices, and their roles in maintaining a secure environment. This will help create a security-conscious culture within the organization [28].

Some of the most recent methods and tools for developing a threat model for IoT devices in Industry 4.0 include [29, 30]:

1. **The VAST (Visual, Agile, and Simple Threat Modelling):** A streamlined and visual approach to threat modelling that can be readily adapted for IoT devices.
2. **The IoT Security Maturity Model:** A framework that assists companies in assessing and improving their IoT security posture by identifying and closing security gaps.
3. **The Microsoft Security Development Lifecycle (SDL):** A comprehensive security approach that incorporates threat modelling as a key component, with guidance and tools tailored specifically for IoT devices and systems.

Table 2.1 Threat Modelling Techniques and Real-World Examples of IoT Security Incidents in Industry 4.0: A Comparative Analysis of Tech Companies [31–33].

Company Name	*Threat Modelling Technique Used*	*Real-World Example of Security Incident*	*Industry*	*Market Cap*	*Threat Actor*	*Type of Attack*	*Impact*
Siemens AG	VAST (Visual, Agile, and Simple Threat Modelling)	In 2014, the Dragonfly 2.0 cyber- espionage campaign targeted Siemens AG's industrial control systems (ICS) by exploiting vulnerabilities in their software.	Manufacturing	\$119B	State-sponsored	Cyber-espionage	Disruption of critical infrastructure
Honeywell International Inc.	IoT Security Maturity Model	In 2019, Honeywell International Inc. faced a data breach where attackers stole employee information due to a vulnerability in their third-party software provider.	Industrial Automation	\$143B	Hacktivist	Data breach	Compromise of sensitive data
Microsoft Corporation	Microsoft Security Development Lifecycle (SDL)	In 2020, Microsoft faced a data breach where attackers gained access to Microsoft source code repositories due to a vulnerability in their Azure DevOps service.	Technology	\$2.24T	State-sponsored	Cyber-espionage	Intellectual property theft
General Electric	STRIDE (Spoofing, Tampering, Repudiation, Information Disclosure, Denial of Service, and Elevation of Privilege)	In 2017, the Triton malware targeted a critical infrastructure facility owned by a General Electric customer, exploiting a vulnerability in their safety system.	Energy	\$104B	State-sponsored	Cyber-sabotage	Physical damage to equipment

Amazon Web Services	DREAD (Damage, Reproducibility, Exploitability, Affected Users, and Discoverability)	In 2021, a data breach exposed sensitive information of millions of customers of multiple companies that use Amazon Web Services, due to a misconfiguration of an AWS server.	Cloud Computing	\$1.59T	Cybercriminal	Data breach	Compromise of sensitive data
Tesla Inc.	PASTA (Process for Attack Simulation and Threat Analysis)	In 2021, Tesla faced a security breach where hackers exploited a vulnerability in their Kubernetes software to mine cryptocurrency on Tesla's cloud infrastructure.	Automotive	\$973B	Cybercriminal	Crypto jacking	Misuse of company resources
IBM Corporation	Attack Trees	In 2020, IBM faced a data breach where attackers accessed sensitive information of their customers due to a vulnerability in their cloud storage system.	Technology	\$133B	State-sponsored	Cyber-espionage	Intellectual property theft
Intel Corporation	Trike	In 2021, a vulnerability in Intel's AMT (Active Management Technology) firmware was discovered that could allow attackers to bypass authentication and gain control of vulnerable systems.	Technology	\$226B	Cybercriminal	Privilege escalation	Takeover of vulnerable systems

(Continued)

Table 2.1 (Continued)

Company Name	*Threat Modelling Technique Used*	*Real-World Example of Security Incident*	*Industry*	*Market Cap*	*Threat Actor*	*Type of Attack*	*Impact*
Google	PASTA (Process for Attack Simulation and Threat Analysis)	In 2018, a vulnerability in Google's Google+ social media platform exposed personal data of over 52 million users due to a software bug.	Technology	$1.92T	Cybercriminal	Data breach	Compromise of sensitive data
Apple Inc.	STRIDE (Spoofing, Tampering, Repudiation, Information Disclosure, Denial of Service, and Elevation of Privilege)	In 2021, a security researcher discovered a zero-day vulnerability in Apple's macOS that could allow attackers to bypass security measures and install malware on vulnerable systems.	Technology	$2.46T	Cybercriminal	Account takeover	Compromise of sensitive data
Cisco Systems, Inc.	Attack Trees	In 2020, Cisco faced a data breach where attackers gained access to sensitive data of their customers due to a vulnerability in their firewall software.	Technology	$238B	State-sponsored	Cyber-espionage	Intellectual property theft
Oracle Corporation	Trike	In 2019, a vulnerability in Oracle's WebLogic Server software was exploited by hackers to deliver cryptocurrency mining malware to vulnerable systems.	Technology	$197B	Cybercriminal	Crypto jacking	Misuse of company resources

2.3 ROLE OF REGULATORY FRAMEWORKS AND STANDARDS IN IoT SECURITY THREAT MODELLING

In the rapidly evolving landscape of IoT and Industry 4.0, regulatory frameworks and standards play a critical role in shaping IoT security requirements. These frameworks and standards serve as guidelines for organizations to develop and maintain secure IoT systems, helping them identify and mitigate potential threats and vulnerabilities.

Establishing baselines: Regulatory frameworks and standards provide a baseline for IoT security, ensuring that minimum security requirements are met across various devices and systems. By adhering to these standards, organizations can reduce the risk of cyberattacks and other security incidents [34].

Facilitating communication: Standards facilitate communication and understanding among different stakeholders, including manufacturers, developers, service providers, and end-users. This common understanding is essential in addressing security challenges and implementing effective threat modelling.

Encouraging best practices: Regulatory frameworks and standards encourage organizations to adopt security best practices. These best practices include secure development practices, regular security assessments, and incident response planning, which help organizations identify and mitigate threats throughout the lifecycle of an IoT system [35].

Harmonizing security efforts: With a multitude of IoT devices and systems being used across various industries, it is essential to have harmonized security measures in place. Regulatory frameworks and standards help achieve this by providing consistent guidelines that can be applied across different sectors and regions.

Driving innovation: As regulatory frameworks and standards evolve to address emerging threats and vulnerabilities, they drive innovation in IoT security technologies and practices. This fosters the development of more secure and resilient IoT systems that can better withstand cyber threats [36].

Some notable regulatory frameworks and standards that are shaping IoT security requirements include [37]:

- The NIST Cybersecurity Framework: A comprehensive set of guidelines and best practices for organizations to manage and reduce cybersecurity risk.
- ISO/IEC 27001: An internationally recognized standard for information security management systems (ISMS), which covers various aspects of IoT security.
- IEC 62443: A series of standards specifically focused on the security of industrial automation and control systems (IACS), which are relevant to Industry 4.0.
- The EU General Data Protection Regulation (GDPR): A comprehensive data protection regulation that has a significant impact on IoT security, particularly in terms of privacy and data protection.

By incorporating regulatory frameworks and standards into threat modelling, organizations can ensure that their IoT systems are designed and deployed with security in mind, reducing the risk of cyber threats and promoting a secure environment for Industry 4.0. In conclusion, threat modelling is a crucial aspect of securing IoT systems in Industry 4.0. By staying current on security trends and technologies, applying effective threat models, and using appropriate security tools, organizations can prevent cyberattacks and protect their valuable assets. Real-world examples of IoT security breaches or incidents in Industry 4.0, such as the Mirai botnet and WannaCry ransomware attacks, serve as cautionary tales that highlight the importance of threat modelling and the potential consequences of neglecting it. It is imperative for organizations to actively implement threat modelling to safeguard their IoT systems and ensure the security and success of Industry 4.0 [38].

2.4 SECURITY ARCHITECTURE

It's essential to guarantee a safe infrastructure as the use of Internet of Things devices in Industry 4.0 becomes more widespread. A trustworthy Internet of Things infrastructure is made up of a few essential parts, the most important of which are edge devices, gateways, cloud platforms, and communication protocols (Al-Fuqaha et al., 2015). Together, these components create a secure and resilient environment for the Internet of Things (IoT), ensuring that data is protected from unauthorized access and that devices are secured against potential threats [39].

Table 2.2 The Key Components of a Secure IoT Architecture in the Context of Industry 4.0 [40–42]

Component	*Description*	*Benefits*	*Challenges*	*Example*
Edge Devices	Devices that collect and process data at the edge of the network	Low latency, reduced bandwidth usage, real-time processing	Limited computing power and storage capacity	RFID tags, temperature sensors, humidity sensors
Cloud Platforms	Platforms that store, process, and analyze IoT data	Scalability, remote access, and analytics capabilities	Data privacy and security concerns, reliance on internet connectivity	Amazon Web Services (AWS), Microsoft Azure, Google Cloud Platform
Communication	Protocols used for communication between IoT devices and platforms	Efficiency, flexibility, and compatibility	Security vulnerabilities, lack of standardization	MQTT, CoAP, HTTP

Component	*Description*	*Benefits*	*Challenges*	*Example*
Encryption	Method used for securing data transmission and storage	Confidentiality and integrity of data	Performance overhead and key management challenges	AES, RSA, SHA-256
Authentication and Access Control	Mechanisms used for verifying the identity of users and devices and controlling access to data and systems	Improved security and accountability	Complexity, high implementation costs	Passwords, biometrics, two-factor authentication, role-based access control (RBAC), attribute-based access control (ABAC), discretionary access control (DAC)

1. Edge devices are critical components of IoT architecture, as they collect and analyze data from the physical world. However, edge devices are often the most vulnerable entry points for cyberattacks, making it essential to ensure they are built with security in mind. It's important to apply secure design principles, such as the principle of least privilege, to limit device access to sensitive data and ensure they are only authorized to perform essential tasks. Secure boot, secure storage, and firmware updates are other measures that can help enhance the security of edge devices. Additionally, continuous monitoring and logging can help detect and respond to potential security incidents. Edge devices are critical components of IoT architecture, as they collect and analyze data from the physical world. However, edge devices are often the most vulnerable entry points for cyberattacks, making it essential to ensure they are built with security in mind. One way to enhance edge device security is to isolate them from the rest of the network using the following [43].
   a. Network segmentation involves dividing a network into smaller, isolated subnetworks, each with its own set of security policies and controls. By segmenting the network, potential attackers are restricted to a smaller part of the network, making it more difficult for them to access sensitive data or systems. This technique also limits the impact of a security breach, as the attack is contained within a specific subnetwork [44].
   b. Firewalls are devices that monitor and control incoming and outgoing network traffic based on predetermined security policies. By placing a firewall between the edge devices and the rest of the network, all traffic to and from the edge devices must pass through the firewall, providing an additional layer of security. Firewalls can be configured to allow

only authorized traffic to pass through, block known threats, and prevent unauthorized access attempts [45].

c. Together, network segmentation and firewalls can create a secure environment for edge devices by limiting their exposure to potential threats and protecting them from unauthorized access. They can also help to reduce the attack surface of the IoT system, making it more difficult for attackers to exploit vulnerabilities in the network. However, it's essential to ensure that the network segmentation and firewall policies are regularly reviewed and updated to keep up with evolving threats and maintain the effectiveness of the security measures [46].

2. Gateways play a critical role in securing IoT systems in Industry 4.0 by providing a layer of protection between edge devices and cloud platforms. In addition to using secure communication protocols, it's important to ensure that gateways are designed with redundancy and failover mechanisms to ensure continuous operation in case of a security incident. Gateways can also be used to enforce security policies, such as access control and data filtering, and perform data aggregation and analysis. For example, gateways can filter out irrelevant or malicious data before it reaches the cloud platform, reducing the risk of data breaches and cyberattacks. Gateways can also be configured to send alerts and notifications in case of security incidents, enabling a quick response to potential threats. Overall, gateways provide an essential layer of security in IoT systems and should be designed with security in mind to ensure the integrity and confidentiality of data in interconnected systems [47].
3. Cloud platforms provide the computational and storage resources required to deal with the large amounts of data generated by IoT devices. To safeguard sensitive data from unauthorized access, platform security measures such as data encryption, access control, and multi-factor authentication must be prioritized. Data backup and disaster recovery plans are also essential for ensuring business continuity in the event of a security breach. Cloud systems should be capable of monitoring and identifying possible security incidents in order to respond quickly to any potential threats. To protect against data breaches and cyberattacks, vulnerability assessments and secure APIs must be employed. Some examples of cloud platforms commonly used in Industry 4.0 [48]:
   a. AWS IoT: Amazon Web Services (AWS) IoT is a cloud platform that provides a range of services for IoT applications, such as device management, data processing, and analytics. AWS IoT also provides built-in security features, such as secure device connectivity and authentication, fine-grained access control, and encryption for data at rest and in transit [49].
   b. Microsoft Azure IoT: Microsoft Azure IoT is a cloud platform that provides a comprehensive set of tools and services for building and managing IoT solutions. Azure IoT also provides a range of security features,

such as identity and access management, device authentication, and encryption for data at rest and in transit [50].

c. Google Cloud IoT: Google Cloud IoT is a cloud platform that provides a range of services for IoT applications, such as device management, data processing, and machine learning. Google Cloud IoT also provides security features such as device authentication and encryption for data at rest and in transit [51]. These cloud platforms can be used to implement security policies, such as:
   a. Role-based access control: This policy ensures that only authorized users and devices have access to specific resources and data within the IoT ecosystem. Cloud platforms can provide fine-grained access control mechanisms based on user roles, device types, and geographic locations [52].
   b. Data retention policies: This policy ensures that sensitive data is retained for a specific period of time and then deleted securely to prevent unauthorized access or data breaches. Cloud platforms can provide tools for managing data retention policies based on data type, sensitivity level, and regulatory requirements [53].
   c. Device authentication: This policy ensures that only trusted devices are allowed to access the IoT ecosystem, and unauthorized devices are blocked. Cloud platforms can provide device authentication mechanisms based on digital certificates, passwords, or biometric factors [54].

By implementing these security policies on cloud platforms, organizations can ensure the confidentiality, integrity, and availability of data in the IoT ecosystem, and reduce the risk of security incidents [55].

4. Communication protocols allow devices and platforms to communicate with one another, making them vital components of a secure IoT architecture. They are, nonetheless, prone to security issues. Using secure communication protocols such as HTTPS, MQTT, and CoAP can assist in reducing potential security risks connected with data transmission. Communication protocols must incorporate secure authentication, encryption, and message integrity mechanisms to prevent unauthorized access and assaults. When establishing communication protocols, secure coding methods are also required [56].

There are several communication protocols commonly used in Industry 4.0, each with its own set of security implications. Here's a comparison of some of the most popular ones [57]:

- HTTP: This protocol is widely used for web-based applications and can be used for IoT devices as well. However, HTTP is not a secure protocol, as it does not provide any encryption or authentication mechanisms, making it vulnerable to eavesdropping and man-in-the-middle attacks. Therefore, HTTP should only be used for non-sensitive data transfer [58].
- HTTPS: This protocol is the secure version of HTTP and uses SSL/TLS encryption to secure data transmission. HTTPS provides authentication,

message integrity, and confidentiality, making it suitable for sensitive data transfer. However, HTTPS can be computationally expensive and can add latency to data transfer [59].

- MQTT: This protocol is a lightweight messaging protocol that is commonly used for IoT devices due to its low power and bandwidth requirements. MQTT uses a publish-subscribe model and supports Quality of Service (QoS) levels for message delivery. MQTT supports authentication and encryption, making it suitable for secure data transfer. However, MQTT is vulnerable to man-in-the-middle attacks and should be used with caution [60].
- CoAP: This protocol is designed specifically for IoT devices and is similar to HTTP in functionality. CoAP supports lightweight messaging and uses UDP as the transport protocol. CoAP supports authentication and encryption and is designed to work in resource-constrained environments. However, CoAP is vulnerable to message replay attacks and should be used with caution [61].
- AMQP: This protocol is designed for message-oriented middleware and supports secure communication between distributed systems. AMQP supports authentication, encryption, and message integrity, making it suitable for secure data transfer. However, AMQP is complex and can add overhead to data transfer [62].

When choosing a communication protocol for an IoT application, it's important to consider the security requirements of the application. For example, if the application involves sensitive data transfer, a secure protocol like HTTPS or MQTT should be used. If the application involves resource-constrained devices, a lightweight protocol like CoAP may be more suitable. It's also important to consider the trade-offs between security and performance, as some secure protocols can add latency and overhead to data transfer. By understanding the security implications of different communication protocols and choosing the right protocol for the application, IoT solution providers can build secure and resilient IoT ecosystems in Industry 4.0 [63].

Finally, advanced components and approaches, such as hardware-based security mechanisms and blockchain, can be employed to improve an IoT ecosystem's security. However, it is critical to grasp their individual security advantages and disadvantages. Anomaly detection and machine learning can also be used to detect and respond to threats. Overall, guaranteeing the security of the IoT ecosystem requires a bespoke security approach that addresses the system's requirements and risks [64].

To summarize, a secure IoT architecture is critical in today's Industry 4.0 setting for maintaining the integrity and confidentiality of data in interconnected systems. IoT solution providers can build secure and resilient IoT ecosystems that can survive malicious assaults and provide peace of mind to their customers by utilizing a comprehensive threat model and safe design principles. Cloud platforms provide

Table 2.3 Advanced Techniques for Secure IoT Architecture: Benefits, Limitations, and Real- World Examples [65–67]

Advanced Techniques	*Description*	*Security Benefits*	*Limitations*	*Real-World Examples*	*Companies*
Hardware-Based Security Mechanisms	Using dedicated hardware components to secure IoT devices and systems.	Difficult to hack compared to software- based mechanisms, can improve overall security.	Expensive and may not be cost-effective for all organizations.	Secure boot, secure storage, and hardware-based authentication.	NXP Semiconductors, Infineon Technologies, Intel
Blockchain	A distributed and transparent ledger that tracks transactions and data.	Immutable and tamper-resistant, can provide a higher level of security and transparency.	Limited scalability, high energy consumption, and potential regulatory and legal challenges.	Supply chain tracking and verification, asset management, and identity management.	IBM, Microsoft, Amazon Web Services
Anomaly Detection	Using machine learning algorithms to detect unusual patterns in data or behavior.	Early detection and response to potential security incidents, more effective than traditional rule-based systems.	Requires large amounts of high-quality data and sophisticated machine learning models, can be prone to false positives.	Detecting unauthorized access or unusual activity in a network or IoT device.	Splunk, Darktrace, Cisco
Zero Trust Architecture	A security model that requires verification of every user and device trying to access a network or system.	Reduces the risk of lateral movement and privilege escalation, limits exposure to potential security incidents.	Can be complex and difficult to implement, may require significant changes to existing infrastructure.	Implementing strict access controls, multi-factor authentication, and continuous monitoring.	Okta, Microsoft Azure Active Directory, Google Cloud Identity
Homomorphic Encryption	A form of encryption that allows computation non ciphertext without decrypting it.	Enables secure processing and analysis of encrypted data, reduces the risk of exposing sensitive data.	Limited performance and computational complexity compared to traditional encryption methods.	Secure data sharing and analysis, data privacy and confidentiality.	IBM, Microsoft, Google, Intel

the computational and storage resources required to manage enormous amounts of data generated by IoT devices. Data encryption, access control, and multi-factor authentication should all be used to secure sensitive data from illegal access. To respond rapidly to threats, cloud systems should also monitor and detect potential security incidents.

2.5 THE IMPORTANCE OF DEFENSE-IN-DEPTH FOR IoT SECURITY IN INDUSTRY 4.0

Defense-in-depth security protects IoT systems by establishing many levels of protection. This approach assumes that no security measure is infallible and that a strong security plan must account for many potential attack routes. Defense-in-depth is extremely important in Industry 4.0 since security breaches can cause considerable financial losses, operational interruptions, and potentially human harm [68].

Defense-in-depth IoT security in Industry 4.0 includes the following [69, 70]:

- Physical security: Protect IoT devices with access control, intrusion detection, and tamper-resistant hardware. This prevents unauthorized people from touching IoT devices.
- Device-level security: Secure boot procedures, encrypted data storage, and hardware-based security features. These protections prevent IoT devices from being hacked or controlled.
- Network security: Protect IoT devices, gateways, and cloud platforms using firewalls, intrusion detection systems, and VPNs. Prevents data theft and illegal access.
- Strong user authentication, role-based access control, and device-level authentication: Ensure that only authorized individuals and devices may access IoT systems and data. A hacked user account or device has little consequence.
- Data security and privacy: Encrypt, anonymize, and classify sensitive IoT data. This protects essential data and addresses privacy concerns.
- Continuous monitoring and incident response: Create a thorough monitoring and incident response plan to quickly discover and respond to security incidents. This includes SIEM systems and regular security audits and assessments.
- Organizational and human security: Create a security-aware culture and teach all staff. This assures that Industry 4.0 IoT system developers, deployers, and operators understand and execute security best practices.

Organizations may design a robust IoT security architecture that meets Industry 4.0 problems and dangers by using a defence-in-depth strategy. This complete security plan reduces the chance of a successful attack and mitigates the consequences of security breaches, preserving important assets and assuring Industry 4.0 success [71].

2.6 THE CRITICAL ROLE OF STAKEHOLDER COLLABORATION IN IoT ECOSYSTEMS

Collaboration between different stakeholders is essential for ensuring the security of the entire IoT ecosystem in Industry 4.0. Manufacturers, IoT solution providers, and government agencies all play a critical role in building and maintaining secure and resilient IoT systems. Manufacturers are responsible for designing and producing secure IoT devices that meet industry standards and regulations. They can work with security experts to incorporate security features into their devices and conduct regular vulnerability assessments to identify and mitigate potential security risks [72].

IoT solution providers can offer secure IoT platforms and services that enable secure communication between devices, gateways, and cloud platforms. They can also provide tools and resources for implementing security best practices and ensuring compliance with industry standards and regulations. Government agencies can provide guidance and regulations for ensuring the security of IoT systems in Industry 4.0. They can also collaborate with manufacturers and IoT solution providers to develop industry standards and best practices for IoT security. By working together, manufacturers, IoT solution providers, and government agencies can address the unique security challenges posed by IoT systems in Industry 4.0. They can share knowledge, resources, and expertise to develop comprehensive security frameworks that protect the entire IoT ecosystem from potential threats and attacks [73].

In conclusion, collaboration between different stakeholders is crucial for ensuring the security of the entire IoT ecosystem in Industry 4.0. It requires a coordinated effort from manufacturers, IoT solution providers, and government agencies to build and maintain secure and resilient IoT systems that can withstand potential threats and attacks [74].

2.7 SECURITY TECHNOLOGIES

As IoT systems become more prevalent in Industry 4.0, their security becomes a critical issue. Because IoT devices are frequently employed in harsh and uncontrolled settings, they are vulnerable to unauthorized access, data breaches, and cyberattacks. A security breach in an IoT device can have serious repercussions ranging from data theft to physical damage or harm. Therefore, to secure IoT systems in Industry 4.0, appropriate security technologies must be implemented. We will look at the key security technologies used to safeguard Industry 4.0 IoT systems, such as encryption, authentication, access control, and blockchain. We will describe how these technologies work, their benefits, and drawbacks, and provide real-world examples of their application [75].

2.7.1 Encryption and Its Importance in IoT Security

Encryption is a basic security technology that entails using mathematical algorithms and keys to convert plain text into a coded message. The encoded data can only be viewed by those who have the decryption key, making it extremely secure. Encryption is critical in the context of IoT for securing the massive amounts of data produced by connected devices and machines. Because of the delicate nature of IoT data and the highly interconnected nature of IoT systems, encryption is a critical safeguard against cyber threats [76].

2.7.2 Authentication and Its Importance in IoT Security [78]

Authentication verifies a user or devices and restricts IoT system and data access. IoT authentication prevents cyberattacks and illegal access to devices and data. Types of authentications used in IoT security:

There are several types of authentication methods used in IoT security, including password-based authentication: This is the most common authentication method and involves users entering a password or passphrase to verify their identity. To verify a user's passcode, a database is used.

- **Certificate-based authentication:** Devices or users are verified by certificates. Certificates from a trusted authority can identify devices or users without passwords.
- **Biometric authentication:** Fingerprints, facial recognition, and vocal recognition are used for biometric authentication. Due to its security and usability, biometric identification is growing in IoT security.
- **Multi-factor authentication:** This secure method uses multiple login methods. An IoT system may demand a password and fingerprint scan.

Examples of Authentication in IoT Security

Authentication is used in various IoT applications to ensure secure access and protect against cyberattacks. Some examples of authentication in IoT security are shown in the following table.

Table 2.4 From Smart Homes to Quantum IoT: Encryption Solutions for Today's Connected Devices [77]

Application	*Type of Encryption*	*Example Algorithm*	*Advantages*	*Disadvantages*	*Real-World Implementations*	*Companies Using this*	*Benefits from Attacks*
Smart Home Security	Symmetric Encryption	Advanced Encryption Standard (AES)	Fast and efficient, widely used and supported	Same key used for encryption and decryption; key distribution is a challenge	Smart door locks and security cameras use AES encryption to protect data transmitted over a network	ADT, Vivint, SimpliSafe	Protects customer data, prevents unauthorized access to security systems
Industrial IoT	Asymmetric Encryption	Rivest–Shamir–Adleman (RSA)	Strong security, effective for digital signatures and key exchange	Slower than symmetric encryption, requires more processing power	Industrial IoT devices, such as sensors and controllers, use RSA encryption for secure firmware updates	Siemens, General Electric, ABB	Prevents tampering with firmware and ensures data integrity, protects critical infrastructure
Healthca re IoT	SSL/TLS Encryption	Transport Layer Security (TLS)	Trusted and widely used, efficient and flexible	Vulnerable to man-in-the-middle attacks, can be difficult to configure	Medical devices and systems, such as patient monitors and electronic health records, use SSL/TLS encryption for secure data transmission	Philips, GE Healthcare, Medtronic	Protects sensitive patient data, prevents unauthorized access to healthcare systems
Automotive IoT	Stream Cipher Encryption	Salsa20 or ChaCha2 0	Fast and efficient, widely supported	Vulnerable to certain types of attacks, key management can be challenging	Connected cars and other automotive IoT devices use stream cipher encryption for secure firmware updates and data transmission	Tesla, General Motors, BMW	Protects customer data, prevents unauthorized access to vehicle systems

(Continued)

Table 2.4 (Continued)

Application	*Type of Encryption*	*Example Algorithm*	*Advantages*	*Disadvantages*	*Real-World Implementations*	*Companies Using this*	*Benefits from Attacks*
Energy IoT	Elliptic Curve Cryptography (ECC)	Elliptic Curve Digital Signature Algorithm (ECDSA)	Strong security, efficient and fast	Requires special hardware for implementation, not as widely supported as other encryption algorithms	Smart meters and other energy IoT devices use ECC encryption for secure data transmission and management	GE, Siemens, ABB	Prevents unauthorized access to the Power grid and ensures data integrity
Financial IoT	Homomorphic Encryption	Paillier Encryption	Allows computation on encrypted data without decrypting it, preserves privacy	Slower than other encryption methods, requires specialized hardware and software	Homomorphic encryption is used in financial IoT applications, such as secure data sharing and processing	Goldman Sachs, JPMorgan Chase, Bank of America	Protects sensitive financial data, prevents unauthorized access to financial systems
Quantum IoT	Quantum Key Distribution (QKD)	BB84 protocol	Provably secure, immune to eavesdropping	Limited distance and transmission rates, expensive and complex hardware	QKD is used in quantum IoT applications, such as secure communication between quantum devices and quantum key generation	Toshiba, ID Quantique, QuintessenceLabs	Protects quantum data, prevents eavesdropping and unauthorized access to quantum systems

Table 2.5 From Smart Homes to Transportation: How Authentication Protects IoT Applications [79]

IoT Application	*Authentication Mechanism*	*Implementation Details*	*Challenges Faced*	*Benefits Gained*	*Successful Implementation Metrics*	*Future Trends*
Smart homes	Facial recognition	Cameras installed at entry points to authenticate users	Ensuring accuracy of facial recognition	Improved home security and access control	Reduced break-in attempts	Integration with voice assistants and other smart home devices
Industrial Control	Certificate-based authentication	Certificates issued to devices to verify identity	Managing certificate revocation lists	Secure access to critical infrastructure	Reduced downtime	Use of blockchain technology for improved security
Healthcare IoT	Biometric authentication	Fingerprint scanning to authenticate healthcare workers	Cost of biometric identification	Secure patient data and access control	Improved patient outcomes	Use of wearable technology for continuous authentication
Smart cities	Certificate-based authentication	Certificates issued to devices to authenticate access	Ensuring proper device identification	Protecting city infrastructure from cyberattacks	Improved traffic management	Integration with public safety systems and emergency response
Automotive IoT	Password and biometric	Users authenticate with a password and facial recognition	Ensuring accuracy of facial recognition	Secure access to vehicle systems and data	Reduced car thefts	Use of blockchain technology for secure vehicle data sharing
Financial IoT	Multi- factor authentication	Users authenticate with a password and fingerprint scan	Ensuring proper management of passwords	Secure access to sensitive financial data	Reduced fraud and data breaches	Use of machine learning for continuous risk assessment
Retail IoT	Certificate-based authentication	Certificates issued to devices to authenticate access	Ensuring proper device identification	Secure access to retail systems and point- of-sale data	Improved inventory management	Use of IoT sensors for personalized customer experiences

(Continued)

Table 2.5 (Continued)

IoT Application	*Authentication Mechanism*	*Implementation Details*	*Challenges Faced*	*Benefits Gained*	*Successful Implementation Metrics*	*Future Trends*
Energy IoT	Multi- factor authentication	Users authenticate with a password and certificate	Ensuring proper certificate management	Protecting critical infrastructure from cyberattacks	Improved energy efficiency	Use of AI for predictive maintenance and anomaly detection
Transportation IoT	Biometric authentication	Passengers authenticate with facial recognition	Ensuring accuracy of facial recognition	Secure access to transportation systems and data	Reduced ticket fraud	Use of blockchain technology for secure and transparent transactions

Oauth and OpenID Connect are authentication protocols that are commonly used in IoT security. Here's a brief explanation of how these protocols work in simple terms [80]:

Oauth is an authentication protocol that allows users to grant third-party applications access to their resources without sharing their login credentials. For example, if you use a mobile app to log in to your Google account, the app uses Oauth to obtain a token from Google, which it can use to access your Google Drive files or other resources, without knowing your Google account password. Oauth works by using a series of redirects and tokens between the user, the third-party application, and the authentication server, to ensure that only authorized access is granted.

OpenID Connect is a more advanced authentication protocol that builds upon Oauth. OpenID Connect allows users to authenticate with an identity provider, such as Google or Facebook, and then share their identity information with third-party applications. This allows users to log in to multiple applications with a single set of credentials, while maintaining control over their identity information. OpenID Connect works by using JSON Web Tokens (JWTs) to exchange identity information between the user, the identity provider, and the application. This ensures that only authorized access is granted and that the user's identity information is protected.

Overall, both Oauth and OpenID Connect provide secure and convenient authentication mechanisms for IoT devices and applications, allowing users to easily grant access to their resources while maintaining control over their identity information.

2.8 EMERGING SECURITY TECHNOLOGIES AND RESEARCH: QUANTUM COMPUTING AND CRYPTOGRAPHIC INNOVATIONS FOR IoT DEVICES

Quantum computing is a new technology that has the potential to transform several disciplines, including cybersecurity. Quantum computers handle complicated tasks much quicker than ordinary computers by exploiting quantum physics concepts. This section will go over the possible influence of quantum computing on encryption as well as the development of novel cryptographic approaches tailored particularly for IoT devices.

Quantum Computing's Impact on Encryption:

Traditional encryption techniques, such as RSA and elliptic curve cryptography (ECC), which are frequently employed to safeguard data flows in IoT devices, are under attack from quantum computers. Once strong enough, quantum computers will be able to crack these encryption algorithms in seconds, rendering them worthless for safeguarding critical information.

To combat this threat, researchers are hard at work on post-quantum cryptography (PQC), a new family of encryption algorithms thought to be immune to quantum computer assaults. PQC techniques, such as lattice-based cryptography, code-based cryptography, and multivariate polynomial cryptography, are being researched and standardized to replace present encryption systems and safeguard IoT devices from prospective quantum attacks.

Cryptographic Innovations for IoT Devices:

IoT devices have unique constraints, such as limited processing power, storage capacity, and energy resources, which make it challenging to implement traditional cryptographic techniques. To address these challenges, researchers are developing new cryptographic methods tailored for IoT devices.

- Lightweight cryptography: These algorithms are designed to minimize computational and memory requirements while still providing adequate security. Examples of lightweight cryptographic algorithms include PRESENT, LED, and SIMON/SPECK, which have been specifically developed to work efficiently on resource-constrained devices.
- Physically Unclonable Functions (PUFs): PUFs are hardware security primitives that generate unique device fingerprints based on the inherent variability of their physical properties. They can be used for secure key storage, device authentication, and other security applications in IoT devices without the need for additional hardware or memory resources.
- Secure device onboarding: Secure device onboarding techniques, such as zero-touch provisioning, enable IoT devices to securely join networks and establish trust relationships without human intervention. These techniques can simplify the deployment of secure IoT devices, particularly in large-scale Industry 4.0 environments.
- Privacy-preserving cryptographic techniques: IoT devices often collect and process sensitive data, making privacy a critical concern. Researchers are developing cryptographic techniques, such as homomorphic encryption and secure multi-party computation, which allow data to be processed without revealing its contents, thus preserving user privacy.

By incorporating emerging security technologies and research, the IoT ecosystem in Industry 4.0 can be better prepared for potential threats posed by quantum computing and other future challenges. These innovative cryptographic techniques will play a crucial role in ensuring the security and privacy of IoT devices while accommodating their unique constraints.

Table 2.6 Exploring the Latest Authentication Techniques for IoT: From Multi-Factor to Brainwave Authentication

Authentication Techniques	*Description*	*Benefits*	*Limitations*	*Potential Use Cases*	*In Use*	*Future Scope*
Multi- factor authentication (MFA)	Requires users to provide two or more forms of authentication, such as a password and a fingerprint scan.	Enhances security, reduces the risk of unauthorized access.	Can be inconvenient for users, requires additional setup and configuration.	Smart homes, healthcare IoT, automotive IoT.	Widely used.	Integration with emerging technologies such as AI and machine learning.
Certificate-based authentication	Involves issuing digital certificates to devices or users, which are then used to authenticate them.	More secure than passwords, provides a scalable and manageable way to authenticate devices and users.	Requires careful certificate management, can be complex to set up.	Industrial control systems, financial IoT, energy IoT.	Widely used.	Integration with emerging quantum technologies.
Biometric authentication	Uses physical or behavioral characteristics, such as fingerprints, facial recognition, or voice recognition, to authenticate users.	Provides a high level of security, can be convenient for users.	Can be costly to deploy, requires specialized hardware and software.	Healthcare IoT, automotive IoT, smart cities.	Increasingly used.	Integration with emerging biometric technologies, such as gait analysis.
Token- based authentication	Involves issuing a token to a user or device, which is then used to authenticate them.	Provides a scalable and secure way to authenticate devices and users, tokens can be easily revoked if compromised.	Requires careful token management, can be complex to set up.	Retail IoT, transportation IoT, smart homes.	Increasingly used.	Integration with emerging blockchain technologies.

(Continued)

Table 2.6 (Continued)

Authentication Techniques	*Description*	*Benefits*	*Limitations*	*Potential Use Cases*	*In Use*	*Future Scope*
Behavioral biometrics	Analyzes a user's behavior patterns, such as typing speed or mouse movements, to authenticate them.	Provides continuous authentication, can enhance security in IoT applications.	Can be difficult to accurately measure and analyze behavior patterns.	Financial IoT, healthcare IoT, smart cities.	Emerging use.	Integration with emerging AI and machine learning technologies.
Contextual authentication	Uses context, such as location or time of day, to determine whether to grant access to a device or system.	Provides an additional layer of security, can be convenient for users.	Can be complex to set up, may require additional hardware or software.	Smart homes, transportation IoT, energy IoT.	Emerging use.	Integration with emerging context- aware technologies.
Quantum authentication	Involves using quantum encryption and communication techniques to authenticate users or devices.	Provides a high level of security, is not susceptible to hacking or eavesdropping.	Requires specialized hardware and software, is still in the early stages of development, still in the early stages of development.	Industrial control systems, financial IoT, energy IoT.	Emerging use.	Integration with emerging quantum technologies.
Brainwave authentication	Analyzes a user's brainwave patterns to authenticate them.	Provides a high level of security, can be convenient for users.	Requires specialized hardware and software, is still in the early stages of development.	Healthcare IoT, transportation IoT, smart homes.	Emerging use.	Integration with emerging brainwave analysis technologies.

2.9 BENEFITS AND LIMITATIONS OF AUTHENTICATION IN IoT SECURITY

Authentication improves security and protects against unauthorized entry and data breaches in IoT security. IoT systems can ensure that only authorized people have access by verifying the identity of users and devices, stopping cyberattacks and data breaches.

Authentication, on the other hand, has some limitations. Password verification is susceptible to brute-force or guessing attacks. Biometric identification is costly to deploy and may not be appropriate for all applications. To avoid unauthorized access, certificate-based authentication necessitates careful certificate management. Authentication is an important aspect of IoT protection. IoT systems can enhance security and protect against cyberattacks and data breaches by utilizing appropriate authentication methods.

2.9.1 Access Control and Its Importance in IoT Security

Access control is a security mechanism that allows only authorized people or devices to access resources or systems. To guarantee that only authorized devices and users can access and interact with IoT networks, access control is critical in IoT security.

Unauthorized entry can lead to data breaches, system failure, and other security concerns. Access control contributes to the security of IoT systems by providing a layered security strategy that limits resource access based on user or device permissions.

A comparison of the pros and cons of encryption, authentication, access control, and blockchain, as well as how to combine them into a complete security framework:

> In Industry 4.0, IoT systems are kept safe with key security technologies like encryption, identification, access control, and the blockchain. Each of these technologies has its own advantages and disadvantages, and which one to use will rely on what the IoT application needs.
>
> Encryption is a simple form of security that is used in many IoT systems. It gives a high level of protection by changing the data into a format that can't be read and can only be read with the right key. Encryption is strong because it can keep data safe even if it is read by people who shouldn't be able to. But encryption can also add extra work to IoT devices, which can be bad in places where resources are limited.
>
> IoT systems also use authentication, which is another important security tool. It makes sure that only people or devices with permission can get into the system or data. Authentication is strong because it adds another layer of security that can help stop cyberattacks and illegal access. Authentication,

on the other hand, can be hard to set up in large-scale IoT systems and may take more resources to keep up.

Access control is a security measure that makes sure only approved users or devices can use certain system resources or functions. It adds another layer of security by limiting access based on rules that have already been set. Access control's strength is that it lets IoT systems be split up and managed to stop people from getting in without permission. Access control can be hard to set up, though, and if it's not set up right, it can leave security holes.

Blockchain is a type of distributed ledger technology that lets you store and handle data in a way that is safe and open to everyone. It is especially useful when data needs to be shared between multiple parties while keeping the data private and safe. The best thing about blockchain is that it can keep a record of events that can't be changed. But blockchain can also make IoT systems more complicated and add extra work, which can be bad in places where resources are limited.

For organizations to use these technologies as part of a complete security framework, they need to carefully look at their security needs and choose the best technologies. For example, a system that needs a high level of data privacy might benefit from using encryption and blockchain, while a system that needs strict control over user access might put identity and access control at the top of its list. It is also important to think about how the chosen security tools can be scaled up, maintained, and used together. Organizations should make sure that the chosen technologies can be quickly added to the infrastructure they already have and that they can be maintained and updated over time to deal with new threats.

2.10 POTENTIAL USE CASES

1. A real-world use case of encryption and blockchain securing IoT systems:

   Imagine a smart city using IoT to improve traffic flow and eliminate congestion. Road, traffic signal, and vehicle sensors gather and analyze data in real time. The smart city uses encryption and blockchain to protect data.

   Encryption secures IoT device data. Vehicle sensor data is secured using a symmetric encryption method like AES, and the encryption keys are securely held on the devices. This prevents unauthorized parties from reading data intercepted during transmission.

   The smart city uses blockchain to protect data. Tamper-proof and transparent distributed ledgers store IoT device data. Each blockchain block contains a cryptographic hash of the preceding block to identify data tampering. The blockchain guarantees data privacy and security by dispersing data control.

   Encryption and blockchain provide a safe and scalable IoT-based traffic control system in the smart city. Encryption protects data secrecy, while blockchain assures data integrity and transparency. This use case shows how several security solutions may be used to secure IoT systems in Industry 4.0.

2. A smart home security system to demonstrate how encryption, authentication, access control, and blockchain may function together.

   Encryption: Smart home security systems may encrypt user passwords, sensor data, and video recordings. For instance, home security cameras can encrypt and store video footage on a cloud server that only authorized individuals with the decryption key can access.

   Authentication: Only authorized users can access the smart home security system. To unlock the front door or disable the alarm, the system might demand a password or biometric verification like face recognition or fingerprint scanning.

   Access control: Access control can restrict smart home security system devices and functionality. The security camera feed and alarm system can be restricted to approved users.

   Blockchain: Blockchain can protect and tamper-proof smart home security system events. A blockchain ledger may record when people enter and depart the property, when the alarm system is armed or disengaged, and when security cameras record footage. This can give a reliable, unalterable record of occurrences.

   The smart home security system protects against cyberattacks and unwanted access by combining various technologies into a complete security architecture. Encryption, authentication, access control, and blockchain can safeguard and transparently record events and protect critical data.

3. Smart manufacturing might utilize large-scale encryption, authentication, access control, and blockchain in IoT systems. Smart manufacturing uses IoT devices and data to optimize production, save costs, and increase quality. The IoT ecosystem in smart manufacturing must be secure as linked devices and data increase. Smart manufacturing systems may safeguard data and restrict access to approved devices and users by using encryption, authentication, access control, and blockchain. In a smart manufacturing facility, IoT devices can be secured to protect important data and functionality. Authentication can restrict system access to authorized users, while access control can limit functionality by user role.

   In smart manufacturing, blockchain technology can store and manage IoT device data securely and transparently. Blockchain technology can provide a tamper-proof distributed ledger of all smart manufacturing ecosystem interactions. This improves data integrity and prevents harmful cyberattacks.

   Encryption, authentication, access control, and blockchain can make smart manufacturing IoT systems safe, transparent, and scalable. Smart manufacturing systems may increase operational efficiency, cost, quality control, and data security by offering a complete security framework.

In summary, encryption, authentication, access control, and blockchain are essential for Industry 4.0 IoT system security. Comprehensive security is essential as IoT devices grow more important in numerous businesses. Organizations must carefully assess their security demands and use complementary security technologies to support the success and resilience of IoT installations in Industry 4.0. Businesses can build a safe, scalable IoT ecosystem that can survive new threats by understanding these technologies' pros and cons and learning from real-world use cases. In Industry 4.0, proactive and adaptive IoT security requires continual development and keeping up with security advances.

2.11 BEST PRACTICES

IoT security is a crucial consideration in Industry 4.0, where an increasing number of interconnected devices generate vast amounts of data. Proper implementation of IoT security best practices is vital to safeguard against cyberattacks and ensure the confidentiality, integrity, and availability of data.

A checklist that serves as quick references for implementing the best practices in IoT security:

1. Secure firmware updates:
   - Authenticate the update source.
   - Verify the update's integrity using digital signatures.
   - Encrypt the update.
   - Check device compatibility and storage space.
2. Data protection:
   - Encrypt sensitive data in transit and at rest.
   - Implement access controls.
   - Regularly audit data access logs.
3. Physical security:
   - Implement locks, alarms, and security cameras.
   - Use tamper-evident seals or sensors.
   - Conduct regular physical security audits.
4. Supply chain security:
   - Verify security controls of suppliers and vendors.
   - Conduct regular security audits and risk assessments.
   - Implement strict security controls for components and software.
5. Continuous monitoring and risk management:
   - Implement security monitoring tools.
   - Conduct regular vulnerability assessments and penetration testing.
   - Create and maintain an incident response plan.
6. Security by design:
   - Integrate security considerations into the entire device lifecycle.
   - Use secure coding practices and perform code reviews.

7. Network segmentation:
   - Isolate devices using network segmentation.
   - Implement access controls and firewalls between segments.
8. Device authentication and authorization:
   - Assign unique, strong credentials for each IoT device.
   - Implement role-based access control (RBAC).
9. Regular software patching and upgrades:
   - Keep device software up to date.
   - Monitor for vulnerabilities and exploits.
10. Employee training and awareness:
    - Provide regular training and awareness programs.
    - Educate employees on IoT security and their role in protecting assets.

Table 2.7 Best Practices in IoT Security

Best Practice	*Steps*	*Purpose/Objective*	*Tools/Technologies*
Secure Firmware Updates	1. Authenticate the update source	Ensure authenticity and integrity of updates	Digital signatures, cryptographic algorithms
	2. Verify the update's integrity using digital signatures		
	3. Encrypt the update	Protect update from unauthorized access or tampering	Encryption algorithms (e.g., AES)
	4. Check device compatibility and storage space	Confirm update compatibility and prevent failures	Device management tools
Data Protection	1. Encrypt sensitive data in transit and at rest	Safeguard sensitive data	Encryption algorithms (e.g., TLS, AES)
	2. Implement access controls	Limit access to authorized personnel only	Access control mechanisms (e.g., RBAC)
	3. Regularly audit data access logs	Detect potential security breaches	Log analysis tools
Physical Security	1. Implement locks, alarms, and security cameras	Protect devices from physical theft or tampering	Physical security devices
	2. Use tamper-evident seals or sensors	Detect physical tampering	Tamper-evident seals, sensors
	3. Conduct regular physical security audits	Identify vulnerabilities	Security audit tools, guidelines

(Continued)

Table 2.7 (Continued)

Best Practice	*Steps*	*Purpose/Objective*	*Tools/Technologies*
Supply Chain Security	1. Verify security controls of suppliers and vendors	Ensure secure components and software	Vendor risk management tools
	2. Conduct regular security audits and risk assessments	Evaluate supply chain security	Security audit tools, risk assessment methodologies
	3. Implement strict security controls for components and software	Minimize supply chain- related risks	Security control frameworks
Continuous Monitoring and	1. Implement security monitoring tools	Detect and respond to security incidents	Security Information and Event Management (SIEM)
Risk Management	2. Conduct regular vulnerability assessments and penetration testing	Identify and address vulnerabilities	Vulnerability scanners, penetration testing tools
	3. Create and maintain an incident response plan	Quickly respond to security incidents	Incident response frameworks, tools
Security by Design	1. Integrate security considerations into the entire device lifecycle	Build security into devices from the start	Secure design principles, guidelines
	2. Use secure coding practices and perform code reviews	Prevent security vulnerabilities in code	Static and dynamic code analysis tools
Network Segmentation	1. Isolate devices using network segmentation	Limit the impact of a security breach	Network segmentation techniques, firewalls
	2. Implement access controls and firewalls between segments	Control communication between segments	Access control mechanisms, firewalls
Device Authentication and Authorization	1. Assign unique, strong credentials for each IoT device	Prevent unauthorized access	Authentication protocols (e.g., Oauth)
	2. Implement role-based access control (RBAC)	Limit access based on roles and responsibilities	RBAC tools and systems
Regular Software Patching and	1. Keep device software up to date	Address vulnerabilities and maintain device security	Maintenance, OTA updates

Best Practice	*Steps*	*Purpose/Objective*	*Tools/Technologies*
	2. Implement automatic or semi-automatic patch management systems	Ensure timely updates and patches	Patch management tools, device management tools
	3. Monitor security bulletins and advisories	Stay informed of emerging threats and vulnerabilities	Security feeds, vulnerability databases
User Training and Awareness	1. Provide regular security training for employees	Reduce human-related security risks	Security training materials, e-learning platforms
	2. Establish clear security policies and procedures	Communicate expectations and best practices	Policy development tools, templates
	3. Conduct regular security awareness campaigns	Reinforce the importance of security in daily operations	Awareness campaign materials, posters, emails

Real-world examples showcasing the implementation of IoT security best practices in various industries:

1. Microsoft:
   In 2019, Microsoft announced a $5 billion investment in IoT research and development over four years, focusing on areas such as security, innovation, and scalability. By July 2020, the Azure Security Center for IoT, a unified security management and advanced threat protection service, had protected over 10 million IoT and edge devices.
2. IBM:
   IBM has a history of investing in IoT security research and development. In 2017, the company launched the Watson IoT Cybersecurity Center of Excellence, a research hub focused on addressing IoT security challenges. IBM X-Force Red, the company's security testing team, also reported a 400% increase in IoT security testing requests in 2018, highlighting the growing demand for secure IoT solutions.
3. Amazon Web Services (AWS):
   AWS IoT Device Defender, a managed service that helps customers secure their IoT devices, monitors billions of IoT events each month. In 2018, AWS introduced IoT security enhancements such as IoT Device Defender Audit, which continuously audits IoT configurations to ensure they adhere to security best practices.
4. Google Cloud:
   In 2018, Google announced the launch of Cloud IoT Core, a fully managed service that allows users to securely connect, manage, and ingest data

from millions of globally dispersed devices. The company also introduced Edge TPU, a purpose-built ASIC chip designed to run TensorFlow Lite machine learning models on edge devices, enhancing security by processing data locally.

5. Mercedes-Benz:

   Mercedes-Benz, a division of Daimler AG, has been working on enhancing the security of its connected vehicles and manufacturing processes. The company has implemented secure firmware updates, data encryption, and access control measures to protect sensitive information. Moreover, Mercedes-Benz is a founding member of the Automotive Information Sharing and Analysis Center (Auto-ISAC), which aims to improve cybersecurity collaboration within the automotive industry.
6. ARM:

   In 2018, ARM unveiled its Platform Security Architecture (PSA), a comprehensive framework designed to simplify the development of secure IoT devices. By providing a set of threat models, security analyses, and hardware and firmware reference designs, PSA aims to raise the security bar across the IoT ecosystem. ARM's PSA Certified program, which assesses IoT devices for security compliance, has reported a growing number of certified devices since its launch in 2019.
7. Rolls-Royce:

   Rolls-Royce has been focusing on IoT security in its operations and products. In 2018, Rolls-Royce partnered with T-Systems, a subsidiary of Deutsche Telekom, to develop a secure Industrial IoT (IioT) platform. The partnership aimed to enhance the security of connected systems in the aerospace and automotive industries by implementing secure data transmission, access control, and data protection measures.
8. Toyota:

   Toyota has been working on improving IoT security in its connected vehicles and manufacturing processes. In 2018, the company announced a partnership with the Automotive Information Sharing and Analysis Center (Auto-ISAC), aiming to enhance cybersecurity knowledge sharing and collaboration within the automotive industry. Toyota has also established an internal cybersecurity team focused on ensuring the security of its IoT-enabled production processes.
9. Cisco:

   Cisco has made significant strides in IoT security through its SecureX platform launched in 2020. The platform integrates key security technologies such as encryption, authentication, and access control to protect connected devices and systems from cyberthreats. As a result, Cisco has reported a decrease in security incidents and improved customer trust in its IoT solutions.

10. Honeywell:

    In 2020, Honeywell launched its Forge Cybersecurity platform, designed to improve the security of its industrial control systems. The platform integrates several IoT security best practices, such as continuous monitoring, access control, and secure firmware updates. As a result, Honeywell has experienced a decrease in security incidents and improved operational efficiency.
11. ABB:

    ABB, a leading industrial technology company, introduced its Ability™ cybersecurity platform in 2019 to enhance the security of its IoT devices and connected systems. By adopting secure design principles, data encryption, and continuous monitoring, ABB has reported fewer cybersecurity incidents and improved customer trust.
12. Bosch:

    Bosch has been proactive in implementing IoT security best practices for its connected devices and systems. In 2021, the company introduced an IoT security testing service to help its customers assess and improve the security of their IoT products. By following industry standards, such as IEC 62443 and ISO/IEC 27001, Bosch has maintained a strong security posture and protected its devices and systems from cyberattacks.
13. Schneider Electric:

    In 2020, Schneider Electric launched its cybersecurity monitoring and advisory services to help industrial customers improve their security posture. By following IoT security best practices such as secure firmware updates, data protection, and supply chain security, the company has successfully mitigated the risk of cyberattacks and ensured the reliability of its industrial control systems.
14. Philips Healthcare:

    In 2017, Philips Healthcare managed to avoid significant disruption during the WannaCry ransomware attack that affected many organizations across the globe. Their proactive security measures, including security updates, strong encryption, and staff training, helped protect their medical devices and systems from the attack. As a result, they maintained their reputation and avoided potential financial and legal consequences.
15. Johnson Controls:

    Johnson Controls reported a 40% reduction in security incidents affecting their building automation systems after adopting their IoT security strategy. The implementation of encryption, authentication, and supply chain security measures has contributed to this success, enabling them to maintain operational efficiency and customer trust.

Case Studies

1. Title: Real-World IoT Security in the Automotive Industry: The Tesla Case Study Executive Summary

This case study examines Tesla's approach to securing its connected cars, focusing on encryption, authentication, and secure firmware updates. The study demonstrates how Tesla has successfully implemented IoT security best practices in the real world, faced real-world security challenges, and highlights the key lessons learned, which can be applied across various industries.

Background

Tesla, a pioneer in electric and connected vehicles, has recognized the importance of IoT security in the automotive industry. As connected cars face challenges like unauthorized access, data breaches, and firmware vulnerabilities, Tesla has proactively addressed these issues to protect its vehicles and customers. The 2016 hack by Tencent Keen Security Lab researchers, who remotely exploited vulnerabilities in a Tesla Model S, serves as a prime example of the real-world challenges that connected car manufacturers must confront. In 2016, a reported 200 million connected cars were on the road, with this number expected to reach 600 million by 2025, highlighting the growing significance of connected car security (IHS Markit, 2020).

Case Evaluation

The case study focuses on the following aspects of Tesla's IoT security approach:

- Real-world security challenges: Tesla's experience with security incidents, such as the 2016 remote hacking demonstration by Tencent's Keen Security Lab, and the 2018 data breach by a former employee.
- Encryption: Tesla's use of state-of-the-art cryptographic algorithms to protect data transmission between the vehicle and external networks.
- Authentication: Tesla's implementation of strong authentication mechanisms to secure remote connections to the vehicle's systems.
- Secure firmware updates: Tesla's development of a secure over-the-air (OTA) update process to maintain firmware integrity.

Proposed Solutions

Based on the evaluation of Tesla's IoT security approach, the following solutions are recommended for other industries:

- Embrace proactive security measures: Establish bug bounty programs, collaborate with security researchers, and continuously monitor for potential threats.

- Regular updates and maintenance: Develop OTA update systems to enable rapid deployment of security patches and software improvements.
- Use security as a competitive advantage: Build trust with consumers by focusing on security and differentiating from competitors in the market.

Conclusion

Tesla's real-world approach to IoT security in the automotive industry showcases the importance of adopting comprehensive security measures and continuously improving these measures to address emerging threats. The key lessons learned from Tesla's experience, including how they addressed real-world security challenges, can be applied across various industries, emphasizing the importance of security in the era of IoT and Industry 4.0.

Implementation

For industries looking to implement the proposed solutions, the following steps are suggested [87–90]:

- Assess current security practices and identify potential areas for improvement.
- Develop a comprehensive security strategy that incorporates encryption, authentication, and secure firmware updates.
- Collaborate with security experts, suppliers, and other stakeholders to establish a unified security framework.
- Regularly evaluate the effectiveness of security measures and update them as necessary to stay ahead of emerging threats.

2. Title: Smart Manufacturing and IoT Security: Securing the Supply Chain with Blockchain Technology—The IBM and Maersk Case Study

Executive Summary

This case study explores the role of IoT in smart manufacturing, focusing on supply chain integrity and the challenges faced by organizations. Using IBM and Maersk's joint solution, Trade Lens, as a real-world example, we examine how the platform employs blockchain technology to secure the supply chain, address security challenges such as counterfeit components and tampering, and improve overall security and competitiveness in Industry 4.0. We will describe the challenges faced by organizations, the steps they took to address these challenges, and the lessons learned from their experiences, supported by statistical data.

Background

According to a 2018 study by the International Chamber of Commerce, the global economic value of counterfeit and pirated goods is estimated to be $1.9 trillion. This highlights the significance of addressing supply chain security challenges. IoT plays a crucial role in smart manufacturing, enabling enhanced automation, real-time monitoring, and data-driven decision-making. With these benefits come unique security challenges in the supply chain, such as counterfeit components, tampering, and unauthorized access to sensitive data. Securing the supply chain is vital for ensuring the integrity of IoT devices and maintaining trust in smart manufacturing systems.

Case Evaluation

The case study focuses on the following aspects of IBM and Maersk's TradeLens platform:

- Overview of TradeLens: The joint blockchain-based solution developed by IBM and Maersk to improve transparency, traceability, and security in the supply chain.
- Addressing counterfeit components: TradeLens's ability to verify the authenticity of components and prevent counterfeits from entering the supply chain.
- Ensuring supply chain integrity: The use of blockchain in TradeLens to prevent tampering, track components through the supply chain, and maintain a secure record of transactions.
- Lessons learned: The implementation of TradeLens has demonstrated the importance of adopting innovative blockchain solutions to address IoT security challenges, the need for collaboration with suppliers and partners, and the value of continuously evaluating and improving supply chain security measures.

Proposed Solutions

Based on IBM and Maersk's experience with TradeLens, the following solutions are recommended for other manufacturing companies looking to secure their supply chains:

- Adopt blockchain technology: Implement blockchain solutions to enhance supply chain security, transparency, and traceability.
- Collaborate with suppliers and partners: Establish a unified security framework across the supply chain by working closely with suppliers and partners.
- Continuously evaluate and improve: Regularly assess supply chain security measures and adapt them to address new challenges and evolving technologies.

Conclusion

The IBM and Maersk's TradeLens platform demonstrates the potential benefits of adopting innovative blockchain solutions to address IoT security challenges in smart manufacturing. By ensuring the integrity of IoT devices and improving overall supply chain security, companies can enhance their competitiveness in Industry 4.0.

Implementation

For manufacturing companies looking to implement the proposed solutions, the following steps are suggested [91–95]:

- Conduct a thorough assessment of the current supply chain to identify potential security vulnerabilities.
- Research and evaluate suitable blockchain solutions, such as TradeLens, for the specific needs of the company and its supply chain.
- Develop a comprehensive implementation plan, including collaboration with suppliers and partners, employee training, and necessary infrastructure upgrades.
- Regularly review and update supply chain security measures to stay ahead of emerging threats and evolving technologies.

3. Title: Energy Sector and IoT Security: Protecting Critical Infrastructure with AI and Machine Learning—The Darktrace Case Study

Executive Summary

This case study explores the importance of IoT in modernizing energy infrastructure and enabling renewable energy integration while addressing potential risks and security challenges, including cyberattacks on critical infrastructure and the manipulation of energy systems. Using Darktrace, a leading cybersecurity company, as a real-world example, we demonstrate how their AI and machine learning-driven solutions have been utilized by energy companies for anomaly detection and threat prevention in IoT systems, improving resilience and security posture against evolving threats.

Background

IoT has become essential in modernizing energy infrastructure and integrating renewable energy sources. It enables real-time monitoring, remote control, and data-driven decision-making, enhancing the efficiency and reliability of energy systems. However, the increasing reliance on IoT exposes the energy sector to cyberattacks and security challenges that can potentially compromise critical infrastructure and disrupt energy systems. According to the World Energy Council, the energy sector is the second most targeted industry for cyberattacks, accounting for 32% of all cyberattacks on critical infrastructure.

Case Evaluation

The case study focuses on the following aspects of Darktrace's approach to IoT security in the energy sector:

- Importance of AI and machine learning: Darktrace's development of AI and machine learning technologies to enhance security and anomaly detection in IoT systems.
- Anomaly detection and threat prevention: Darktrace's AI-driven solutions, such as their Industrial Immune System, which helps energy companies identify and mitigate potential security risks and cyberattacks on their energy infrastructure.
- Improved resilience and security posture: The impact of Darktrace's AI and machine learning-based approach on the ability of energy companies to protect critical infrastructure and energy systems from evolving threats.

Darktrace's approach to IoT security in energy sector:

- Challenges faced by organizations: Energy companies face challenges such as securing their critical infrastructure from cyberattacks, ensuring the integrity of their IoT devices, and protecting sensitive data from unauthorized access.
- Steps taken to address challenges: Darktrace developed AI and machine learning technologies to enhance security and anomaly detection in IoT systems, offering solutions such as their Industrial Immune System, which helps energy companies identify and mitigate potential security risks and cyberattacks on their energy infrastructure.
- Lessons learned from experiences: Organizations can benefit from adopting AI-driven solutions, establishing a robust cybersecurity framework, and fostering a culture of security awareness.

Proposed Solutions

Based on Darktrace's experience, the following solutions are recommended for other energy companies looking to enhance their IoT security:

- Adopt AI and machine learning technologies: Implement AI-driven solutions like Darktrace's Industrial Immune System for anomaly detection and threat prevention in IoT systems.
- Establish a robust cybersecurity framework: Develop a comprehensive security strategy that includes the continuous monitoring and assessment of risks in the energy infrastructure.
- Foster a culture of security awareness: Train employees and stakeholders to recognize potential security threats and promote a proactive approach to cybersecurity.

Conclusion

Darktrace's real-world approach to IoT security in the energy sector demonstrates the potential benefits of adopting AI and machine learning technologies to protect critical infrastructure and energy systems. By effectively detecting anomalies and preventing threats, energy companies can improve their resilience and security posture in the face of evolving cybersecurity challenges.

Implementation

For energy companies looking to implement the proposed solutions, the following steps are suggested [96–99]:

- Assess the current state of IoT security and identify potential areas for improvement.
- Research and evaluate suitable AI and machine learning technologies, such as Darktrace's Industrial Immune System, for anomaly detection and threat prevention.
- Develop a comprehensive implementation plan, including employee training, infrastructure upgrades, and the establishment of a robust cybersecurity framework.
- Regularly review and update security measures to stay ahead of emerging threats and evolving technologies.

4. Title: Healthcare and IoT Security: Safeguarding Medical IoT Devices and Patient Data—The UPMC Case Study

Executive Summary

This case study examines the role of IoT in healthcare, particularly in remote patient monitoring and telemedicine, while addressing the unique security challenges in healthcare IoT, such as patient data privacy and the safety of medical devices. By analyzing the University of Pittsburgh Medical Center's (UPMC) successful implementation of robust access control, data protection, and continuous monitoring to secure their IoT devices and protect patient data, we illustrate the benefits of this approach for both patient safety and regulatory compliance.

Background

IoT has become increasingly important in healthcare, enabling remote patient monitoring, telemedicine, and improved patient outcomes. However, the reliance on connected medical devices and systems exposes the healthcare sector to security challenges, such as patient data privacy and the safety of medical devices. Ensuring the security of IoT devices in healthcare is critical for patient safety and compliance with regulatory requirements.

Case Evaluation

The case study focuses on the following aspects of UPMC's approach to IoT security in healthcare:

- IoT applications at UPMC: An overview of UPMC's innovative use of IoT for remote patient monitoring, telemedicine, and personalized patient care.
- Security challenges in healthcare IoT: A discussion of unique security concerns, such as patient data privacy and medical device safety, faced by UPMC.
- Implementing robust access control, data protection, and continuous monitoring: UPMC's successful strategy to secure IoT devices and protect patient data, including their response to a cyberattack attempt on their medical devices.

Challenges Faced by UPMC

UPMC faced several challenges when implementing IoT security in healthcare, including:

- Ensuring patient data privacy: Maintaining the confidentiality and integrity of sensitive patient data transmitted between IoT devices and medical systems.
- Protecting medical devices: Ensuring the safety and functionality of connected medical devices to provide accurate and reliable patient care.
- Maintaining regulatory compliance: Adhering to healthcare-specific regulations, such as HIPAA and GDPR, to avoid penalties and protect patients' rights.

Steps Taken to Address Challenges

UPMC took several steps to address these challenges, including:

- Implementing robust access control measures, such as multi-factor authentication and strict access policies, to ensure only authorized personnel could access IoT devices and patient data.
- Encrypting sensitive patient data, both at rest and in transit, to prevent unauthorized access and maintain patient privacy.
- Establishing continuous monitoring of IoT devices and systems for signs of potential security threats and promptly responding to any detected anomalies.

Lessons Learned from UPMC's Experience

- Proactive risk assessment: Conducting thorough risk assessments to identify vulnerabilities and potential security threats to IoT devices and patient data is crucial in preparing for potential cyberattacks.
- Collaborative approach: Engaging with device manufacturers, vendors, and other stakeholders to improve IoT device security and adopt best practices is essential for a robust security posture.
- Ongoing training and awareness: Regularly training employees and raising awareness about cybersecurity best practices and potential threats help in fostering a culture of security within the organization.

Conclusion

UPMC's real-world approach to IoT security in healthcare demonstrates the potential benefits of implementing robust access control, data protection, and continuous monitoring. This approach enhances patient safety, protects sensitive patient data, and ensures compliance with regulatory requirements.

Implementation

For healthcare organizations looking to implement the proposed solutions, the following steps are suggested [100–103]:

- Conduct a thorough risk assessment to identify vulnerabilities and potential security threats to IoT devices and patient data.
- Develop a comprehensive implementation plan that covers network architecture, employee training, and the procurement of necessary hardware and software for robust access control and data protection measures.
- Implement incident response and disaster recovery plans to ensure quick and effective response to detected anomalies or cyberattacks.
- Periodically conduct audits, penetration testing, and vulnerability assessments to evaluate the effectiveness of the security measures in place and stay ahead of emerging threats and evolving technologies.

5. Title: Automotive Industry and Secure Vehicle-to-Everything (V2X) Communication: The BMW Case Study

Executive Summary

This case study examines the adoption of IoT and Industry 4.0 principles in the automotive industry by focusing on BMW, a leading automotive manufacturer that has incorporated secure vehicle-to-everything (V2X) communication into the design and production of its vehicles. The case study discusses the security challenges related to V2X communication and presents the security measures implemented by BMW, including strong encryption, secure authentication, and continuous monitoring for anomalous behavior. It also highlights the increased safety, improved traffic flow, and other benefits achieved through secure V2X communication in the automotive industry.

Background

BMW, a leading automotive manufacturer, has embraced IoT and Industry 4.0 principles in its vehicle design and production processes. A key aspect of this transformation is the implementation of secure vehicle-to-everything (V2X) communication, which allows vehicles to communicate with other vehicles, infrastructure, and devices to enhance safety, efficiency, and overall driving experience.

Challenges

V2X communication presents several security challenges, including:

- Protecting against cyberattacks: Ensuring the integrity of communication between vehicles and infrastructure is critical to prevent accidents and disruptions.
- Ensuring data privacy: Safeguarding the privacy of driver and vehicle data from unauthorized access and misuse.
- Maintaining secure communication: Ensuring that communication between vehicles and infrastructure is secure and free from interference.

Lessons Learned

The following lessons can be drawn from BMW's experience with secure V2X communication:

- Prioritize security from the outset: Designing vehicles and infrastructure with security in mind from the beginning is crucial to address potential risks and vulnerabilities.
- Collaborate with stakeholders: Collaboration between manufacturers, suppliers, infrastructure providers, and regulatory authorities is essential to establish industry-wide security standards and best practices.
- Invest in employee training: Educating employees on the importance of IoT security and providing them with the necessary training helps to ensure the successful implementation and maintenance of secure V2X communication.

Solutions [103–105]

BMW has implemented various security measures to address these challenges:

- Strong encryption: BMW uses advanced encryption techniques to secure data transmitted between vehicles and infrastructure, preventing unauthorized access and ensuring data integrity.
- Secure authentication: Vehicles and infrastructure components utilize secure authentication methods to verify the identity of communication partners and prevent unauthorized access.
- Continuous monitoring: BMW employs continuous monitoring systems to detect and respond to anomalous behavior, preventing potential cyberattacks and maintaining secure V2X communication.

Results [106, 107]

Through the implementation of secure V2X communication, BMW has achieved significant benefits, including:

- Increased safety: According to a 2020 study by the National Highway Traffic Safety Administration (NHTSA), V2X communication has the potential to prevent or mitigate up to 80% of non-impaired collisions, significantly enhancing road safety.
- Improved traffic flow: V2X communication enables better traffic management and coordination, reducing congestion and improving overall traffic flow. In a study conducted by the University of Michigan Transportation Research Institute, V2X technology was found to reduce travel times by up to 19%.
- Enhanced driving experience: Secure V2X communication allows for features such as real-time traffic updates, dynamic route guidance, and vehicle diagnostics, resulting in an improved driving experience for BMW customers.

Conclusion

BMW's successful incorporation of secure V2X communication into its vehicle design and production processes demonstrates the potential of IoT and Industry 4.0 principles in the automotive industry. By addressing the security challenges associated with V2X communication, BMW has achieved increased safety, improved traffic flow, and an enhanced driving experience for its customers.

Future Trends in IoT Security: Challenges, Opportunities, and Emerging Technologies

As we venture further into the age of Industry 4.0, it is crucial to recognize and anticipate emerging trends and challenges in IoT security. This section will discuss some key developments that are poised to shape the future landscape of IoT security in Industry 4.0.

AI and machine learning: The use of artificial intelligence (AI) and machine learning (ML) in IoT security is becoming increasingly prevalent. These technologies can help identify patterns, detect anomalies, and predict threats in real time, significantly enhancing the security and resilience of IoT systems. For example, AI-powered intrusion detection systems can quickly identify and respond to potential threats, reducing the risk of data breaches and system failures.

Edge computing: The rise of edge computing has led to a shift in data processing and storage from centralized cloud servers to IoT devices themselves. This decentralized approach can help reduce latency, optimize bandwidth usage, and improve overall system performance. However, it also introduces new security challenges, as edge devices become more vulnerable to attacks. Securing edge devices and ensuring data privacy will be crucial as edge computing continues to gain prominence in Industry 4.0.

Data privacy and regulatory frameworks: As IoT devices continue to proliferate and collect vast amounts of data, the importance of data privacy is increasing. Stricter data protection regulations, such as the European Union's General

Data Protection Regulation (GDPR) and California's Consumer Privacy Act (CCPA), are being implemented worldwide. Adhering to these regulatory frameworks and ensuring data privacy will be crucial for IoT security in Industry 4.0.

Industry 5.0 shift: As Industry 4.0 evolves towards Industry 5.0, the focus will shift from automation and efficiency towards human-centric collaboration and personalization. In this new paradigm, IoT security will need to adapt to ensure the protection of both human and machine interactions.

Technologies like biometric authentication, secure human-machine interfaces, and context-aware access controls will play a crucial role in safeguarding Industry 5.0 ecosystems.

In Industry 5.0, human-machine intelligence will present new security concerns, such as protecting sensitive data from human-machine interactions. IoT systems will also need to adapt to growing customization and personalization, creating new vulnerabilities and attack vectors. To combat these new risks, IoT security must be more flexible and stronger.

Organizations must create thorough Industry 5.0-specific security strategies to meet these issues. This may include adopting security-by-design principles, deploying advanced encryption and authentication mechanisms, and developing explicit policies and processes for managing human-machine interactions. Industry 5.0 security requires continual monitoring and threat intelligence.

Collaboration amongst industry players, such as manufacturers, technology providers, and regulators, will be vital in creating and implementing comprehensive security standards for Industry 5.0. Organizations can handle IoT security and maximize human-centric collaboration and personalization in Industry 5.0 by increasing security awareness and best practices.

Ethical implications: With the increasing complexity and interconnectedness of IoT systems in Industry 4.0, the ethical implications of IoT security become more significant. Ensuring the responsible use of data, safeguarding user privacy, and addressing potential biases in AI and ML algorithms will be essential to maintain public trust and ensure the long-term success of IoT-enabled industrial processes.

Trust and value: As highlighted by McKinsey, trust is a crucial factor in unlocking the value of IoT deployments. Organizations must ensure that their IoT infrastructure is secure and reliable to build trust with customers and stakeholders, which in turn will enable new business models and revenue streams.

IoT-specific security solutions: As the IoT landscape becomes more complex, specialized security solutions tailored to the unique needs of IoT devices will become increasingly important. These solutions should address device authentication, data encryption, and secure communication protocols.

Privacy-aware IoT: Ensuring user privacy is a critical aspect of IoT security. Organizations must adopt privacy-by-design principles and implement privacy-enhancing technologies to protect user data and comply with data protection regulations.

Securing automotive IoT: The rapid growth of connected cars and smart transportation systems is transforming the automotive industry and creating new security challenges. As vehicles become increasingly connected, they become vulnerable to a wide range of threats, such as remote hacking, unauthorized access to sensitive data, and privacy breaches. According to TechTarget, some key security implications for automotive IoT include ensuring secure data storage, secure communications, and protection against unauthorized code execution.

To address these risks, industry stakeholders are developing advanced security measures, such as intrusion detection and prevention systems, secure over-the-air updates, and cryptographic communication protocols. Collaboration between automakers, technology providers, and regulators is essential in developing and implementing comprehensive security standards to protect vehicles from potential threats. As the automotive IoT landscape continues to evolve, it is crucial for all stakeholders to prioritize security and remain vigilant against emerging threats.

The impact of 5G on IoT security: The advent of 5G technology will significantly affect the IoT landscape, offering faster data transmission, lower latency, and higher bandwidth, leading to an acceleration in IoT adoption. However, as noted by TechTarget, the benefits of 5G also come with new security challenges. The massive number of connected devices in a 5G network increases the attack surface and necessitates advanced security measures to protect data and ensure network integrity.

Edge computing and network slicing are two approaches that can help mitigate these risks by providing localized security and isolating critical network functions. Edge computing brings data processing and storage closer to the IoT devices, reducing latency and enhancing security by limiting the exposure of sensitive data. Network slicing enables the creation of virtual networks with dedicated resources and security measures, ensuring the protection of critical network functions and data.

Moreover, as highlighted by Telit, security issues related to 5G IoT include the need for end-to-end encryption, the protection of network elements, and enhanced authentication mechanisms. Fortinet also emphasizes the importance of adopting a security-driven networking approach, integrating security measures into the network infrastructure to address the complex security challenges posed by 5G and IoT.

As the IoT ecosystem evolves with the integration of 5G technology, it is vital for organizations to stay informed about the latest security developments and implement robust security measures to safeguard their IoT infrastructure from potential threats.

Hardware firewalls and cybersecurity solutions for consumer IoT devices: As consumer IoT devices such as smart home appliances and wearables become more prevalent, the need for robust security solutions grows. Hardware firewalls and other cybersecurity solutions can help protect these devices from unauthorized access, malware, and data breaches. Manufacturers must prioritize security in the design process, and consumers should be educated on the importance of updating firmware, using strong passwords, and taking other precautions to safeguard their devices.

Implementing Root of Trust (RoT) and Threat Modelling for IoT Devices: Establishing a Root of Trust (RoT) is essential for ensuring the security and integrity of IoT devices. RoT provides a foundation for secure boot, cryptographic key management, and device authentication. In addition, threat modelling is a crucial step in identifying potential security risks and developing appropriate mitigation strategies. Industry standards and best practices for RoT and threat modelling will be critical in fostering a secure IoT ecosystem.

Innovations in encryption and network visibility for IoT devices: Emerging technologies, such as homomorphic encryption and quantum-resistant algorithms, promise to enhance the security of IoT devices by enabling secure data processing and protecting against future threats. Furthermore, advanced network visibility tools can help organizations monitor and manage the security of their Io networks, detecting potential vulnerabilities and threats in real time. These innovations will play a vital role in securing the IoT landscape as it continues to grow and evolve.

Cybersecurity skills gap: The demand for skilled cybersecurity professionals is predicted to grow, with a significant gap between the number of available positions and the qualified workforce required to fill them. Addressing this skills gap will be crucial to ensure the effective implementation of IoT security measures across industries.

Evolving cyber threats: As technology advances, cybercriminals will also adapt and develop new attack vectors, making it crucial for organizations and governments to stay ahead of emerging threats. Continuous learning, attending conferences, and participating in industry forums can help security practitioners stay ahead of the curve and ensure their IoT systems remain secure in the rapidly evolving landscape of Industry 4.0 and beyond.

Regulatory landscape: Governments worldwide are expected to implement new cybersecurity regulations and standards to protect critical infrastructure and sensitive information. Staying up to date with these evolving regulations, such as the European Union's General Data Protection Regulation (GDPR) and the United States' IoT Cybersecurity Improvement Act, will be vital for businesses to maintain compliance and avoid potential fines or penalties.

Quantum computing: The development of quantum computing technologies presents both opportunities and challenges for cybersecurity. While quantum computers could potentially break current encryption methods, they may also enable the development of new cryptographic algorithms resistant to quantum attacks.

The advancement of post-quantum cryptography is a critical breakthrough impacting the future of IoT security. The capacity of quantum computers to undermine various existing encryption algorithms poses a significant threat to the security and confidentiality of data communicated and stored by IoT devices as their arrival draws near. Post-quantum cryptography is the development of cryptographic algorithms that are resistant to the formidable computing powers of quantum computers, safeguarding the long-term security and privacy of IoT systems.

Several post-quantum cryptographic methods are now being researched and standardized, including lattice-based cryptography, code-based cryptography, and multivariate cryptography. To enable wider implementation, the National Institute of Standards and Technology (NIST) is leading the assessment and standardization of these algorithms. Organizations can address fundamental security aspects such as data protection, device authentication, key management, secure firmware updates, and long-term security by incorporating post-quantum cryptographic algorithms into IoT security solutions, ultimately establishing a solid foundation for the future of IoT.

In conclusion, to stay informed and adapt to these emerging trends, IoT security professionals must engage in continuous learning and research. Leveraging real-world data, attending conferences, and participating in industry forums can help security practitioners stay ahead of the curve and ensure their IoT systems remain secure in the rapidly evolving landscape of Industry 4.0 and beyond. The future of IoT security will be shaped by the challenges and opportunities presented by emerging technologies and trends, such as automotive IoT, 5G networks, consumer device security, IoT implementation, and advances in encryption and network visibility. By staying informed about these developments and implementing best practices, organizations can effectively protect their IoT infrastructure and ensure the responsible use of connected devices in an increasingly interconnected world.

Conclusion

In conclusion, it is critical for researchers and practitioners alike to actively engage in security practices and contribute to the continued growth of this discipline within the context of Industry 4.0. The ever-changing IoT and Industry 4.0 world needs professionals staying informed and current on emerging dangers, best practices, and technology breakthroughs.

Furthermore, as the global industrial landscape evolves toward Industry 5.0, IoT security will become increasingly more critical. Industry 5.0 stresses collaboration between humans and intelligent technologies, emphasizing the need for secure IoT settings to support worker safety, efficiency, and productivity.

As a result, we encourage our valued readers to accept this challenge and take ownership of their learning path. Practitioners may play a critical role in developing a more secure and robust IoT ecosystem inside Industry 4.0 and beyond by utilizing the knowledge and insights obtained from this chapter in real-world scenarios.

To stay on top of IoT security, professionals should attend industry conferences, participate in online forums, and interact with other specialists to share their experiences and skills. Review and update security practices on a regular basis, and keep an eye out for new advances in the industry.

Looking ahead, IoT security will continue to improve, fueled by advancements in technologies such as AI, machine learning, and edge computing, as well as the shift to Industry 5.0. Researchers and practitioners may help to create a safer, more resilient, and efficient industrial landscape for future generations by staying informed and actively engaged in this field. The scientific community's combined cooperation is essential for the continuous advancement and success of IoT security in Industry 4.0 and the next era of Industry 5.0.

REFERENCES

[1] Dorri, A., Kanhere, S. S., Jurdak, R., & Gauravaram, P. (2017). Blockchain for IoT security and privacy: The case study of a smart home. In 2017 IEEE International Conference on PervasiveComputing and Communications Workshops (PerCom Workshops) (pp. 618–623). IEEE.

[2] Atzori, L., Iera, A., & Morabito, G. (2010). The Internet of Things: A survey. Computer Networks, 54(15), 2787–2805. doi: 10.1016/j.comnet.2010.05.010

[3] Kang, H., Lee, J. H., Kim, Y., Lee, J., & Kim, J. H. (2018). Security architecture for industrial Internet of Things. Future Generation Computer Systems, 88, 388–399.

[4] Li, Z., Han, X., Wang, Q., Zhang, T., & Xia, F. (2019). Development of threat modeling for industrial Internet of Things. Journal of Ambient Intelligence and Humanized Computing, 10(9), 3435–3447.

[5] Miorandi, D., Sicari, S., De Pellegrini, F., & Chlamtac, I. (2012). Internet of Things: Vision, applications and research challenges. Ad Hoc Networks, 10(7), 1497–1516.

[6] Kang, H., Kim, H., & Kim, H. (2020). Threat modeling for Internet of Things (IoT) security. Journal of Information Processing Systems, 16(5), 1155–1169.

[7] Li, M., Liu, P., Xiao, Y., & Li, X. (2021). A threat modeling approach for securing IoT systems. Journal of Parallel and Distributed Computing, 149, 3–14.

[8] Panda, P., Mohapatra, S. K., & Mohanty, S. P. (2020). Threat modeling and vulnerability assessment in Internet of Things (IoT) based industrial control systems (ICS). Journal of Ambient Intelligence and Humanized Computing, 11(3), 1403–1418.

[9] Patel, K., & Patel, K. (2017). A comprehensive survey of IoT threats, vulnerabilities and security solutions. Journal of Network and Computer Applications, 84, 1–25. doi: 10.1016/j.jnca.2016.11.008

[10] Siddique, S. B., Hussain, S., & Khan, M. A. (2019). Security threats and vulnerabilities in IoT-based networks: A review. Journal of Ambient Intelligence and Humanized Computing, 10(2), 569–585. doi: 10.1007/s12652-018-1009-6

[11] Abbas, R., Saleem, Z., & Al-Fuqaha, A. S. (1945). Cybersecurity for industrial Internet of Things: Recent advances and future challenges. IEEE Internet of Things Journal, 5(3), 1930–1945. doi: 10.1109/JIOT.2018.2822005

[12] Pereira, C. N. B., Gomes, A. G., & de Oliveira, A. L. A. (2018). Industrial Internet of Things: A review of the literature. IEEE Transactions on Engineering Management, 65(3), 373–384. doi: 10.1109/TEM.2018.2807520

[13] Liu, J., Luo, W., Zhang, Z., & Li, M. (2020). Threat modeling for industrial Internet of Things: A systematic review. Journal of Ambient Intelligence and Humanized Computing, 11(1), 13–27. doi: 10.1007/s12652-018-1009-6

[14] Dey, K., Sarkar, P., & Biswas, S. (2019). Security and privacy issues in the industrial Internet of Things: A review. IEEE Access, 7, 147634–147646. doi: 10.1109/ACCESS.2019.2949034

[15] Bojinov, H., Buchegger, S., & Ginzboorg, P. (2018). IoT attack trees: Mapping and testing attacks on smart home ecosystems. IEEE Pervasive Computing, 17(1), 68–76.

[16] Bhatt, C., Joshi, A., & Chaudhari, N. S. (2020). Machine learning-based security analysis for smart homes. IEEE Internet of Things Journal, 7(4), 3113–3124.

[17] Elkhodr, M., Shahrestani, S., Cheung, W. K., & Hassanien, A. E. (2016). A framework for reliable and secure IoT-based smart city applications. IEEE Access, 4, 217–227.

[18] Howard, M., & Longstaff, T. A. (2010). Take only what you need: Leveraging the principle of least privilege to manage data exposure. IEEE Security & Privacy, 8(3), 44–51.

[19] Wang, Y., Wang, J., Li, Q., Li, X., & Li, G. (2021). Secure data storage and processing in the IoT-based cloud computing environment. IEEE Transactions on Industrial Informatics, 17(6), 4296–4307.

[20] Garcia-Morchon, O., Kumar, S., & Sethi, M. (2017). Security architecture for the Internet of Things: A survey. IEEE Communications Surveys & Tutorials, 20(1), 559–590. doi: 10.1109/comst.2017.2771523

[21] Shafagh, H., Hithnawi, A., & Hummen, R. (2018). Secure communication for the Internet of Things–a survey. IEEE Internet of Things Journal, 5(1), 1–28. doi: 10.1109/jiot.2017.2758384

[22] Kim, H., Shin, H., & Kim, Y. (2019). Edge computing for secure IoT-based smart city: A survey. IEEE Communications Magazine, 57(5), 34–39. doi: 10.1109/mcom.2019.1800257

[23] Kshetri, N. (2018). Blockchain's roles in meeting key supply chain management objectives. International Journal of Information Management, 39, 80–89. doi: 10.1016/j.ijinfomgt.2017.12.007

[24] Deng, Y., Yao, L., & Duan, Y. (2019). A security architecture for the Internet of Things based on fog computing. Security and Communication Networks, 2019, 1–12. doi: 10.1155/2019/4698609

[25] Zhu, L., Wang, Y., & Liu, D. (2020). Security and privacy protection for edge computing enabled IoT systems: A survey. Journal of Network and Computer Applications, 156, 102547. doi: 10.1016/j.jnca.2020.102

[26] NIST. "Securing the IoT Ecosystem." www.nist.gov/system/files/documents/2018/04/27/securing_the_iot_ecosystem_final.pdf

[27] OWASP. "IoT Security Best Practices." https://owasp.org/www-pdf-archive/OWASP_Internet_of_Things_Project_Top_10–2018.pdf

[28] McKinsey & Company. "Security Architecture for the Internet of Things." www.mckinsey.com/business-functions/digital-mckinsey/our-insights/security-architecture-for-the-internet-of-things

[29] "IoT Security Foundation." https://iotsecurityfoundation.org/what-we-do/best-practice-guidelines/

[30] Intel Secure Device Onboard (SDO). www.intel.com/content/www/us/en/internet-of-things/secure-device-onboard.html
[31] VeChain. www.vechain.com/
[32] Darktrace. www.darktrace.com/
[33] Nokia Secure Service Enablement. www.nokia.com/networks/portfolio/secure-service-enablement/
[34] Google Cloud Identity-Aware Proxy (IAP). https://cloud.google.com/iap/
[35] Trusted Computing Group (TCG) Trusted Platform Module (TPM). https://trustedcomputinggroup.org/
[36] AWS Identity and Access Management (IAM). https://aws.amazon.com/iam/
[37] Microsoft Azure IoT Hub Device Management. https://azure.microsoft.com/en-us/services/iot-hub/device-management/
[38] Forescout CounterACT. www.forescout.com/platform/counteract/
[39] Symantec Endpoint Protection. www.symantec.com/products/endpoint-protection
[40] Samsung Knox. www.samsungknox.com/
[41] IBM Watson. www.ibm.com/watson
[42] Cisco IoT. www.cisco.com/c/en/us/products/internet-of-things/iot-security/index.html
[43] Google Cloud Security Command Center. https://cloud.google.com/security-command-center
[44] Azure IoT Hub Device Management. https://azure.microsoft.com/en-us/services/iot-hub/device-management/
[45] Apple Secure Enclave. https://developer.apple.com/documentation/security/cryptographic_services/secure_enclave
[46] Palo Alto Networks Cortex XDR. www.paloaltonetworks.com/cortex/cortex-xdr
[47] CrowdStrike Falcon Intelligence. www.crowdstrike.com/solutions/falcon-intelligence/
[48] Al-Fuqaha et al. (2015). "A Survey of Internet of Things Security Challenges and Solutions."
[49] Ray et al. (2018). "Internet of Things Security: A Review."
[50] Ahmed et al. (2020). "IoT Security: Review, Blockchain Solutions, and Open Challenges."
[51] Zeadally et al. (2019). "Blockchain-Based Security Solutions for the Internet of Things: A Review."
[52] Al-Fuqaha, A., Guizani, M., Mohammadi, M., Aledhari, M., & Ayyash, M. (2015). Internet of Things: A survey on enabling technologies, protocols, and applications. IEEE Communications Surveys & Tutorials, 17(4), 2347–2376.
[53] Ray, P. P. (2018). A review on Internet of Things (IoT) for defense and public safety. Journal of Network and Computer Applications, 107, 1–13.
[54] Microsoft. "Symmetric Key Cryptography." https://docs.microsoft.com/en-us/windows/win32/seccrypto/symmetric-key-cryptography
[55] Study.com. "Symmetric Encryption: Definition & Examples." https://study.com/academy/lesson/symmetric-encryption-definition-examples.html
[56] Asymmetric Encryption: "Public Key Cryptography." RSA. www.rsa.com/en-us/products/public-key-cryptography
[57] "Asymmetric Encryption: Definition & Examples." Study.com. https://study.com/academy/lesson/asymmetric-encryption-definition-examples.html

[58] Hashing: "Hash Functions." NIST. https://nvlpubs.nist.gov/nistpubs/FIPS/NIST.FIPS.180-4.pdf

[59] "Hashing: Definition & Example." Investopedia. www.investopedia.com/terms/h/hashing.as

[60] "Homomorphic Encryption." IBM. www.ibm.com/cloud/learn/homomorphic-encryption

[61] "Homomorphic Encryption: Definition & Examples." Study.com. https://study.com/academy/lesson/homomorphic-encryption-definition-examples.html

[62] Quantum Key Distribution (QKD): "Quantum Cryptography." IBM. www.ibm.com/cloud/learn/quantum-cryptography

[63] "Quantum key distribution: Definition & example." Investopedia. www.investopedia.com/terms/q/quantum-key-distribution.asp

[64] Tsai, C. W., & Lai, C. F. (2019). Authentication mechanisms for Internet of Things: A survey. IEEE Communications Surveys & Tutorials, 21(1), 660–674.

[65] Jøsang, A., & Pope, S. (2005). User authentication for IoT devices. Proceedings of the 3rd International Conference on Mobile Computing and Ubiquitous Networking, 2005, 366–375.

[66] Manickam, S., & Hsu, C. H. (2019). Authentication and access control for the Internet of Things: A survey. IEEE Internet of Things Journal, 6(4), 6444–6464.

[67] Fan, K., & Wang, S. (2017). A survey on IoT security: Application areas, security threats, and solution architectures. IEEE Access, 5, 3543–3552.

[68] Hong, S. S., & Park, J. H. (2020). A survey of authentication methods for the Internet of Things. Electronics, 9(2), 247.

[69] Access cntrl. "Smart padlock flaw exposes millions of homes to hackers." BBC News. www.bbc.com/news/technology-51148622

[70] "Fitness tracking company exposed personal data of users." BBC News. www.bbc.com/news/technology-57120714

[71] "Researchers find flaw in smartwatch authentication process." ZDNet. www.zdnet.com/article/researchers-find-flaw-in-smartwatch-authentication-process/

[72] "US car manufacturer reports security breach, customer data exposed." ZDNet. www.zdnet.com/article/us-car-manufacturer-reports-security-breach-customer-data-exposed/

[73] "Security researcher finds vulnerability in US hospital access control system." Threatpost. https://threatpost.com/us-hospital-access-control-vulnerability/156326/

[74] "Ransomware attack on US IT software provider leads to shutdown of businesses worldwide." CNN Business. www.cnn.com/2021/07/02/tech/kaseya-ransomware-attack-explained/index.html

[75] "An analysis of blockchain security in IoT." Sensors, 2020. www.mdpi.com/1424-8220/20/18/5154

[76] "IoT security: A comprehensive survey." IEEE Communications Surveys & Tutorials, 2019. https://ieeexplore.ieee.org/abstract/document/8677113

[77] "IoT security: Review, blockchain solutions, and open challenges." Future Internet, 2020. www.mdpi.com/1999-5903/12/6/103

[78] "Authentication and authorization for constrained IoT devices: A survey." IEEE Communications Surveys & Tutorials, 2018. https://ieeexplore.ieee.org/abstract/document/8362587

[79] "Access control in IoT systems: A survey." ACM Computing Surveys, 2018. https://dl.acm.org/doi/abs/10.1145/3178876

[80] "Encryption for IoT security: A survey." IEEE Internet of Things Journal, 2019. https://ieeexplore.ieee.org/abstract/document/8663854
[81] Gartner. "Magic Quadrant for IoT Security, 2022."
[82] Market research future. "IoT security market research report—Global forecast till 20250."
[83] The 5 worst examples of IoT hacking and vulnerabilities in history. iotforall.com
[84] The ongoing rise in IoT attacks: What we're seeing in 2023. Security Boulevard.
[85] Symantec. (2014, July 29). Dragonfly: Western energy companies under sabotagethreat.
[86] New York Times. (2019, March 1). Honeywell says it's latest victim of Chinese cybertheft.
[87] Koscher, K., Czeskis, A., Roesner, F., Patel, S., Kohno, T., Checkoway, S.,... Savage, S. (2010). Experimental security analysis of a modern automobile. In 2010 IEEE Symposium on Security and Privacy (pp. 447–462). IEEE. https://ieeexplore.ieee.org/document/5504804.
[88] Tesla, Inc. (2021). Vehicle security. www.tesla.com/about/security
[89] Zetter, K. (2016). Hackers remotely kill a jeep on the highway—With me in it. www.wired.com/2015/07/hackers-remotely-kill-jeep-highway/. www.blackhatethicalhacking.com/articles/free-access/iot-hacking/
[90] Greenberg, A. (2018). Tesla model 3 can be hacked and stolen in seconds using this key fob attack. https://electrek.co/2022/09/13/tesla-vehicles-stolen-relay-attack-caveat/. www.autoblog.com/2023/03/29/tesla-model-3-hacked/
[91] IBM. (2021). TradeLens: A blockchain-enabled, global trade digitization solution. www.ibm.com/blockchain/solutions/tradelens
[92] Kayıkcı, Y., Chaudhuri, A., & Subramanian, N. (2020). Blockchain and Supply Chain Logistics: Evolutionary Case Studies. Springer International Publishing.
[93] Darktrace. (2021). The Industrial Immune System: Real-Time Threat Detection for Industrial Control Systems. https://darktrace.com/news/darktrace-launches-industrial-immune-system-for-critical-infrastructure
[94] Smith, D. C. (2021). Cybersecurity in the energy sector: Are we really prepared? Journal of Energy & Natural Resources Law, 39(3), 265–270. doi: 10.1080/02646811.2021.1943935
[95] www.tandfonline.com/action/showCitFormats?doi=10.1080%2F02646811.2021.1943 935
[96] Giotis, G., Giannoutsou, N., & Giannopoulos, G. (2017). A survey of IoT key security issues. In 2017 6th International Conference on Modern Circuits and Systems Technologies (MOCAST).
[97] Pantelopoulos, A., & Bourbakis, N. G. (2010). A survey on wearable sensor-based systems for health monitoring and prognosis. IEEE Transactions on Systems, Man, and Cybernetics, Part C (Applications and Reviews), 40(1), 1–12.
[98] UPMC (University of Pittsburgh Medical Center). (2021). About UPMC. www.upmc.com/about
[99] National Highway Traffic Safety Administration (NHTSA). (2020). Vehicle-to-everything (V2X) communications. www.nhtsa.gov/technology-innovation/vehicle-vehicle-communication
[100] Campolo, C., Molinaro, A., & Paratore, S. Y. (2017). 5G network slicing for vehicle-to-everything services. IEEE Wireless Communications, 24(6), 38–45. https://ieeexplore.ieee.org/document/8459911

[101] University of Michigan Transportation Research Institute. (2018). The impact of V2X communication on traffic flow efficiency. https://deepblue.lib.umich.edu/bitstream/handle/2027.42/145688/104238.pdf

[102] BMW Group. (2021). BMW ConnectedDrive: Digital services. www.bmw.com/en/innovation/connected-car.html

[103] "Enhanced intelligent driver model to access the impact of driving strategies on traffic capacity." www.researchgate.net/publication/46158245_Enhanced_Intelligent_Driver_Model_to_Access_the_Impact_of_Driving_Strategies_on_Traffic_Capacity

[104] Kruse, C. S., Frederick, B., Jacobson, T., & Monticone, D. K. (2017). Cybersecurity in healthcare: A systematic review of modern threats and trends. Technology and Health Care, 25(1), 1–10.

[105] Chinthapalli, K. (2020). The cybersecurity of medical devices. The Lancet Digital Health, 2(9), e431–e432.

[106] Dargahi, T., Dehghantanha, A., Conti, M., & Bianchi, S. (2020). A review on security of IoT in healthcare: A systematic literature review. Computers & Security, 97.

[107] Kshetri, N. (2018). Blockchain's roles in meeting key supply chain management objectives. International Journal of Information Management, 39, 80–89, www.sciencedirect.com/science/article/abs/pii/S0268401217305248

Chapter 3

Smart Home Environment for Society

An Overview of IoT-Enabled Air Conditioners

Shalom Akhai

3.1 INTRODUCTION

Smart homes are revolutionizing the way we live, allowing us to create more integrated, automated, and convenient living environments [1–5]. The air conditioning system is a critical component of a smart home, and it plays a vital role in ensuring a comfortable indoor climate [6–8]. Studies have shown that human comfort and productivity are heavily influenced by the indoor environment and thermal comfort conditions, which are typically managed by air conditioning systems in modern buildings. As a result, almost all types of buildings, including residential, offices, hospitals, and hotels, now have air conditioning systems installed to provide comfort to people [9–14].

Figure 3.1 illustrates an IoT-based air conditioning remote control, which is just one example of the various types of IoT-enabled air conditioning systems currently available on the market. These systems use sensors and machine learning algorithms to adjust temperature and humidity levels, and optimize energy efficiency. One of the primary advantages of IoT-enabled air conditioning systems is the real-time monitoring and control they provide, allowing users to adjust temperature settings and monitor energy usage remotely. However, the use of sensitive data transmission also poses risks, such as data breaches and hacking, which manufacturers need to address through robust security measures like encryption and authentication protocols to protect user data. So to mitigate these risks, it is important for manufacturers to implement robust security measures, such as encryption and authentication protocols, to protect user data [15–21].

Despite these challenges, the demand for IoT-enabled air conditioning systems is expected to continue to grow in the coming years. In addition to providing real-time monitoring and control, the integration of artificial intelligence and big data analytics is also expected to play an increasingly important role in the development of smart air conditioning systems. For example, machine learning algorithms can be used to analyze energy usage patterns and adjust the air conditioning system to optimize energy efficiency, while big data analytics can provide valuable

DOI: 10.1201/9781003480181-3

Figure 3.1 IoT-based remote control.

insights into usage patterns and help users make informed decisions about how to conserve energy and save money [22–26].

Thus, IoT-enabled air conditioning systems are an exciting development in the field of smart homes, providing real-time monitoring, control, and optimization of air conditioning systems. While there are challenges associated with data privacy and security, the benefits of these systems are significant, including improved energy efficiency, cost savings, and increased comfort and convenience for homeowners. As technology continues to evolve, we can expect to see even more advanced and sophisticated IoT-enabled air conditioning systems that will continue to revolutionize the way we live in our homes.

3.2 LITERATURE REVIEW OF IOT-ENABLED AIR CONDITIONING SYSTEMS

There are several kinds of air conditioning systems available in the market and available in literature that are enabled by IoT technology and offer distinct features and benefits. Here is an overview of the various types of IoT-enabled air conditioning systems. These systems have different types, each with their exclusive advantages and features. The three most popular IoT-enabled air conditioning systems available on the market with unique features and benefits, as mentioned in Table 3.1:

Table 3.1 IoT-Enabled Air Conditioning Systems Comparison

Type of IoT-Enabled Air Conditioning System	*Description*	*Features*	*Control*	*References*
Smart Thermostats	Standalone devices installed on existing HVAC systems.	Touchscreen interface. Remote control via mobile app. Sensors to detect temperature and humidity changes. Adjusts system for comfortable indoor climate.	Remote control via mobile app.	[27–29]
Integrated Air Conditioning Systems	Built into home's infrastructure, works with other smart home devices.	Sensors for temperature, humidity, and air quality changes. Machine learning algorithms for energy optimization. Seamless integration with smart home devices.	Integrated with smart home devices.	[30–32]
Portable Air Conditioners	Standalone units that can be moved from room to room.	Remote control. Mobile app control. Sensors for temperature and humidity changes. Adjusts system for comfortable indoor climate.	Remote control via mobile app.	[33–36]

- **Smart Thermostats**—Smart thermostats are standalone devices that can be easily installed on an existing HVAC system. They typically feature a touchscreen interface and can be controlled remotely via a mobile app. Smart thermostats use sensors to detect changes in temperature and humidity levels, adjusting the air conditioning system accordingly to maintain a comfortable indoor climate. Smart thermostats typically feature a touchscreen interface that allows users to easily adjust temperature settings and view energy usage data. They can also be controlled remotely via a mobile app, allowing users to monitor and control their air conditioning system from anywhere. Smart thermostats can help homeowners save money on energy bills by ensuring that their air conditioning system is operating at peak efficiency [27–29].
- **Integrated Air Conditioning Systems**—Integrated air conditioning systems are built into the home's infrastructure and are designed to work seamlessly with other smart home devices. These systems typically include sensors that can detect changes in temperature, humidity, and air quality,

as well as machine learning algorithms that can analyze usage patterns and adjust the air conditioning system to optimize energy efficiency. Integrated air conditioning systems are designed to work seamlessly with other smart home devices, allowing users to create a fully integrated and automated living environment. These systems typically feature sensors that can detect changes in temperature, humidity, and air quality, as well as machine learning algorithms that can analyze usage patterns and adjust the air conditioning system to optimize energy efficiency. Integrated air conditioning systems can help homeowners save money on energy bills by providing valuable insights into usage patterns and optimizing energy usage accordingly [30–32].

- **Portable Air Conditioners**—Portable air conditioners are standalone units that can be easily moved from room to room. Portable air conditioners use sensors to detect changes in temperature and humidity levels, adjusting the air conditioning system accordingly to maintain a comfortable indoor climate. Portable air conditioners are ideal for homeowners who want the flexibility of being able to move their air conditioning system from room to room. These units typically feature a remote control and can be controlled remotely via a mobile app, allowing users to monitor and control their air conditioning system from anywhere. Portable air conditioners can help homeowners save money on energy bills by ensuring that they only cool the rooms that are being used, rather than cooling the entire home unnecessarily [33–36].

Thus, IoT-enabled air conditioning systems come in different types, each with its own unique features and benefits. Smart thermostats are standalone devices that can be easily installed on an existing HVAC system, while integrated air conditioning systems are built into the home's infrastructure and are designed to work seamlessly with other smart home devices. Portable air conditioners are standalone units that can be easily moved from room to room. By providing real-time monitoring, control, and optimization of air conditioning systems, IoT-enabled air conditioning systems can help homeowners save money on energy bills, while also providing increased comfort and convenience.

3.3 BENEFITS OF IoT-ENABLED SOLUTIONS

IoT-enabled solutions for air conditioning systems in smart homes offer several benefits that enhance energy efficiency, real-time monitoring and control, and overall comfort and convenience.

3.3.1 Increased Energy Efficiency

One of the most significant benefits of IoT-enabled air conditioning systems is increased energy efficiency. By providing real-time monitoring and analysis of usage patterns, these systems can help homeowners optimize their energy usage and reduce energy waste. For example, sensors in the air conditioning system can detect when no one is in a room and adjust the temperature accordingly, ensuring that energy is not wasted cooling an empty room. Additionally, machine learning algorithms can analyze usage patterns and make adjustments to the air conditioning system to optimize energy efficiency [37–40].

3.3.2 Real-Time Monitoring and Control

IoT-enabled air conditioning systems provide real-time monitoring and control capabilities that give homeowners greater control over their indoor climate. Through the use of sensors and mobile apps, homeowners can monitor the temperature, humidity, and air quality of their home from anywhere, and make adjustments as needed. This level of control ensures that the indoor climate is always at the desired level of comfort, regardless of whether or not someone is home [41–45].

3.3.3 Improved Comfort and Convenience

IoT-enabled air conditioning systems provide improved comfort and convenience in several ways. For example, homeowners can set schedules for their air conditioning system to ensure that their home is at the desired temperature when they arrive home from work. They can also adjust the temperature of individual rooms, providing greater flexibility in controlling the indoor climate. Additionally, these systems can provide valuable insights into usage patterns and provide alerts when maintenance is required, ensuring that the air conditioning system is always operating at peak efficiency [46–48].

Thus, it is clear that IoT-enabled solutions for air conditioning systems in smart homes provide several benefits, including increased energy efficiency, real-time monitoring and control, and improved comfort and convenience. By providing real-time monitoring, control, and optimization of air conditioning systems, these solutions can help homeowners save money on energy bills, while also providing increased comfort and convenience. These benefits make IoT-enabled air conditioning systems an ideal choice for homeowners who want to create a comfortable and energy-efficient living environment.

3.4 CHALLENGES OF IoT-ENABLED SOLUTIONS

IoT-enabled solutions for air conditioning systems in smart homes offer numerous benefits, but there are also challenges that must be addressed. These challenges include security and privacy concerns, integration with existing systems, and cost and technical challenges.

3.4.1 Security and Privacy Concerns

One of the most significant challenges of IoT-enabled air conditioning systems is security and privacy. These systems collect sensitive data such as temperature preferences and usage patterns, which must be protected from unauthorized access. If a hacker gains access to this data, they can use it to gain access to other parts of the smart home system or steal personal information. To address these concerns, manufacturers of IoT-enabled air conditioning systems must implement robust security protocols to protect user data. This includes the use of encryption and authentication methods to ensure that only authorized users can access the data. Additionally, manufacturers must provide regular security updates to address any vulnerabilities that are discovered [49–52].

3.4.2 Integration with Existing Systems

Another challenge of IoT-enabled air conditioning systems is integration with existing systems. Many homes have existing air conditioning systems that are not compatible with IoT-enabled systems, which can make installation and integration more challenging.

To address this challenge, manufacturers must design their systems to be compatible with a wide range of existing air conditioning systems. They must also provide clear instructions and support to help users with the installation and integration process [53–55].

3.4.3 Cost and Technical Challenges

IoT-enabled air conditioning systems can be more expensive than traditional air conditioning systems, which can be a barrier to adoption. Additionally, these systems require technical expertise to install and maintain, which can be a challenge for some homeowners.

To address these challenges, manufacturers must provide affordable options for IoT-enabled air conditioning systems, while also providing clear and easy-to-follow instructions for installation and maintenance. They must also provide ongoing support to help users with any technical issues that arise [56–64].

Thus IoT-enabled solutions for air conditioning systems in smart homes offer numerous benefits, but there are also challenges that must be addressed. These challenges include security and privacy concerns, integration with existing systems, and cost and technical challenges. Manufacturers of IoT-enabled air conditioning systems must work to address these challenges to ensure that their systems are accessible, affordable, and secure. By doing so, they can help homeowners create a comfortable and energy-efficient living environment, while also providing peace of mind when it comes to security and privacy concerns

3.5 EMERGING INNOVATIONS

The integration of air conditioning systems with smart home environments has already made significant progress with IoT-enabled solutions as discussed. However, there is still room for improvement, and the latest trends and future directions in this area include the integration of artificial intelligence, the use of big data, and the development of energy-efficient solutions.

3.5.1 Integration of Artificial Intelligence

The integration of artificial intelligence (AI) into air conditioning systems can offer several benefits, such as improved energy efficiency, increased comfort, and personalized temperature control. AI algorithms can learn from user preferences and usage patterns to optimize temperature settings and reduce energy consumption.

One of the most promising applications of AI in air conditioning systems is predictive maintenance. By analyzing data from sensors and other sources, AI algorithms can predict when maintenance is needed and alert homeowners to potential issues before they become a problem.

3.5.2 Use of Big Data

The use of big data in air conditioning systems can help manufacturers and homeowners to identify patterns and optimize performance. By collecting data from a wide range of sources, such as weather forecasts, occupancy sensors, and energy usage data, manufacturers can develop more efficient and effective air conditioning systems.

For homeowners, the use of big data can help them to optimize their air conditioning usage patterns and reduce energy consumption. For example, by analyzing data from occupancy sensors, homeowners can determine when rooms are most frequently used and adjust their air conditioning settings accordingly.

3.5.3 Development of Energy-Efficient Solutions

Energy efficiency is one of the most critical factors in air conditioning systems, and the development of energy-efficient solutions is a top priority for manufacturers. One of the most promising developments in this area is the use of renewable energy sources, such as solar power, to power air conditioning systems.

Additionally, manufacturers are developing more efficient components, such as compressors and fans, to reduce energy consumption. They are also developing more advanced controls, such as variable refrigerant flow (VRF) systems, to optimize temperature settings and reduce energy usage.

3.5.4 IoT-Enabled AC Solutions in Smart Homes

In conclusion, the integration of air conditioning systems with smart home environments through IoT-enabled solutions has brought numerous benefits, such as increased energy efficiency, real-time monitoring and control, and improved comfort and convenience. However, there are also several challenges that need to be addressed, such as security and privacy concerns, integration with existing systems, and cost and technical challenges. To overcome these challenges, manufacturers and developers are exploring the latest trends and future directions in this area, such as the integration of artificial intelligence, the use of big data, and the development of energy-efficient solutions. These developments have the potential to make air conditioning systems more efficient, effective, and personalized, and can have a significant impact on reducing our environmental footprint.

Thus, the latest trends and future directions in IoT-enabled air conditioning systems include the integration of artificial intelligence, the use of big data, and the development of energy-efficient solutions. These developments have the potential to revolutionize the way we use air conditioning systems in our homes, making them more efficient, effective, and personalized. As technology continues to evolve, we can expect to see even more exciting developments in this area, improving the quality of life for homeowners and helping to reduce our impact on the environment.

3.6 CONCLUSIONS

- IoT-enabled air conditioning systems are now a reality and provide real-time monitoring, control, and optimization of air conditioning systems, allowing for increased energy efficiency, improved comfort, and convenience. These systems come in different types, each with its own unique features and benefits, including smart thermostats, remote control systems, and centralized systems.
- The integration of artificial intelligence and big data analytics has the potential to make air conditioning systems more efficient and personalized, reducing energy consumption while adapting to the user's preferences. Developing energy-efficient solutions is also crucial in reducing our environmental footprint.
- Despite the benefits of these systems, there are challenges that need to be addressed, such as security and privacy concerns, integration with existing systems, and cost and technical challenges. Manufacturers and developers are exploring the latest trends and future directions in this area to overcome these challenges, such as integrating AI, using big data, and developing energy-efficient solutions.
- IoT-enabled air conditioning systems offer several benefits, including increased energy efficiency, real-time monitoring and control, and improved comfort and convenience, making them an ideal choice for homeowners who want to create a comfortable and energy-efficient living environment.
- The latest trends and future directions in IoT-enabled air conditioning systems include integrating AI, using big data, and developing energy-efficient solutions, which have the potential to revolutionize the way we use air conditioning systems in our homes, making them more efficient, effective, and personalized.

- However, there are also challenges that must be addressed, including security and privacy concerns, integration with existing systems, and cost and technical challenges. Manufacturers must work to address these challenges to ensure that their systems are accessible, affordable, and secure.

Overall, IoT-enabled air conditioning systems are a promising technology that can create more integrated, automated, and convenient living environments in our homes. As technology continues to evolve, we can expect to see even more exciting developments that have the potential to improve our quality of life and reduce our impact on the environment.

3.7 FUTURE RESEARCH SIGNIFICANCE

IoT-enabled solutions for society air conditioners in smart home environments have become a reality with the advent of the Internet of Things (IoT). These solutions provide real-time monitoring, control, and optimization of air conditioning systems, allowing for increased energy efficiency, improved comfort, and convenience. There are different types of IoT-enabled air conditioning systems available in the market, such as smart thermostats, remote control systems, and centralized systems, each with their own unique features.

Despite the benefits of these systems, there are also several challenges that need to be addressed, such as security and privacy concerns, integration with existing systems, and cost and technical challenges. Manufacturers and developers are exploring the latest trends and future directions in this area, such as the integration of artificial intelligence, the use of big data, and the development of energy-efficient solutions to overcome these challenges.

The integration of artificial intelligence and big data analytics has the potential to make air conditioning systems more efficient and personalized, allowing them to adapt to the user's preferences and provide optimal comfort while reducing energy consumption. The development of energy-efficient solutions is also crucial in reducing our environmental footprint.

Thus, IoT-enabled air conditioning systems are a promising technology that can create more integrated, automated, and convenient living environments in our homes. With the latest trends and future directions in this area, we can expect to see even more exciting developments that have the potential to improve our quality of life and reduce our impact on the environment.

REFERENCES

1. Hammi, B., Zeadally, S., Khatoun, R., & Nebhen, J. (**2022**). Survey on smart homes: Vulnerabilities, risks, and countermeasures. *Computers & Security*, *117*, 102677.

2. Verma, R., Mishra, P. K., Nagar, V., & Mahapatra, S. (**2021**). Internet of Things and Smart Homes: A Review. In *Wireless Sensor Networks and the Internet of Things* (pp. 111–128). Apple Academic Press.
3. Aliero, M. S., Qureshi, K. N., Pasha, M. F., & Jeon, G. (**2021**). Smart home energy management systems in Internet of Things networks for green cities demands and services. *Environmental Technology & Innovation, 22*, 101443.
4. Orosz, D. (**2021**). Examining the contribution of smart homes to the smart performance of cities. *Theory Methodology Practice: Club of Economics in Miskolc, 17*(SI), 23–30.
5. Rottinghaus, A. R. (**2021**). Smart homes and the new white futurism. *Journal of Futures Studies, 25*(4), 45–56.
6. Kumar, P., & Akhai, S. (**2022**). Effective energy management in smart buildings using VRV/VRF systems. *Additive Manufacturing in Industry, 4.0*, 27–35.
7. Yu, L., Xie, W., Xie, D., Zou, Y., Zhang, D., Sun, Z.,... Jiang, T. (**2019**). Deep reinforcement learning for smart home energy management. *IEEE Internet of Things Journal, 7*(4), 2751–2762.
8. Yao, Y., & Shekhar, D. K. (**2021**). State of the art review on model predictive control (MPC) in heating ventilation and air-conditioning (HVAC) field. *Building and Environment, 200*, 107952.
9. Akhai, S., Mala, S., & Jerin, A. A. (**2021**). Understanding whether air filtration from air conditioners reduces the probability of virus transmission in the environment. *Journal of Advanced Research in Medical Science & Technology, 8*(1), 36–41.
10. Akhai, S., Mala, S., & Jerin, A. A. (**2020**). Apprehending air conditioning systems in context to COVID-19 and human health: A brief communication. *International Journal of Healthcare Education & Medical Informatics, 7*(1 & 2), 28–30.
11. Akhai, S., Singh, V. P., & John, S. (**2016**). Investigating indoor air quality for the split-type air conditioners in an office environment and its effect on human performance. *Journal of Mechanical Civil Engineering, 13*(6), 113–118.
12. Tanwar, N., & Akhai, S. (**2017**). Survey analysis for quality control comfort management in air conditioned classroom. *Journal of Advanced Research in Civil and Environmental Engineering, 4*(1 & 2), 20–23.
13. Akhai, S., Thareja, P., & Singh, V. P. (**2017**). Assessment of indoor environment health sustenance in air conditioned class rooms. *Journal of Advanced Research in Civil and Environmental Engineering, 4*(1 & 2), 1–9.
14. Akhai, S., Singh, V. P., & John, S. (**2016**). Human performance in industrial design centers with small unit air conditioning systems. *Journal of Advanced Research in Production Industrial Engineering, 3*(2), 5–11.
15. Philip, A., Islam, S. N., Phillips, N., & Anwar, A. (**2022**). Optimum energy management for air conditioners in IoT-enabled smart home. *Sensors, 22*(19), 7102.
16. Ali, A. M., Shukor, S. A., Rahim, N. A., Razlan, Z. M., Jamal, Z. A. Z., & Kohlhof, K. (**2019**, June). IoT-based smart air conditioning control for thermal comfort. In *2019 IEEE International Conference on Automatic Control and Intelligent Systems (I2CACIS)* (pp. 289–294). IEEE.
17. Rocha, F., Dantas, L. C., Santos, L. F., Ferreira, S., Soares, B., Fernandes, A.,... Batista, T. (**2020**). Energy efficiency in smart buildings: An IoT-based air conditioning control system. In *Internet of Things. A Confluence of Many Disciplines: Second IFIP International Cross-Domain Conference, IFIPIoT 2019, Tampa, FL,*

USA, October 31–November 1, 2019, Revised Selected Papers 2 (pp. 21–35). Springer International Publishing.

18. Hossein Motlagh, N., Mohammadrezaei, M., Hunt, J., & Zakeri, B. (**2020**). Internet of Things (IoT) and the energy sector. *Energies*, *13*(2), 494.
19. Islam, F. B., Nwakanma, C. I., Kim, D. S., & Lee, J. M. (**2020**, October). IoT-based HVAC monitoring system for smart factory. In *2020 International Conference on Information and Communication Technology Convergence (ICTC)* (pp. 701–704). IEEE.
20. Yaïci, W., Entchev, E., & Longo, M. (**2022**, June). Internet of Things (IoT)-based system for smart home heating and cooling control. In *2022 IEEE International Conference on Environment and Electrical Engineering and 2022 IEEE Industrial and Commercial Power Systems Europe (EEEIC/I&CPS Europe)* (pp. 1–6). IEEE.
21. Alves, A. A., Monteiro, V., Pinto, J. G., Afonso, J. L., & Afonso, J. A. (**2020**). Development of an Internet of Things system for smart home HVAC monitoring and control. In *Sustainable Energy for Smart Cities: First EAI International Conference, SESC 2019, Braga, Portugal, December 4–6, 2019, Proceedings 1* (pp. 197–208). Springer International Publishing.
22. Cheng, C. C., & Lee, D. (**2019**). Artificial intelligence-assisted heating ventilation and air conditioning control and the unmet demand for sensors: Part 1. Problem formulation and the hypothesis. *Sensors*, *19*(5), 1131.
23. Bae, Y., Bhattacharya, S., Cui, B., Lee, S., Li, Y., Zhang, L.,... Kuruganti, T. (**2021**). Sensor impacts on building and HVAC controls: A critical review for building energy performance. *Advances in Applied Energy*, *4*, 100068.
24. Png, E., Srinivasan, S., Bekiroglu, K., Chaoyang, J., Su, R., & Poolla, K. (**2019**). An Internet of Things upgrade for smart and scalable heating, ventilation and air-conditioning control in commercial buildings. *Applied Energy*, *239*, 408–424.
25. Yang, S., Wan, M. P., Chen, W., Ng, B. F., & Dubey, S. (**2020**). Model predictive control with adaptive machine-learning-based model for building energy efficiency and comfort optimization. *Applied Energy*, *271*, 115147.
26. Manjarres, D., Mera, A., Perea, E., Lejarazu, A., & Gil-Lopez, S. (**2017**). An energy-efficient predictive control for HVAC systems applied to tertiary buildings based on regression techniques. *Energy and Buildings*, *152*, 409–417.
27. Soudari, M., Kaparin, V., Srinivasan, S., Seshadhri, S., & Kotta, Ü. (**2018**). Predictive smart thermostat controller for heating, ventilation, and air-conditioning systems. *Proceedings of the Estonian Academy of Sciences*, *67*(3), 291–299.
28. Guo, F., & Rasmussen, B. (**2023**). Performance benchmarking of residential air conditioning systems using smart thermostat data. *Applied Thermal Engineering*, *225*, 120195.
29. Aibin, M. (**2020**). The weather impact on heating and air conditioning with smart thermostats. *Canadian Journal of Electrical and Computer Engineering*, *43*(3), 190–194.
30. Majdi, A., Alrubaie, A. J., Al-Wardy, A. H., Baili, J., & Panchal, H. (**2022**). A novel method for indoor air quality control of smart homes using a machine learning model. *Advances in Engineering Software*, *173*, 103253.
31. Yang, T., Zhao, L., Li, W., Wu, J., & Zomaya, A. Y. (**2021**). Towards healthy and cost-effective indoor environment management in smart homes: A deep reinforcement learning approach. *Applied Energy*, *300*, 117335.

32. Valladares, W., Galindo, M., Gutiérrez, J., Wu, W. C., Liao, K. K., Liao, J. C.,... Wang, C. C. (**2019**). Energy optimization associated with thermal comfort and indoor air control via a deep reinforcement learning algorithm. *Building and Environment, 155*, 105–117.
33. Taleb, T., Dutta, S., Ksentini, A., Iqbal, M., & Flinck, H. (**2017**). Mobile edge computing potential in making cities smarter. *IEEE Communications Magazine*, *55*(3), 38–43.
34. Taştan, M., & Gökozan, H. (**2018**). An Internet of Things based air conditioning and lighting control system for smart home. *American Scientific Research Journal for Engineering, Technology, and Sciences (ASRJETS)*, *50*(1), 181–189.
35. Adiono, T., Fathany, M. Y., Fuada, S., Purwanda, I. G., & Anindya, S. F. (**2018**, April). A portable node of humidity and temperature sensor for indoor environment monitoring. In *201[8] 3rd International Conference on Intelligent Green Building and Smart Grid (IGBSG)* (pp. 1–5). IEEE.
36. Hanggoro, A., Putra, M. A., Reynaldo, R., & Sari, R. F. (**2013**, June). Green house monitoring and controlling using Android mobile application. In *2013 International Conference on QiR* (pp. 79–85). IEEE.
37. Ni, J., & Bai, X. (**2017**). A review of air conditioning energy performance in data centers. *Renewable and Sustainable Energy Reviews*, *67*, 625–640.
38. Alsamhi, S. H., Ma, O., Ansari, M. S., & Meng, Q. (**2019**). Greening Internet of Things for greener and smarter cities: A survey and future prospects. *Telecommunication Systems*, *72*, 609–632.
39. Paredes-Valverde, M. A., Alor-Hernández, G., García-Alcaráz, J. L., Salas-Zárate, M. D. P., Colombo-Mendoza, L. O., & Sánchez-Cervantes, J. L. (**2020**). Intelli-Home: An Internet of Things-based system for electrical energy saving in smart home environment. *Computational Intelligence*, *36*(1), 203–224.
40. Lohan, V., & Singh, R. P. (**2019**). Home automation using Internet of Things. In *Advances in Data and Information Sciences: Proceedings of ICDIS 2017, Volume 2* (pp. 293–301). Springer.
41. Dhanalakshmi, S., Poongothai, M., & Sharma, K. (**2020**). IoT based indoor air quality and smart energy management for HVAC system. *Procedia Computer Science*, *171*, 1800–1809.
42. Chung, J. J., & Kim, H. J. (**2020**). An automobile environment detection system based on deep neural network and its implementation using IoT-enabled in-vehicle air quality sensors. *Sustainability*, *12*(6), 2475.
43. Lavanya, A., Jeevitha, M., & Bhagyaveni, M. A. (**2019**). IoT-enabled green campus energy management system. *International Journal of Embedded Systems and Applications*, *9*(2), 21–35.
44. Arya, A. K., Chanana, S., & Kumar, A. (**2018**, December). Smart energy controller for energy management using IoT with demand response. In *2018 IEE[E] 8th Power India International Conference (PIICON)* (pp. 1–6). IEEE.
45. Hariadi, R. R., Yuniarti, A., Kuswardayan, I., Herumurti, D., Arifiani, S., & Yunanto, A. A. (**2019**, July). Termo: Smart air conditioner controller integrated with temperature and humidity sensor. In *2019 [1]2th International Conference on Information & Communication Technology and System (ICTS)* (pp. 312–315). IEEE.
46. Rochd, A., Benazzouz, A., Abdelmoula, I. A., Raihani, A., Ghennioui, A., Naimi, Z., & Ikken, B. (**2021**). Design and implementation of an AI-based &IoT-enabled

home energy management system: A case study in Benguerir—Morocco. *Energy Reports*, *7*, 699–719.

47. Shreenidhi, H. S., & Ramaiah, N. S. (**2022**). A two-stage deep convolutional model for demand response energy management system in IoT-enabled smart grid. *Sustainable Energy, Grids and Networks*, *30*, 100630.
48. Ling, J., Zehtabian, S., Bacanli, S. S., Boloni, L., & Turgut, D. (**2019**, December). Predicting the temperature dynamics of scaled model and real-world IoT-enabled smart homes. In *2019 IEEE Global Communications Conference (GLOBECOM)* (pp. 1–6). IEEE.
49. Goyal, P., Sahoo, A. K., Sharma, T. K., & Singh, P. K. (**2021**). Internet of Things: Applications, security and privacy: A survey. *Materials Today: Proceedings*, *34*, 752–759.
50. Khawla, M., & Tomader, M. (2018, October). A survey on the security of smart homes: Issues and solutions. In *Proceedings of the 2nd International Conference on Smart Digital Environment* (pp. 81–87). ACM Digital Library.
51. Saputra, D. I. S., Suarnatha, I. P. D., Mahardika, F., Wijanarko, A., & Handani, S. W. (**2023**). IoT-based smart air conditioner as a preventive in the post-COVID-19 era: A review. *Journal of Robotics and Control (JRC)*, *4*(1), 60–70.
52. Srivastava, A., Gupta, S., Quamara, M., Chaudhary, P., & Aski, V. J. (**2020**). Future IoT-enabled threats and vulnerabilities: State of the art, challenges, and future prospects. *International Journal of Communication Systems*, *33*(12), e4443.
53. Zahid, H., Elmansoury, O., & Yaagoubi, R. (**2021**). Dynamic predicted mean vote: An IoT-BIM integrated approach for indoor thermal comfort optimization. *Automation in Construction*, *129*, 103805.
54. Tedeschi, S., Emmanouilidis, C., Farnsworth, M., Mehnen, J., & Roy, R. (**2017**). New threats for old manufacturing problems: Secure IoT-enabled monitoring of legacy production machinery. In *Advances in Production Management Systems. The Path to Intelligent, Collaborative and Sustainable Manufacturing: IFIP WG 5.7 International Conference, APMS 2017, Hamburg, Germany, September 3–7, 2017, Proceedings, Part I* (pp. 391–398). Springer International Publishing.
55. Gladence, L. M., Anu, V. M., Rathna, R., & Brumancia, E. (**2020**). Recommender system for home automation using IoT and artificial intelligence. *Journal of Ambient Intelligence and Humanized Computing*, 1–9.
56. Compare, M., Baraldi, P., & Zio, E. (**2019**). Challenges to IoT-enabled predictive maintenance for industry 4.0. *IEEE Internet of Things Journal*, *7*(5), 4585–4597.
57. Çınar, Z. M., Abdussalam Nuhu, A., Zeeshan, Q., Korhan, O., Asmael, M., & Safaei, B. (**2020**). Machine learning in predictive maintenance towards sustainable smart manufacturing in industry 4.0. *Sustainability*, *12*(19), 8211.
58. Bouabdallaoui, Y., Lafhaj, Z., Yim, P., Ducoulombier, L., & Bennadji, B. (**2021**). Predictive maintenance in building facilities: A machine learning-based approach. *Sensors*, *21*(4), 1044.
59. Maraveas, C., Piromalis, D., Arvanitis, K. G., Bartzanas, T., & Loukatos, D. (**2022**). Applications of IoT for optimized greenhouse environment and resources management. *Computers and Electronics in Agriculture*, *198*, 106993.
60. Akhai, S. (**2023**). Navigating the potential applications and challenges of intelligent and sustainable manufacturing for a greener future. *Evergreen, 10*(4), 2237–2243.

61. Singh, S. P., Singh, B., & Gupta, O. P. (**2011**). Performance evaluation of DNS based load balancing techniques for web servers. *International Journal of Computer Science and Technology*, *2*(1), 166–169.
62. Akhai, S. (**2023**). From Black Boxes to Transparent Machines: The Quest for Explainable AI. Available at SSRN: https://ssrn.com/abstract=4390887.
63. Bhanupriya & Singh, S. P. (**2016**). Review on data mining in cloud computing. *International Journal of Computer & IT*, *4*, 1–4.
64. Martín-Lopo, M. M., Boal, J., & Sánchez-Miralles, Á. (**2020**). A literature review of IoT energy platforms aimed at end users. *Computer Networks*, *171*, 107101.

Chapter 4

Intelligent Traffic Management and Identification of Emergency Vehicles

T. Tirupal, B. Uday Kiran Reddy, K. Sai Teja, Uday Kiran Dhane, M. Siva Prasad, and A.O. Salau

4.1 INTRODUCTION

Because of the increase in automobiles, countries all over the world are experiencing traffic problems. Despite the increase in the number of cars using the roads, the static street foundation is essentially unchanged, making it impossible for it to respond to changes without being obstructed. Static roadway dividers [1, 2] have a fixed number of lanes on either side of the street, which is a concern. This necessitates making greater use of already-existing resources, such as the variety of available pathways. To solve this problem, we could build a mobile roadway divider that moves in response to the flow of traffic. The microcontroller can operate the sensors to retrieve the data from them and update it to the web using a Wi-Fi module attached to it. The Internet of Things collects ongoing data from vehicle traffic to identify current traffic activities and traffic stream conditions. The Internet of Things [3–5] will be linked to every object of the traffic, such as roadways and dividers, using infrared sensors.

Typically, it is observed that there is a significant disparity in traffic volume between one side of a street and the opposite side, with one side experiencing heavy traffic while the other side remains relatively free of traffic. In such a scenario, it is possible to automate the adjustment of the wall position, thereby mitigating traffic issues. The implementation of a movable street divider facilitates the organisation of the street boundary, with the aim of maximising the efficiency of roadway utilisation on the existing street. Through the implementation of a divider, it becomes possible to grant traffic clearance for emergency vehicles as and when it is necessary. One concern regarding static street dividers pertains to the fixed allocation of lanes on either side of the road. Due to limited resources and a corresponding increase in the number of vehicles per household, there has been a significant rise in the overall volume of vehicles on roadways.

This necessitates the improved utilisation of available resources, such as the various means of accessibility. The primary objective of this task is to advance the field of traffic control. The objective of this project is to reduce travel time during peak hours, alleviate traffic congestion, and provide a more efficient and effective solution to the aforementioned traffic issues. Our proposed solution involves

 DOI: 10.1201/9781003480181-4

the development of a mobile street divider that is capable of adjusting its position based on the flow of traffic. The parking lot collects and stores ongoing data regarding vehicular traffic, allowing for the monitoring and analysis of current traffic activity and flow conditions. The Internet of Things (IoT) will be interconnected with each individual element of traffic infrastructure, such as streets and dividers, through the utilisation of infrared sensors. In general, it is observed that there is a significant disparity in traffic volume between the two sides of a street divider. In such situations, it is possible to strategically adjust the position of the divider to mitigate traffic congestion issues. Similarly, by employing the construction of a barrier, we can facilitate unimpeded passage for emergency vehicles as necessary. The motivation of the proposed method includes:

- The cumulative effect of the time and fuel savings resulting from the implementation of an additional traffic congestion control measure.
- Furthermore, we aim to eliminate the reliance on manual intervention and manual traffic coordination in order to achieve a more intelligent traffic system throughout the entire city.
- The proposed concept involves the development of an automated and portable road divider instrument capable of maneuvering roadways, thereby facilitating an increasing number of lanes during periods of high traffic congestion.

4.2 LITERATURE REVIEW

The most recent developments in smart transportation research are covered in this area, including models for traffic analysis, predictions of traffic congestion, and the usage of roadside units to transmit messages. Some recent advancement in the field of traffic management is covered in the sections that follow.

In the literature review this methodology proposes structure aids for lowering the likelihood of traffic congestion and, to some extent, provides room for emergency vehicles on the road. With this suggested chapter, we hope to clear the traffic according to the priority lists. This method has been shown to use morphological filtering and blob analysis to find traffic density [6]. The top priority road (one with a lot of traffic) is cleared first. The suggested approach focuses mostly on automobiles. Emergency cars are identified by using image processing. The proposed method performs well in congested traffic based on these factors and in accordance with time intervals. It is operating manually in this document.

Another methodology proposed a structure that helps to systematically reduce the likelihood of traffic congestion caused by high red light delays and successfully gives clearance for emergency vehicles [7]. In this case, we tend to design the system with the goal of clearing priority traffic as directed. We determine the traffic density in this system. Morphological filtering and blob analysis are two techniques. First, the road with the highest priority is cleared. The motorcar is also given importance by the proposed system. If an ambulance is waiting at a red light, the reallane is given priority over other vehicles and this lane is emptied. Image processing identifies emergency vehicles. When an emergency vehicle enters a lane, a small controller is notified. The vehicle is identified using morphological filtering and blob analysis on a camera image. The small controller prioritises the lane with the emergency vehicle and frees that lane. Because it does not operate automatically, this chapter necessitates manual assistance.

Another methodology proposed a structure that serves to some extent to lessen the likelihood of traffic jams and to provide road clearance for emergency vehicles [8]. We want to clear the traffic in accordance with priority with this planned effort. It will assist in lowering highway traffic. Additionally, the government will benefit from enforcing traffic laws. And everyone will abide by the traffic laws. It will be applicable practically everywhere in the city of Pune. It will be used in crosswalks and traffic zones.

Here in this, the movement of the divider in this chapter is only indicated by the use of an LCD display. Another methodology proposed a structure that helps. Instead of sensing vehicles with sensors, image capture is used. The image processing method can be used to control traffic lights in this system. Traffic signals can be controlled by analysing data. Traffic congestion [9] is expected to be avoided with the help of specific calculations, morphology, and imagery. Here in this, only the density can be measured and the moment of the divider cannot be measured. Another methodology of the adjustable road divider is a method which works on image processing in real time to control traffic [10]. It identifies the number of vehicles in each lane and makes decisions for the movement of the barrier. The road with the highest priority is cleared first. This idea reduces the traffic in peak hours.

Another methodology proposed a structure that helps the road and a cloud be connected, allowing for continuous traffic monitoring and the movement of traffic density to the cloud. The chapter also provides a solution for traffic clearance for government and emergency vehicles. Using RFID, a cloud is created to recognise an emergency vehicle's appearance, and after that, a path is created by shifting the street barrier in a similar manner, specifically for emergency vehicles [11]. As a result, it is assumed that relocating the street divider to widen or narrow the roadway and clear traffic is a viable option for avoiding a bottleneck in one direction. It is also possible to provide a free route for the emergency vehicle [12] that is not affected by the nearby traffic. Here in this chapter, Arduino Uno is utilised in this research, but it has less storage than Arduino Mega.

Druthiya et al. [13] predicted a structure that reduces the likelihood of high red light deferrals turning highways into parking lots and allows the emergency vehicle some leeway. We structure the architecture to clear traffic as needed. Traffic thickness exploits morphological sifting and blob examination. The neediest street gets cleansed first. The car's framework also matters. If an emergency vehicle is stuck in a traffic, the real path is given priority and traffic is cleared.

Swapnil et al. [14] proposed a morphological separation and mass analysis recognise the emergency vehicle and communicate it to the little controller. The little controller prioritises the crisis car path and clears it. The article concludes that structure importance and requirement must be considered before building. Movable road dividers prevent accidents and improve transportation. It helps choose and install the correct divider for fast, safe, and orderly traffic.

Anitha et al. [15] proposed a moveable road divider that may move by counting vehicles with sensors. The movable divider reduces traffic. This implies that the programme provides a free and guaranteed ambulance system that will arrive on time and save lives. The programme should help reduce rush-hour travel times and save gasoline.

Varshitha et al. [16] proposed a movable divider that can handle traffic delays on one side of the road while leaving the other side clear. This suggested system allows ambulances to travel freely, ensuring they arrive on time and save lives. It cuts peak-hour travel time and gasoline. It is viable, secure, and requires fewer wires, reducing system maintenance costs.

Aditya et al. [17] proposed and developed Movable Road Dividers for Vehicular Traffic Control, and a demo unit was manufactured with satisfactory results. Since it's a demo module, we just examined Wi-Fi transmission. In real applications, the same approach can be used for many directions based on traffic congestion. For efficient and cost-effective traffic control, this system integrates Wi-Fi modem (IoT) with Ultrasonic Sensor. The technique improves traffic congestion databases at each station. We examined IoT implementation solutions. This system will reduce peak-hour traffic. It automatically reports traffic density to the traffic department.

The objective of the chapter includes:

- To enhance the efficiency of traffic flow. Additionally, there is a need to minimize the complexity associated with junctions.
- To encourage compliance with traffic regulations among individuals.
- To make smart arrangements of the roads in the city.

4.3 PROPOSED METHOD

Every employee in the IT sector is working from morning until evening. To reduce this traffic, these methods are in practice. Need to get to their offices on time so that everyone will arrive at the right time to their office. The proposed system is made of an Arduino Mega microcontroller, which manages the entire process, as shown in Figure 4.1. The transceiver, also referred to as an IR sensor [18], transmits light to count the density. The term "transducer" is also used to describe an ultrasonic sensor [19]. Before moving the partition, sound waves check to see if there are any vehicles nearby; if so, a buzzer sound may be produced. Emergency vehicles like ambulances can be identified using RFID [20] modules, alerting onlookers to let them pass so that the ambulance can be reached and can be updated in the cloud, which allows for remote monitoring.

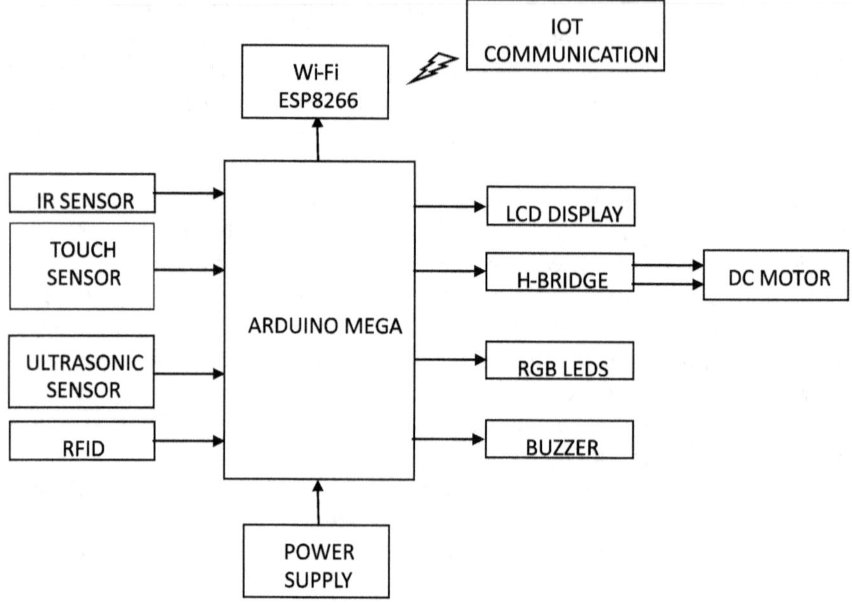

Figure 4.1 Block diagram of the proposed method.

4.3.1 Operation of the Proposed Method

Individuals who experience congestion while commuting and subsequently experience heightened levels of stress may encounter challenges in arriving punctually at their intended destinations, such as their workplace or educational institution. Additionally, this situation may contribute to the presence of diverse forms of pollution, including noise pollution, air pollution, and land pollution. Figure 4.2 depicts the correlation between traffic congestion and its influence on students' academic performance, primarily attributed to the stress induced by traffic delays during their commute to college. Through the implementation of intelligent traffic management systems and the incorporation of emergency vehicle detection technology, traffic flow can be regulated automatically by making necessary adjustments to the dividers [21].

By using this method dividers [22, 23] are moving from one side to another side based on the count of the vehicles. When one side of the road has fewer vehicles and another side has a high number of vehicles the divider [24, 25] will automatically move based on the count of the vehicles as shown in Figure 4.3.

When emergency vehicles [26, 27] like ambulances [28] are stuck in traffic and lives will be in danger in the ambulance [29], to save their lives everyone should give way to the ambulance [30], but there is no way in the road because of the fixed number of lanes. This chapter represents the smart movable road divider. The dividers are moving automatically by controlling the traffic as shown in Figure 4.4.

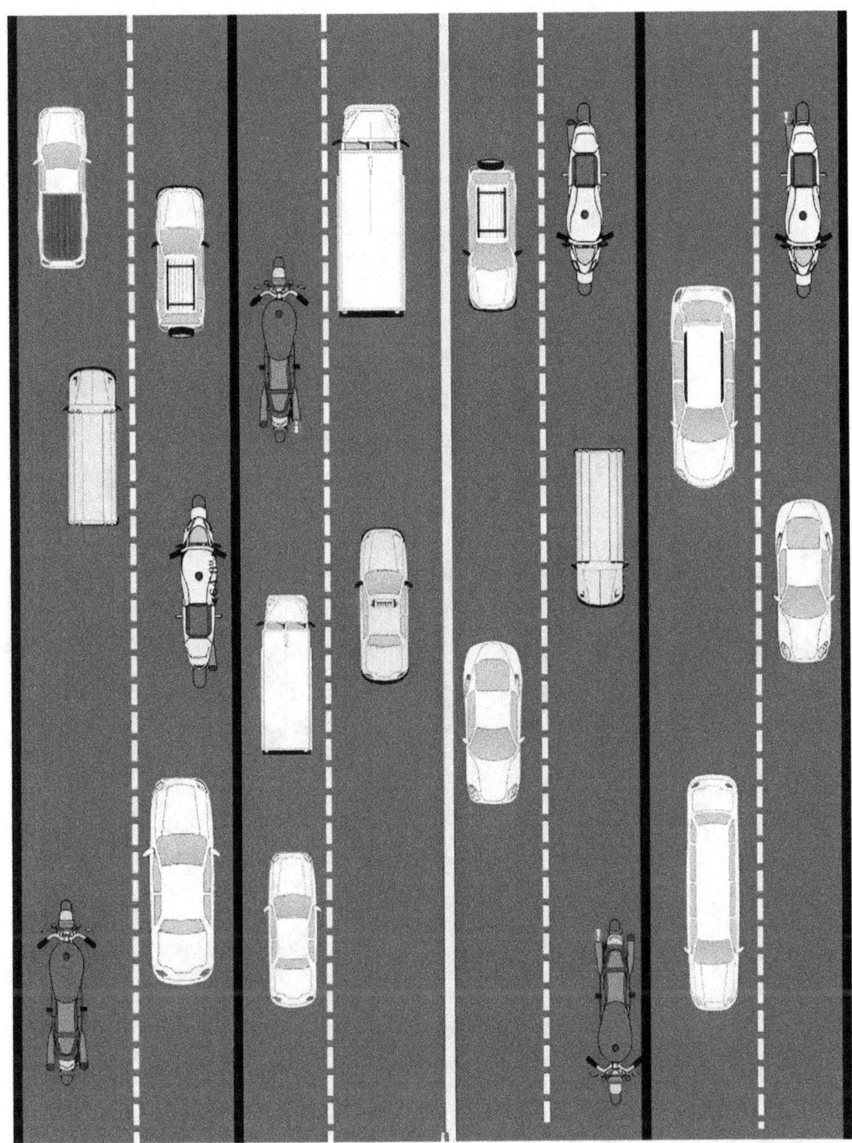

Figure 4.2 Moving the divider based on the vehicles count.

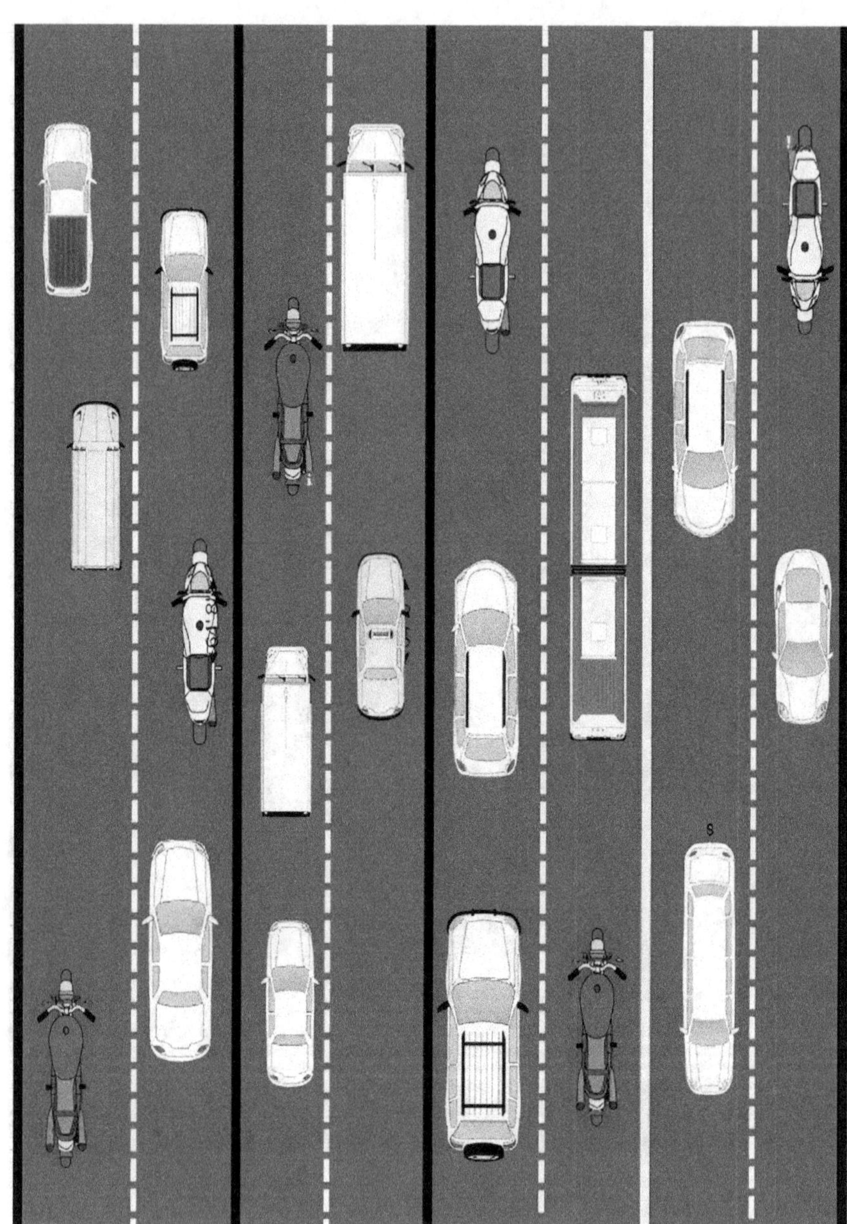

Figure 4.3 The divider's movement towards the lane with fewer vehicles.

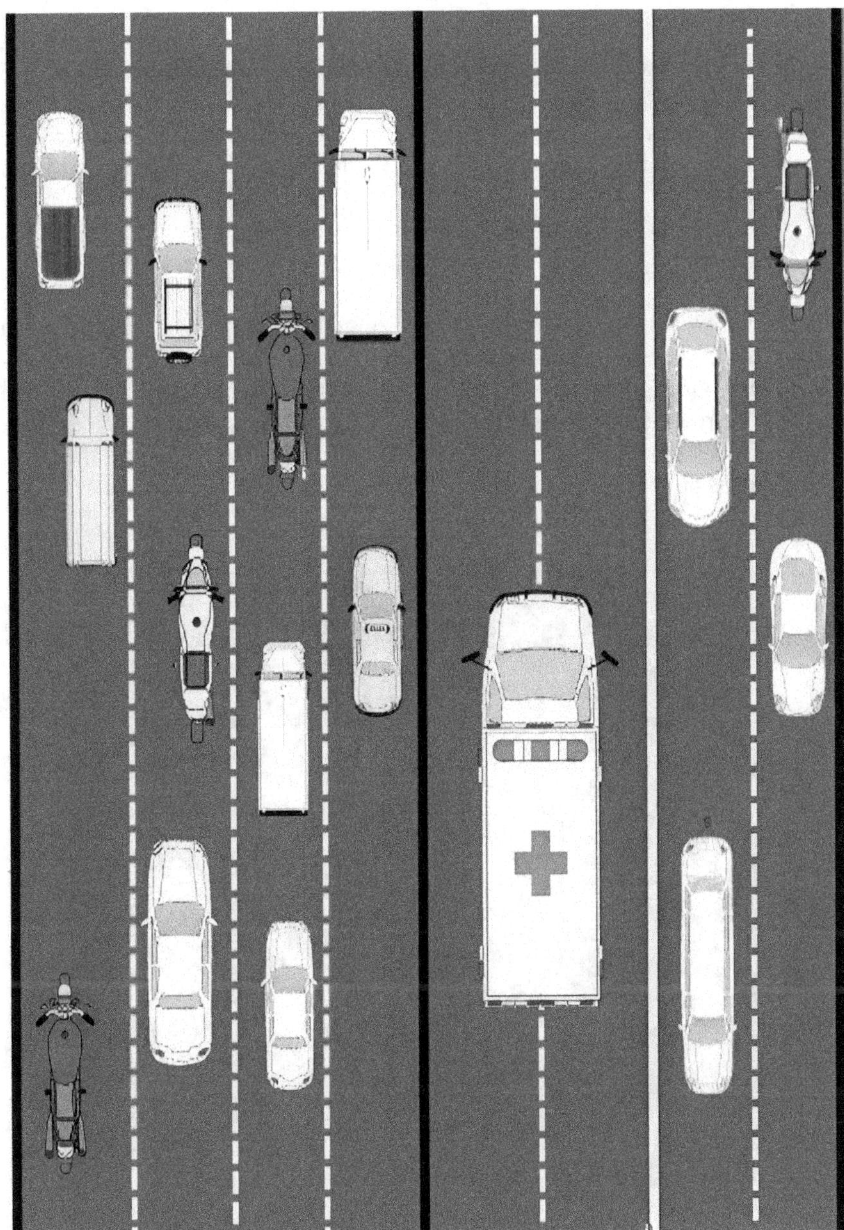

Figure 4.4 Controlling of traffic signals and ambulance detection.

4.3.2 Arduino Mega 2560

Arduino is open-source which includes both hardware and software. It has a 16 MHz artistic resonator, six basic data sources, a USB connection, a force jack, an ICSP header, and a reset button. It has 14 advanced input/output pins. It includes everything needed to support the microcontroller. Figure 4.5 displays an Arduino Mega 2560.

4.3.3 IR Sensor

The IR sensor is an electronic device that emits light to detect an object in the environment. The emitter and receiver are the two components of an infrared sensor. As shown in Figure 4.6, an IR LED serves as the emitter and an IR photodiode serves as the receiver. Both the heat and motion of an object can be measured by an IR sensor. The operating voltage is 5V DC, I/O pins are 3.3V-5V, and the measurement range is up to 20 centimetres.

Figure 4.5 Arduino Mega 2560.

Figure 4.6 IR sensor.

4.3.4 Ultrasonic Sensor

An ultrasonic sensor is a piece of technology that uses ultrasonic sound waves to measure a target object's distance and then turns the sound that is reflected back into an electrical signal. The supply voltage is 5V DC, and the supply current is 15mA shown in Figure 4.7. The distance is measured between 2cm—400cm, with 0.3cm accuracy.

4.3.5 ESP 0.1 Wi-Fi Module

The ESP8266 ESP-01 is a serial to Wi-Fi breakout module with an integrated ARM microprocessor, and 2GPIOs brought out to the header for Serial Wi-Fi Wireless Transceiver Module is a self-contained SOC that allows any microcontroller to connect to your Wi-Fi network. It is equipped with 1MB of Flash Memory, IEEE 802.11 b/g/n Wi-Fi, and 3.3V power. As shown in Figure 4.8, it has a 32bit RISC TensilicaXtensa LX Processor running at 80MHz.

4.3.6 L293D Bridge

An L293D bridge is an electronic circuit. L293D bridge enables the application of a voltage across a load in any direction. The same IC can be utilised to drive a DC motor shown in Figure 4.9. The maximum peak current of the motor is 1.2A. The maximum continuous current of the motor is 600mA. The motor voltage Vcc (Vs) is between 4.5V and 3.6V.

Figure 4.7 Ultrasonic sensor.

Figure 4.8 ESP 0.1 Wi-Fi Module.

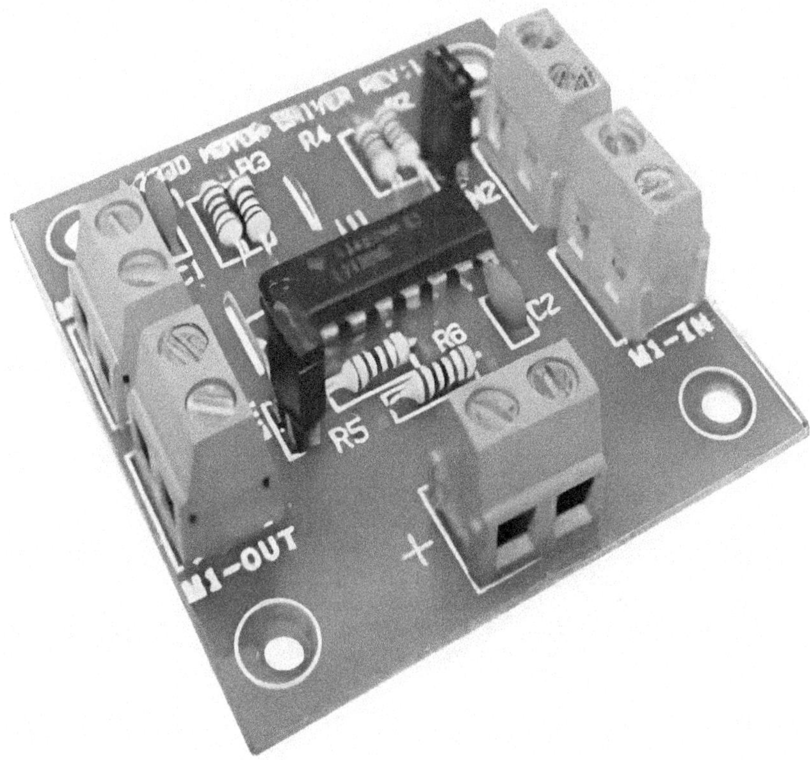

Figure 4.9 L293D bridge.

4.3.7 RFID

RFID is an integrated MF RC52213.56Mhz contactless communication card chip. The RFID tags are small chips that contain a unique id and can be read using an RFID reader. These tags can be embedded in products and documents to automate the tracking process shown in Figure 4.10.

4.3.8 DC Motor

DC Motor transforms direct current electrical power into mechanical power. This DC Motor Voltage rating is 12V DC Motor with 200RPM and can be used in all-terrain robots and a wide range of robotic applications. Figure 4.11 illustrates the better coupling shaft has a hole.

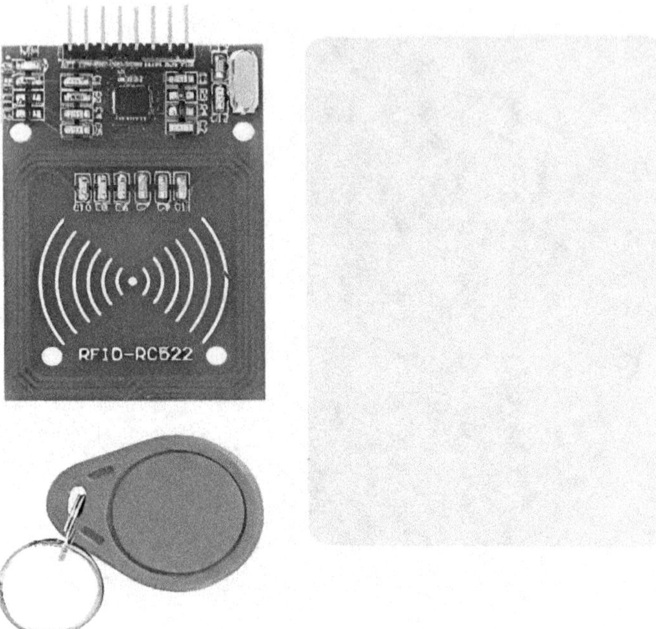

Figure 4.10 RFID.

4.3.9 Liquid Crystal Display

LCD stands for Liquid Crystal Display. The Liquid Crystal Display is a flat panel which uses liquid crystals. As Figure 4.12 illustrates, LCD has two rows and each row produces 16 characters. The display light causes less strain on the eyes. The LCD has operating voltage ranges from 4.7V–5.3V.

4.3.10 Buzzer

A buzzer is a signalling device that emits audible sounds which are mechanical, piezoelectric, or electromechanical. Figure 4.13 shows a buzzer is a simple, effective component to add a sound feature to our chapter. The frequency range is 3,300Hz. Operating voltage ranges from 4V to 9V DC.

4.3.11 Touch Sensor

Electronic sensors that can recognise touch are called touch sensors. When touched, they act as a switch. These sensors are utilised in lighting, mobile touch screens, etc. The user interface provided by touch sensors is simple.

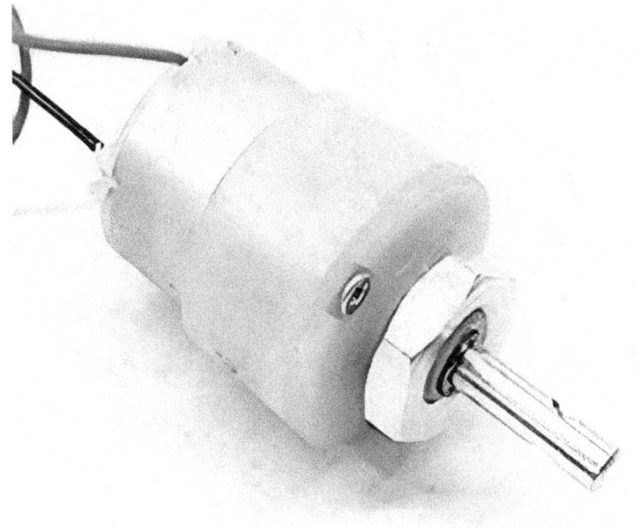

Figure 4.11 DC motor.

Figure 4.12 Liquid Crystal Display.

Figure 4.13 Buzzer.

Figure 4.14 Touch sensor.

4.4 EXPERIMENTAL RESULTS

Upon integration of all the modules, the resulting structure works towards reducing the chances of traffic jams and provides clearance for road vehicles. We have designed and developed a demonstration and work of our chapter. But as this is only a prototype, there is space for improvement when it is implemented as a product type as shown in Figure 4.15.

4.4.1 Comparison of the Existing Model and Proposed Model

In the annals of the past history of vehicle systems and navigation, the serial monitor was the only medium of display of the vehicle count (density of road). After several types of research, we have come up with a unique solution where the vehicle count will be constantly monitored and updated, and additionally, in this chapter the problem of the serial monitor was eradicated and replaced with mobile

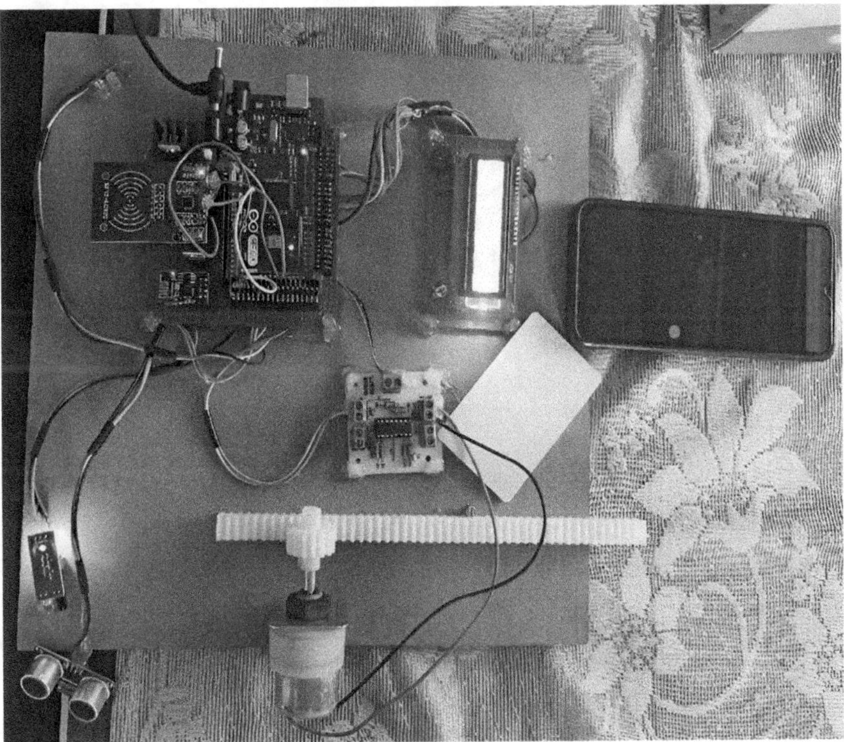

Figure 4.15 Working of the proposed system.

phones. Now, the vehicle count can be constantly viewed on the mobile phone as shown in Figure 4.16.

In earlier times, when this sort of chapter was implemented, there were no proper precautions, which led to many accidents. The main cause was when the bridge was divided into two; there were no proper alert systems and that led to accidents in so many cases. But now in our chapter, our system will notify the control system to clear the vehicles on the bridge before the bridge is getting divided. And moreover, in our chapter, the buzzer is also included which gives a high-frequency beep sound which will alert the drivers of vehicles on the bridge

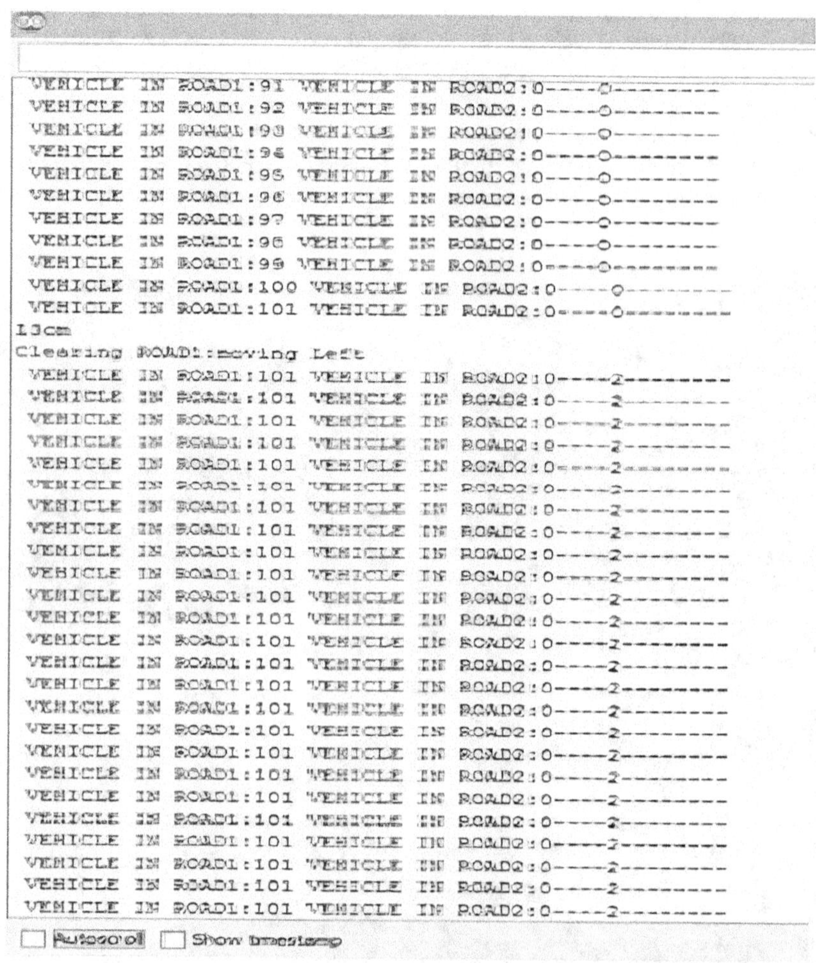

Figure 4.16 (a) In the existing system the density of the traffic [9] is counted by using an IR sensor. (b) Sophisticated, the density of the traffic is counted by using an IR sensor.

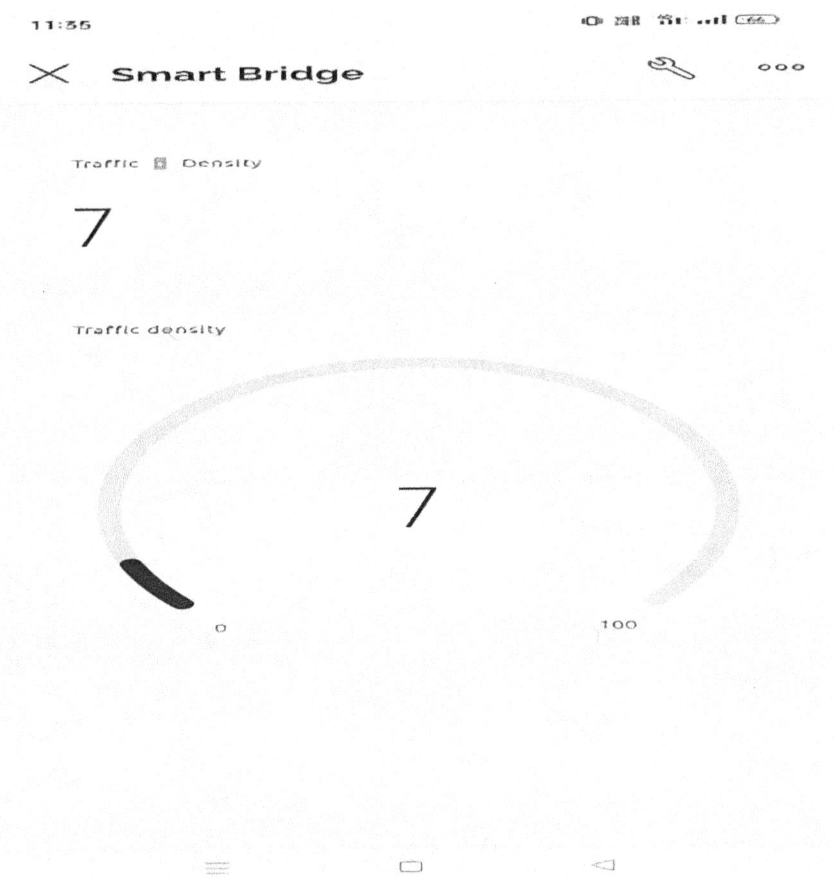

Figure 4.16 (Continued)

to move. This way, our chapter is very effective in terms of handling things and minimizing the dangerous cases as shown in Figure 4.17.

In earlier days if there were an ambulance in traffic there would be no alert system and regardless of whether there were any emergency vehicles in traffic nobody would even get to know about that vehicle. Apparently, it would lead to miscellaneous dangers to the people present in the emergency vehicles or ambulances, and this was a major disadvantage in the previous system. Our chapter consists of a unique solution which will be a remedy for all these disadvantages that were there in the previous system. The solution which is presented in this chapter consists of an idea which will apparently control a traffic signal and makes a way out for the emergency vehicles or ambulances.

There might be a doubt about how the problem of noticing the emergency vehicles and ambulances is done. Our chapter consists of the idea that RFID tags will

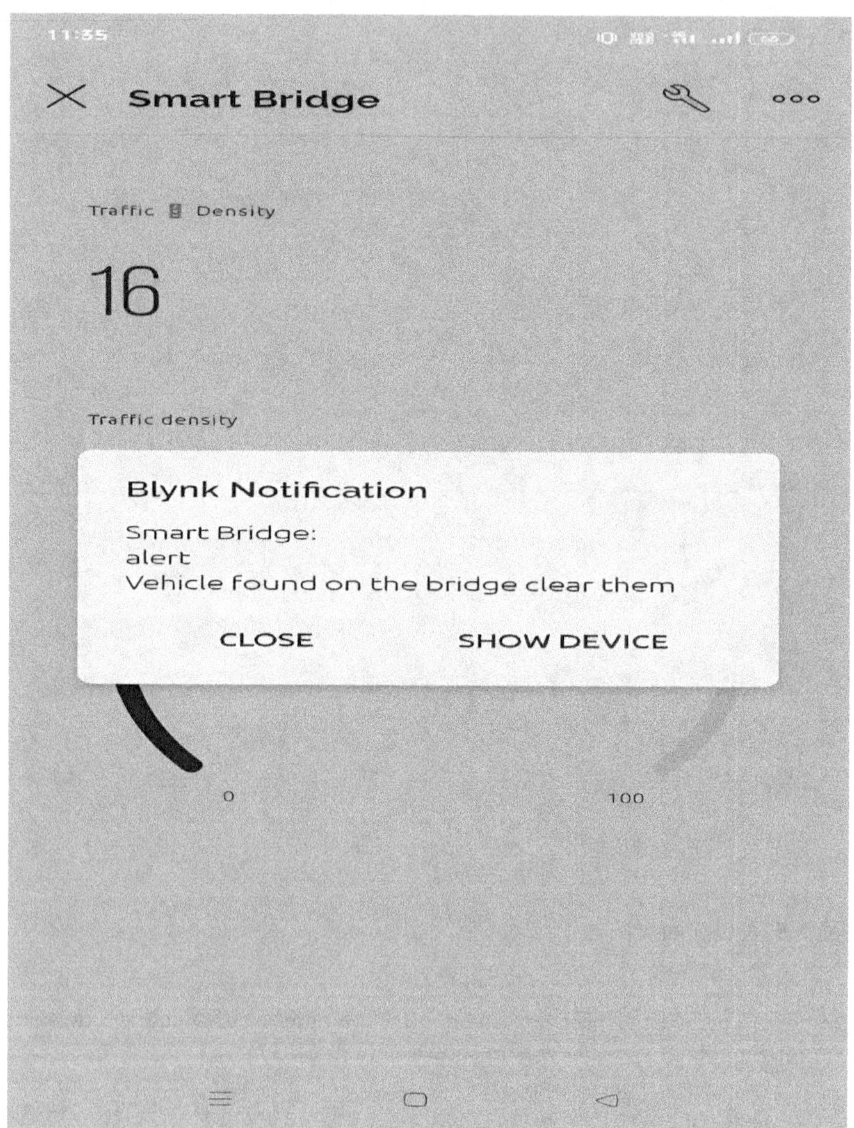

Figure 4.17 Vehicle is detected at the divider using an ultrasonic sensor.

be given to emergency vehicles and ambulances. RFID tags help in noticing them and accordingly vehicles in traffic will make a way for the emergency vehicles and ambulances and even if the vehicles are not responding then there will be a message delivered to the control centre and the control centre personnel ensures that emergency vehicles are available. Then the emergency vehicles gets a separate way which is out of the traffic and accordingly the bridge will be moved and gives the way to the emergency vehicles and ambulances. In this way, there is a lot of time that is being saved for the people in emergency vehicles and there are many lives which are being saved by making a way out for the ambulances in risky traffic. Our chapter gives the best solution which actually saves time for people in emergency vehicles as shown in Figure 4.18.

In the preceding year, individuals attempted to tamper with the divider by physically contacting it. However, the divider touch sensor successfully detected these interactions and promptly transmitted alert signals to the control room. The graphical analysis of intelligent traffic management and identification of emergency

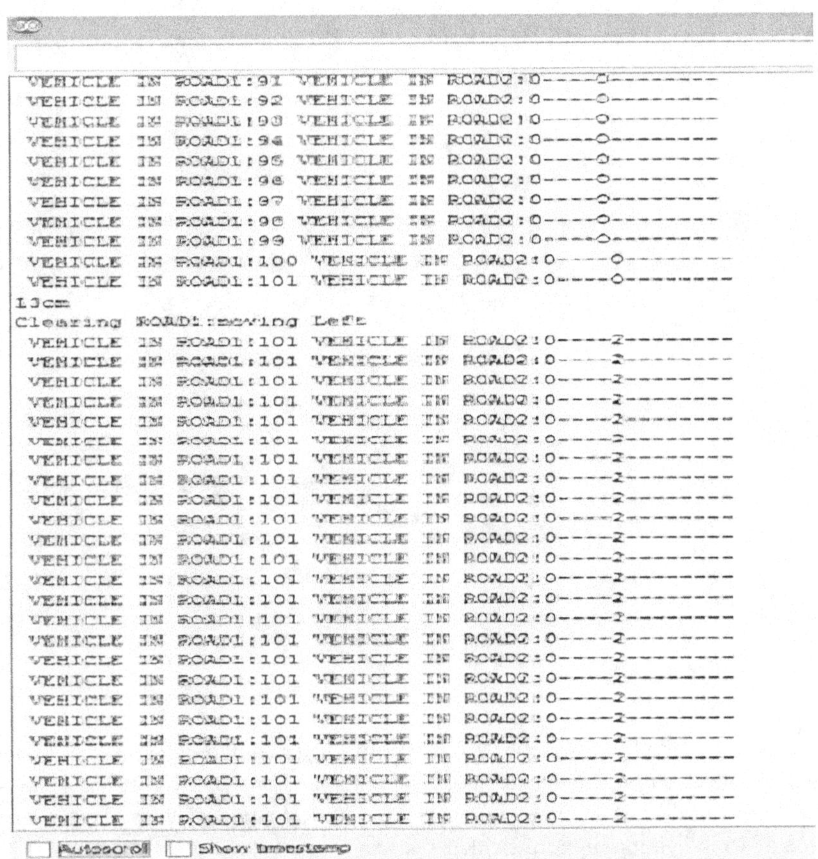

Figure 4.18 (a) In the existing system RFID module is used to give a path to an ambulance [9] and to find the emergency vehicles. (b) Sophisticated RFID module is used to find the emergency vehicles.

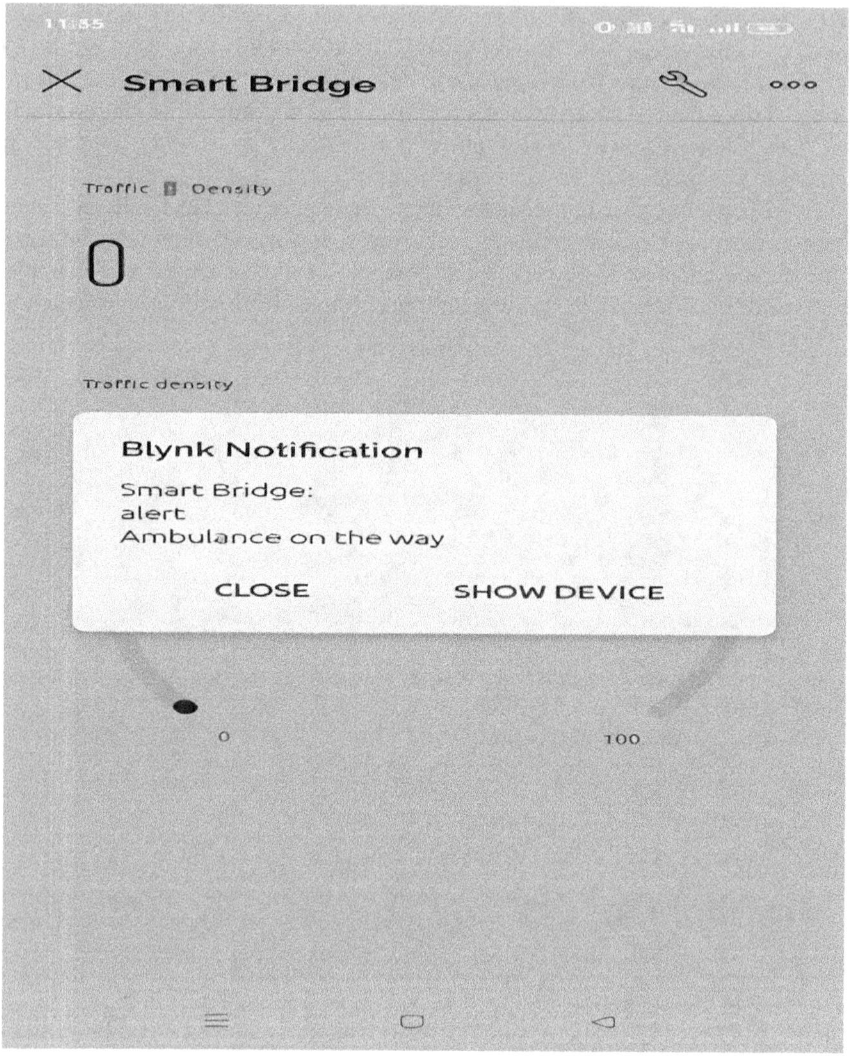

Figure 4.18 (Continued)

vehicles is depicted in Figure 4.19. The graph illustrates the outcomes of implementing intelligent traffic management and emergency vehicle identification systems. The number of vehicles is represented as "situations," with a maximum value of 16 indicating high traffic density.

The current count of vehicles is 16; however, the dividers are not in motion. This indicates the presence of a vehicle in front of the divider, prompting an alert to remove the vehicle obstructing the divider. In the first scenario, the vehicle count is recorded as 16, indicating a high level of traffic. The divider value is

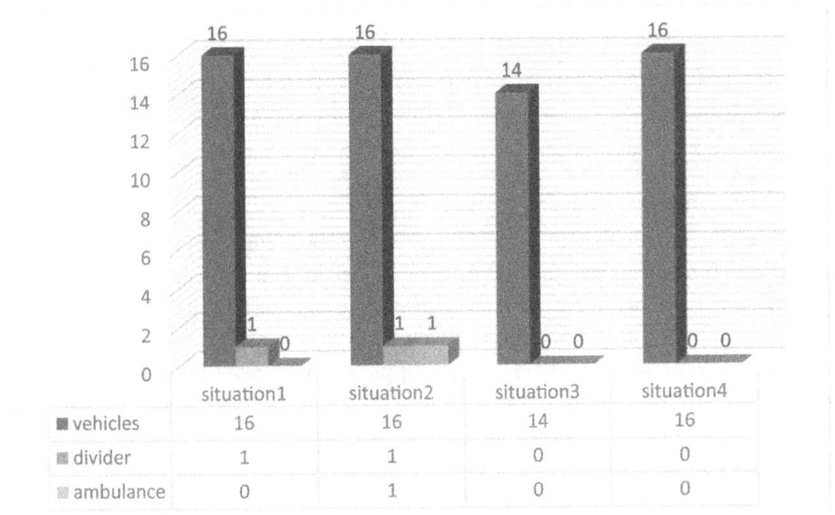

	situation1	situation2	situation3	situation4
vehicles	16	16	14	16
divider	1	1	0	0
ambulance	0	1	0	0

Figure 4.19 Graph of intelligent traffic management and identification of emergency vehicles.

noted as 1, signifying that the divider is in motion. Furthermore, no ambulance is observed in this particular situation, resulting in a corresponding value of 0. There are two vehicles, denoted as Vehicle A and Vehicle B, with a total count of 16.

Additionally, there is a divider, and one ambulance, referred to as Vehicle C, with a count of 1. This implies that the divider is positioned in a manner such that it separates Vehicle A and Vehicle B, while Vehicle C is absent in that particular location. The vehicle count in situation 4 is 14, indicating a lack of traffic and consequently no movement of the divider.

4.4.2 Applications

1. Used for smart traffic management system.

   Smart traffic management systems use integrated sensors like: Radio frequency identification (RFID) tags; automatic identification and data collection (AIDC) tags; temperature sensors; air quality sensors with amount of data that it generates. It's crucial that the system integrates cloud computing and edge processing.
2. Used for rescuing patients.

   Smart traffic management systems are used for rescuing patients, for instance by collecting a patient's details while the patient is in ambulance and admitting the patients in the nearest hospital.
3. Can be used for fire fighting vehicles.

   This chapter is used for fire fighting vehicles, like sending the fire engines to the places effected by fire.

4.5 CONCLUSION

The proposed chapter helps in reducing traffic congestion at the junction and providing clearance of the road for emergency vehicles (such as ambulances and VIP vehicles). It will aid in making smart road arrangements and the traffic density is updated in the cloud via the Blynk application, and the movement of the divider is indicated via buzzer sound and LCD display. These methods adhere to traffic regulations and the future enhancement of the chapter is possible to employ speech alarms to indicate the various controlling devices' operational condition. The introduction of time-controlled gadgets for usage in commercial settings is a further planned development. The forthcoming scope entails the introduction of time-controlled devices designed for utilisation in commercial settings. For instance, the control of a large display in a showroom can be automated between two distinct time intervals, without requiring any user or technician intervention. Additionally, voice alerts can be employed to indicate the operational status of various controlled devices.

REFERENCES

1. Kishore Kumar N, et al., "Smart Automatic Movable Road Divider", Turkish Journal of Computer and Mathematics Education, Vol.12, Issue.11, pp. 6953–6962, 2021.
2. Singh SP, Sharma A, "An Improved Node Selection Algorithm for Routing Protocols in VANET", Communication and Computing Systems, Taylor & Francis Group, pp. 889–894, 2016.
3. Arvind, et al., "Adjustable Road Divider Using IoT", Journal of Science, Computing and Engineering Research, Vol.3, Issue.1, pp. 222–225, 2022.
4. Kansal K, Singh SP, "FIM on Map Reduce Model", International Journal of Computer & IT, pp. 52–55, 2015.
5. Kaur P, Sharma P, Singh SP, "Handoffs in Cellular Systems", International Conference on Research and Innovation in Engineering and Technology (ICRIET-14), pp. 50–52, 2014.
6. Naveen N, Sowmya CN, "IoT Deployed Automatic Movable Smart Road Divider to Avoid Traffic Problems", International Journal of Computer Science Trends and Technology, Vol.7, 2019.
7. Sri BD, Nirosha K, Gouse S, "Design and Implementation of Smart Movable Road Divider Using IOT", International Conference on Intelligent Sustainable Systems (ICISS). doi:10.1109/iss1.2017.8389364.
8. Sowjanya KN, Nikhil Jamadagni HM, Unnimaya K, Bhavana G, "Automatic Movable Smart Road Divider Using IOT", International Advanced Research Journal in Science, Engineering and Technology, Vol.8, Issue.7, pp. 324–329, 2021.
9. Ravish R, Gupta VR, Nagesh KJ, "Software Implementation of an Automatic Movable Road Barrier", International Carnahan Conference on Security Technology, 2019.
10. Agarwal S, Maheshwari P, "Controlling of Smart Movable Road Divider and Clearance Ambulance Path Using IoT Cloud", ICCCI, pp. 1–4, 2021. doi:10.1109/ICCCI50826.2021.9402497.

11. Keerthan AJ, et al., "Automatic Movable Smart Road Dividers-IOT Based Solution to Traffic Congestion Problems", IJISRT, Vol.5, Issue.6, pp. 2456–2165, 2020.
12. Rashmi C, Roopa TN, Samrudh R, Sandhya M, "Movable Road Dividers", IRJET, Vol.7, Issue.6, 2020.
13. Druthiya L, et al., "Smart Automatic Movable Road Divider", JETIR, Vol.6, Issue.5, 2019.
14. Swapnil C, et al., "Movable Road Divider Using IOT", International Journal of Future Generation Communication and Networking, Vol.13, Issue.3s, pp. 321–324, 2020.
15. Anitha G, et al., "Movable Road Divider for Vehicular Traffic Control with Monitoring Over IoT", IJCRT, Vol.11, Issue.4, 2023.
16. Varshitha, et al., "Automatic Movable Divider for Traffic Management", IJCRT, Vol.11, Issue.5, 2023.
17. Aditya T, et al., "Automatic Movable Road Divider", International Research Journal of Modernization in Engineering Technology and Science, Vol.5, Issue.6, 2023.
18. Basavaraju A, Doddigarla S, Naidu N, Malgatti S, "Vehicle Density Sensor System to Manage Traffic", International Journal of Research in Engineering and Technology, pp. 2319–1163, March 2014.
19. Jo Y, Choi J, Jung I. "Traffic Information Acquisition System With Ultrasonic Sensors in Wireless Sensor Networks", International Journal of Distributed Sensor Networks, Vol.10, Issue.5, May 2014.
20. Inserra D, Wen G, "Compact Crossed Dipole Antenna With Meandered Series Power Divider for UHF RFID Tag and Handheld Reader Devices", IEEE Transactions on Antennas and Propagation, Vol.67, Issue.6, pp. 4195–4199, March 2019.
21. Arvind DS, et al., "Adjustable Road Divider Using IoT", Journal of Science, Computing and Engineering Research, Vol.3, Issue.1, pp. 222–225, 2022.
22. Siva RV, Srinivasa Rao M, Pushpa Rani K, "Shrewd Street Dividers Driven by IoT Technology", International Journal of Civil Engineering and Technology, Vol.8, Issue.7, pp. 385–389, 2017.
23. Agrawal S, Maheshwari P, "Controlling of Smart Movable Road Divider and Clearance Ambulance Path Using IoT Cloud", IEEE International Conference on Computer Communication and Informatics, pp. 1–4, 2021.
24. Sai Sri VS, et al., "Automatic Movable Road Divider Using Arduino UNO with Node Micro Controller Unit (MCU)", Materials Today, Vol.80, pp. 1842–1845, 2023.
25. Dinesh M., "Movable Road Dividers for Road Traffic Control with Automated Light Solutions in Embedded", Turkish Journal of Computer and Mathematics Education, Vol.12, Issue.12, pp. 2187–2191, 2021.
26. Immadi G, et al. "Traffic Density Management Using Movable Divider and RFID", International Research Journal of Engineering and Technology, Vol.5, Issue.7, pp. 6130–6134, 2020.
27. Lai YL, Chou YH, Chang LC, "An intelligent IoT Emergency Vehicle Warning System Using RFID and Wi-Fi Technologies for Emergency Medical Services", Technology and Health Care, Vol.26, Issue.1, pp. 43–55, 2018.
28. Amir S, Kamal MS, Khan SS, Salam KA, "PLC-Based Traffic Control System With Emergency Vehicle Detection and Management", IEEE International Conference on Intelligent Computing, Instrumentation and Control Technologies, Vol.6, pp. 1467–1472, 2017.

29. Naik T, Roopalakshmi R, Ravi ND, Jain P, Sowmya BH, "RFID-Based Smart Traffic Control Framework for Emergency Vehicles", IEEE Second International Conference on Inventive Communication and Computational Technologies, pp. 398–401, 2018.
30. Saradha BJ, Vijayshri G, Subha T, "Intelligent Traffic Signal Control System for an Ambulance Using RFID and Cloud", 2nd IEEE International Conference on Computing and Communications Technologies, pp. 90–96, 2017.

Chapter 5

Semantic Based Emotional Voice Coding Using Feed Forward Artificial Neural Network for Smart Home Devices

Firos A

5.1 INTRODUCTION

5.1.1 Artificial Neural Network for Smart Home Devices

Artificial Neural Networks (ANNs) are a type of machine learning model inspired by the human brain's neural networks [1]. They have found widespread applications in various fields, including smart home devices. ANNs are particularly useful for tasks involving pattern recognition, data analysis, and decision-making based on complex input data [2–6].

Here's how an Artificial Neural Network can be applied to smart home devices:

1. **Home Automation and Control:** ANNs can be used to control and automate various smart home devices [7]. For instance, they can analyze data from sensors (e.g., motion sensors, temperature sensors, light sensors) and learn patterns of occupants' behavior to adjust settings automatically. This could include turning on/off lights, adjusting thermostat settings, or controlling smart appliances based on usage patterns.
2. **Energy Management:** ANNs can help optimize energy consumption in a smart home by learning the energy usage patterns of different devices and suggesting ways to reduce energy waste [8]. They can also predict peak energy demand times and automatically adjust devices to reduce energy usage during those periods.
3. **Security and Intrusion Detection:** Neural networks can be used for intelligent security systems. They can analyze data from security cameras, motion sensors, and other sensors to detect unusual activities or potential intrusions [9]. The network can learn to differentiate between regular movements (e.g., family members) and suspicious activities, raising alerts when necessary.
4. **Natural Language Processing (NLP) for Voice Control:** ANNs are the backbone of voice assistants like Amazon Alexa or Google Assistant. NLP algorithms, often based on recurrent neural networks (RNNs), enable smart home devices to understand and respond to voice commands, making interaction with the devices more user-friendly.

DOI: 10.1201/9781003480181-5

5. **Personalized Recommendations:** Artificial Neural Networks can be employed to understand user preferences based on past interactions and suggest personalized content, such as music playlists, movie recommendations, or even preferred lighting settings.
6. **Predictive Maintenance:** Smart home devices, like HVAC systems or smart appliances, can use ANNs to predict maintenance needs based on usage patterns and sensor data. This enables timely maintenance to prevent breakdowns and optimize device lifespan.
7. **Gesture and Emotion Recognition:** ANNs can be employed in smart home devices to recognize gestures or emotional cues from users. For example, a camera-based system could recognize specific hand gestures to control devices, or an emotion recognition system could adapt the home environment based on the user's mood.

To develop an ANN for smart home devices, one would typically follow these steps:

1. **Data Collection:** Gather data from various smart home sensors and devices. This data will be used to train and test the neural network.
2. **Data Preprocessing:** Clean and prepare the data for training. This might involve removing noise, handling missing values, and normalizing the data.
3. **Model Architecture:** Decide on the neural network architecture that suits your specific smart home application. The architecture might involve various layers of neurons, including input, hidden, and output layers.
4. **Training:** Use the prepared data to train the neural network. This process involves feeding the data into the network, adjusting the network's parameters (weights and biases) iteratively to minimize prediction errors.
5. **Validation and Testing:** Validate the trained model on a separate dataset to ensure it generalizes well to unseen data. Fine-tune the model if necessary.
6. **Deployment:** Once the ANN has been trained and validated, integrate it into the smart home system and start using it to make predictions and control devices.

Remember that the success of an ANN in a smart home setting heavily depends on the quality and quantity of data available for training. Additionally, continuous updates and improvements may be required to keep the system effective and secure.

5.1.2 The Feed-Forward Artificial Neural Network (FFANN)

The term "Artificial Neural Network" (ANN) is a broad concept that encompasses various types of neural networks [10]. One of the fundamental types of neural networks is the feed-forward artificial neural network, also known simply as a feed-forward neural network.

An Artificial Neural Network (ANN) is a computational model inspired by the structure and functioning of biological neural networks in the brain [11]. It consists of interconnected nodes, called neurons, organized into layers. Each neuron receives input, processes it, and produces an output that is passed to the next layer of neurons until the final output is generated. ANN can be used for tasks such as classification, regression, pattern recognition, and decision-making based on input data [12].

A Feed-Forward Artificial Neural Network is a specific type of neural network where the data flows in only one direction, from the input layer through one or more hidden layers to the output layer [13, 14]. There are no cycles or loops in the network, meaning the information flows forward and does not form feedback loops. This architecture makes it suitable for many machine learning tasks.

Feed-forward neural networks are typically organized into three types of layers:

1. **Input Layer:** The first layer that receives the input data. Each neuron in this layer represents a feature or attribute of the input data.
2. **Hidden Layers:** These layers come after the input layer and precede the output layer. They are responsible for learning patterns and representations from the input data. A feed-forward neural network can have one or multiple hidden layers, depending on the complexity of the problem.
3. **Output Layer:** The final layer that produces the network's output. The number of neurons in this layer depends on the nature of the task. For instance, in a binary classification problem, there would be one neuron for each class (e.g., one for positive and one for negative).

Feed-forward neural networks are widely used in various applications, including image recognition, natural language processing, and time-series prediction. However, they have certain limitations in handling sequential and time-dependent data, which led to the development of other types of neural networks, such as Recurrent Neural Networks (RNNs) and Long Short-Term Memory (LSTM) networks, to address these specific challenges.

So, a feed-forward artificial neural network is a specific configuration of an artificial neural network where data flows only in one direction, from input to output, making it well-suited for many standard machine learning tasks.

5.1.3 Fuzzy Measures for Smart Home Devices

Fuzzy measures, along with fuzzy logic, are mathematical frameworks that deal with uncertainty and imprecision. They can be applied to various domains, including smart home devices, to enhance decision-making processes and deal with vague and uncertain information.

In the context of smart home devices, fuzzy measures can be used for:

1. **Smart Home Automation:** Fuzzy logic can be applied to automate smart home devices by incorporating linguistic rules that consider various inputs and their

degrees of truth. For example, a rule like "If the temperature is very hot and the humidity is high, then turn on the air conditioner" utilizes fuzzy measures to handle the imprecise nature of linguistic terms like "very hot" and "high."
2. **Context-Aware Systems:** Fuzzy measures can enable smart home devices to adapt to changing contexts and user preferences in a more flexible and adaptive manner. By incorporating fuzzy reasoning, devices can assess various inputs (e.g., time of day, user behavior, outside temperature) and make decisions based on multiple criteria.
3. **Occupant Comfort:** Fuzzy logic can be utilized to assess and optimize occupant comfort by considering multiple factors such as temperature, humidity, and user preferences. Fuzzy control systems can adjust smart thermostats, lighting, and other devices to create comfortable and personalized environments.
4. **Energy Efficiency:** Fuzzy measures can play a role in optimizing energy consumption in smart homes. By taking into account various factors such as occupancy, weather conditions, and appliance usage, fuzzy control systems can manage energy consumption more intelligently.
5. **Fault Detection and Diagnostics:** Fuzzy logic can be applied to identify anomalies and potential faults in smart home devices. By considering multiple sensor inputs and their degrees of deviation from expected values, fuzzy-based fault detection systems can help improve the reliability and safety of smart home devices.
6. **Smart Security Systems:** Fuzzy measures can enhance the decision-making process in smart security systems by considering various inputs, such as motion sensor data, time of day, and occupant behavior, to assess potential threats and trigger appropriate responses.

To implement fuzzy measures in smart home devices, the following steps are typically involved:

1. **Define Linguistic Variables:** Identify the relevant linguistic variables for the smart home application, such as "temperature," "humidity," "brightness," etc. Divide each variable into linguistic terms (e.g., "hot," "cold," "low," "high") and specify their membership functions.
2. **Create Fuzzy Rules:** Formulate fuzzy rules based on expert knowledge or data-driven approaches. These rules establish relationships between the linguistic variables and define how to make decisions based on them.
3. **Fuzzify Inputs:** Convert crisp inputs from sensors into fuzzy values using the membership functions defined for each linguistic term.
4. **Apply Fuzzy Logic Operations:** Utilize fuzzy logic operations (e.g., fuzzy AND, fuzzy OR) to combine multiple inputs and fuzzy rules to obtain fuzzy outputs.
5. **Defuzzify Outputs:** Convert fuzzy outputs into crisp values to command the smart home devices accordingly.

By incorporating fuzzy measures into smart home devices, these systems can handle uncertainties and make more nuanced decisions, leading to improved user experience, energy efficiency, and adaptability to changing conditions.

5.1.4 Fuzzy ANN for Smart Home Devices

A Fuzzy Artificial Neural Network (Fuzzy ANN) is a hybrid approach that combines the benefits of fuzzy logic and artificial neural networks. It integrates the ability of fuzzy logic to handle uncertainty and imprecision with the powerful learning and pattern recognition capabilities of neural networks. In the context of smart home devices, a Fuzzy ANN can be used to enhance decision-making and control processes while considering the inherent uncertainties in the environment.

Here's an outline of how a Fuzzy ANN can be used for smart home devices:

1. **Fuzzy Inference System (FIS):** At the core of a Fuzzy ANN, there is a Fuzzy Inference System (FIS) that contains fuzzy rules and membership functions. The FIS takes linguistic variables as inputs (e.g., temperature, humidity, occupancy level) and converts them into fuzzy sets using membership functions. These fuzzy sets represent the degrees of truth for each linguistic term.
2. **Fuzzification:** The inputs from smart home sensors are fuzzified, i.e., converted into fuzzy values using appropriate membership functions. Fuzzification allows the system to handle imprecise and uncertain sensor readings.
3. **Fuzzy Rule Base:** The FIS **contains** a set of fuzzy rules that encode expert knowledge or learned patterns. These rules establish the relationships between the fuzzy inputs and outputs. For example, a rule might be defined as "IF temperature is high AND humidity is low THEN adjust thermostat to a comfortable level."
4. **Inference Engine:** The inference engine in the FIS applies the fuzzy rules to the fuzzified inputs to determine the fuzzy outputs.
5. **Fuzzy Neural Network (FNN):** In a Fuzzy ANN, the output of the FIS is then used as input to an artificial neural network. The Fuzzy ANN incorporates this hybrid structure to take advantage of both fuzzy logic and neural network capabilities.
6. **Neural Network Training:** The neural network part of the Fuzzy ANN is trained using backpropagation or other learning algorithms to fine-tune its parameters and learn the best mapping between the fuzzy inputs and the desired outputs.
7. **Defuzzification:** Once the Fuzzy ANN produces a crisp output through the neural network, the defuzzification process converts the crisp output back into actionable commands for smart home devices.

5.2 THE BACKGROUND

5.2.1 Smart Home Devices

Smart home devices refer to a wide range of electronic and internet-connected devices that are designed to enhance automation, convenience, security, and energy efficiency within a home. These devices are part of the larger concept of the Internet of Things (IoT), where various objects and devices are connected to the internet and can communicate with each other and with users.

Some common types of smart home devices include:

1. **Smart Device**: A smart device, also known as an adaptive device or multiple-device system, is an advanced device technology used in wireless communication systems to improve the signal quality and increase the efficiency of data transmission and reception. Unlike traditional single-device systems, smart devices use multiple devices and sophisticated signal processing techniques to dynamically adapt their radiation patterns based on the radio frequency (RF) environment and the location of the communicating devices.
2. **Smart Speakers and Voice Assistants:** Devices like Amazon Echo with Alexa, Google Nest Hub with Google Assistant, or Apple HomePod with Siri act as voice-activated assistants that can perform various tasks, answer questions, control other smart devices, and more.
3. **Smart Lighting:** Smart light bulbs, switches, and plugs that can be controlled remotely via a smartphone app or voice command. They often offer features like dimming, color changing, and scheduling.
4. **Smart Thermostats:** These devices allow homeowners to remotely control the heating and cooling of their homes, learn user preferences, and optimize energy usage to save costs.
5. **Smart Home Security Systems:** Including smart doorbells with video cameras, smart locks, motion sensors, and window/door sensors that can enhance home security and provide real-time alerts and monitoring.
6. **Smart Cameras and Baby Monitors:** Internet-connected cameras that enable remote monitoring of your home, pets, or babies through a smartphone app.
7. **Smart Appliances:** Appliances like refrigerators, ovens, washing machines, and robotic vacuums with internet connectivity that allow remote control and scheduling.
8. **Smart Entertainment Systems:** Devices like smart TVs and media streaming devices that can be controlled via smartphone apps and voice commands.
9. **Smart Home Hubs:** Centralized devices that connect and control various smart home devices from different manufacturers, enabling seamless integration and automation.

10. **Smart Smoke Detectors and Carbon Monoxide Detectors:** These detectors can send alerts to your smartphone when smoke or dangerous levels of carbon monoxide are detected.
11. **Smart Water Leak Detectors:** Devices that can detect water leaks and send alerts to prevent water damage.
12. **Smart Plugs and Power Strips:** These devices can turn regular appliances into smart ones, allowing remote control and energy monitoring.
13. **Smart Blinds and Curtains:** Motorized window treatments that can be controlled remotely or programmed to open/close at specific times.

The key benefits of smart home devices include convenience, energy efficiency, enhanced security, and automation. They allow homeowners to control and monitor their homes remotely, save energy and money, and enjoy a more seamless and connected living experience.

5.2.2 Advantages of Fuzzy ANN for Smart Home Devices

The advantages of Fuzzy ANN for smart home devices are:

1. **Handling Uncertainty:** Fuzzy logic allows the system to handle uncertainties in sensor data and user preferences, which are common in a smart home environment.
2. **Interpretability:** Fuzzy rules can be interpreted and understood by humans, making it easier to incorporate domain knowledge into the system.
3. **Adaptability:** Fuzzy ANN can learn and adapt to changing patterns and user behavior through the neural network's learning capabilities.
4. **Efficient Control:** The hybrid approach can lead to more efficient and effective control of smart home devices, considering both linguistic reasoning and learned patterns.

Fuzzy ANN is a promising approach for smart home devices, especially when dealing with imprecise data and complex decision-making processes. It offers a unique way to enhance automation, energy efficiency, and user experience in a smart home environment. It's worth noting that the smart home industry is rapidly evolving, and new devices and technologies are continually being introduced to the market. As the technology advances and becomes more affordable, the adoption of smart home devices is expected to increase, leading to more interconnected and intelligent homes in the future.

5.2.3 Artificial Neural Networks for Automation

Artificial Neural Networks (ANNs) are at the core of many smart devices, providing the intelligence and decision-making capabilities that enable automation and enhanced functionality. The technologies that power ANN-based smart devices include:

1. **Deep Learning:** Deep learning is a subset of machine learning that involves training artificial neural networks with multiple hidden layers. Deep learning techniques, such as convolutional neural networks (CNNs) and recurrent neural networks (RNNs), are widely used in smart devices for tasks like image recognition, natural language processing, and time-series prediction.
2. **Backpropagation:** Backpropagation is a supervised learning algorithm used to train neural networks. It enables the adjustment of the network's weights and biases based on the error between the predicted output and the actual target output. Backpropagation is crucial for the learning process in ANN-based smart devices.
3. **Convolutional Neural Networks (CNNs):** CNNs are specialized neural networks commonly used for image and video processing tasks. They employ convolutional layers to automatically learn relevant features from raw pixel data, making them ideal for tasks like object detection, image classification, and facial recognition in smart cameras and other imaging devices.
4. **Recurrent Neural Networks (RNNs):** RNNs are designed to handle sequential data, making them suitable for time-series analysis and tasks like speech recognition and natural language processing in smart speakers and voice assistants.
5. **Long Short-Term Memory (LSTM) Networks:** LSTM networks are a type of RNN designed to overcome the vanishing gradient problem and effectively handle long-range dependencies in sequential data. They are commonly used in smart devices for tasks like speech recognition, text generation, and time-series prediction.
6. **Natural Language Processing (NLP):** NLP technologies enable smart devices to understand and process human language, allowing for voice-based control and interaction with devices like smart speakers and voice assistants.
7. **Edge Computing:** Edge computing involves performing data processing and analysis closer to the source of data (e.g., smart device or sensor) rather than sending all data to centralized cloud servers. This approach is beneficial for reducing latency, enhancing privacy, and enabling real-time decision-making in smart devices.
8. **Cloud Computing:** While edge computing brings processing closer to the devices, cloud computing plays a crucial role in storing and managing vast amounts of data generated by smart devices. Cloud-based machine learning models can be trained on extensive datasets and deployed to smart devices for local inference.
9. **Internet of Things (IoT) Protocols:** To enable communication and connectivity among smart devices, various IoT protocols and standards like MQTT, CoAP, and WebSocket are used. These protocols facilitate data exchange and control commands between smart devices and cloud platforms.

10. **GPU Acceleration:** Training and inference of deep neural networks can be computationally intensive. Graphics Processing Units (GPUs) and specialized hardware like Tensor Processing Units (TPUs) are used to accelerate neural network computations and improve the performance of smart devices.

These technologies work in synergy to power the intelligence and decision-making capabilities of ANN-based smart devices, enabling them to perform complex tasks and provide valuable services to users in various domains like home automation, healthcare, transportation, and more.

5.2.4 Comparison between Rule-Based Automation and the Machine Learning Automation

Here are two types of home automation algorithms commonly used in the context of smart homes:

5.2.4.1 *Rule-Based Algorithms*

Rule-based algorithms involve setting up specific conditions and corresponding actions. These algorithms follow predefined rules or logic to trigger actions based on certain events or inputs. In the context of home automation, rule-based algorithms can be used to automate tasks according to predefined rules set by the user. For example:

- **If-Then Rules**: If a specific condition is met (e.g., time of day, occupancy status, sensor readings), then a corresponding action is executed (e.g., turn on lights, adjust thermostat).
- **Boolean Logic Rules**: Combining conditions using logical operators (AND, OR, NOT) to create more complex rules. For instance, if it's after sunset AND no one is home, then turn on outdoor lights.
- **Event-Triggered Actions**: Triggering actions based on specific events, such as motion detection, door opening, or a specific device turning on/off.

Rule-based algorithms are relatively straightforward to implement and provide users with a certain level of control over their automation systems. However, they may not handle complex scenarios well and can become cumbersome when dealing with a large number of rules.

5.2.4.2 *Machine Learning Algorithms*

Machine learning algorithms use data-driven approaches to automate tasks and make decisions based on patterns learned from historical data. In the context of home automation, machine learning algorithms can bring a higher level of adaptability and intelligence to the system. Examples include:

- **Energy Optimization**: Machine learning algorithms can learn usage patterns and adjust smart devices like thermostats and lights to optimize energy consumption based on user preferences and changing conditions.
- **Predictive Analysis**: Algorithms can predict user behavior and adjust settings accordingly. For example, anticipating when a user typically arrives home and pre-adjusting the temperature.
- **Anomaly Detection**: Machine learning algorithms can identify unusual patterns or events, such as a sudden spike in energy consumption or unexpected activity, and alert users or take corrective actions.
- **Personalization**: Algorithms can learn user preferences and adjust automation settings to create a more personalized and comfortable environment for each user.

Machine learning algorithms require training on relevant data, and they can adapt to changing conditions and learn from new data over time. However, they might require more computational resources and expertise to implement effectively.

Both types of algorithms have their advantages and use cases. Rule-based algorithms are simple to set up and are suitable for straightforward automation tasks, while machine learning algorithms can provide more adaptive and intelligent automation but may require more initial setup and ongoing maintenance. Often, a combination of both approaches is used to achieve a balance between user control and intelligent automation.

5.2.5 Preference Leveled Evaluation Functions Method to Construct Fuzzy Measures

This section discusses a method to construct parameterized fuzzy measures based on Preference Leveled Evaluation Functions.

The Preference Leveled Evaluation Functions (PLEFs) method is a technique used to construct fuzzy measures, particularly in the context of decision-making and uncertainty management. Fuzzy measures are mathematical functions that assign a degree of "importance" or "coverage" to subsets of a given set, representing uncertainty or vagueness in decision problems.

The PLEFs method is a systematic approach to construct fuzzy measures based on a set of preference statements provided by a decision-maker. These preference statements express the decision-maker's preferences or evaluations of the relative importance of various subsets of the set under consideration. The method seeks to find a fuzzy measure that satisfies these preference statements as closely as possible.

The PLEFs method works in the following way:

The decision-maker provides a set of preference statements that indicate their relative preferences for different subsets of the given set. These preference statements are usually in the form of comparisons between subsets. Based on the provided preference statements, preference levels are constructed for each subset. A preference level represents the degree of preference or importance assigned to a subset relative to other subsets. The preference levels are then normalized to ensure they lie between 0 and 1, representing relative importance on a fuzzy scale. The normalized preference levels are used to construct a fuzzy measure. This measure assigns degrees of importance or coverage to each subset based on the decision-maker's preferences. The constructed fuzzy measure is validated against the provided preference statements to ensure it adequately represents the decision-maker's preferences. If the fuzzy measure does not fully satisfy the preference statements, adjustments can be made to the preference levels or the fuzzy measure. This process may involve iterations to fine-tune the fuzzy measure to better reflect the decision-maker's preferences.

5.2.6 Construction Method Using Preference Leveled Evaluation Functions

A normalized weights function $b = B^{\langle x \rangle}$ having the complementary preference $\hat{\lambda}_b \in [0,1]^x$, and say $\left(F, 2^F, t_X\right)$ be a given fuzzy Frequency measure space where $F = \{x\,(1),\ \ldots,\ x\,(n)\}$. The values, that is the channel values in 2400 MHz range, are given in Table 5.1. There B is the frequency range and $B^{\langle x \rangle}$ is the x-ary aggregate function. T_X the space of all fuzzy measures on X [11, 15].

Let us calculate a parameterized fuzzy value $t_G^{\varphi_1,\ldots\ldots,\varphi_s}$ for parameterized fuzzy measure space $(\mathrm{G}, 2^{\mathrm{G}}, t_G^{\varphi_1,\ldots,\varphi_s})$ where $\mathrm{G} = \{1,2,3,\ldots,x\}$ [11].

We denote by $\left([0,1]^{\mathrm{n}}\right)^{\mathrm{G}}$ the space of all mappings [16],

$$\varphi{:}G \rightarrow \left([0,1]^n\right) \tag{5.1}$$

We denote parameterized fuzzy measure $t_G^{(b,\varphi)} : 2^G \rightarrow [0,1]$ such that for any $C \in 2^G (C \neq \varphi)$, we have

$$t_G^{(b,\varphi)}(C) = t_F(\{f \in F | \left(\vee_{g \in C}\, \varphi(g)\right)(f) \geq \hat{\lambda}_b\left(|C|\right)\}) \tag{5.2}$$

and still define $t_G^{(b,\varphi)}(\varphi) = 0$ for all b and φ. This is adjustability conferring to an anticipated preference value b.

Different complementary preference value $\hat{\lambda}_b$ aid as diverse thresholds to be surpassed by union assessments $\vee_{g \in C}\, \varphi(g)$.

Also, when $C = G$, it follows $\left\{f \in F \mid \left(\vee_{g \in C} \varphi(g)\right)(f) \geq \hat{\lambda}_b\left(|C|\right)\right\} = F$,

$$\text{then } t_G^{(b,\varphi)}(G) = 1, \text{ always.} \tag{5.3}$$

Usually, the greater conservativeness, the lesser the parameterized fuzzy measure $t_G^{(b,\Theta)}$, but it is not always the case. Usually, we will get the expected results, if b and b', the two preferences satisfy some special relation.

$\left(B^{\langle x \rangle}, <\right)$ gives a whole lattice with ordering $<$ demarcated such that for any two $b,\ b' \in B^{\langle x \rangle}$, $b < b' \Leftrightarrow \lambda_b < \lambda_{b'}$ [15].

For any two $b,\ b' \in B^{\langle x \rangle}$, if $b < b'$, then $\lambda_b(j) = \sum_{m=1}^{j} b(m) \leq \sum_{m=1}^{j} b'(m) = \lambda_{b'}(j)$ for all $j \in \{1, \ldots\ldots, x\}$. So, the relation about conservativeness [17], *cn*.

$$cn(b) = \sum_{j=1}^{x} b(j).\frac{j-1}{n-1} = \frac{1}{x-1}\sum_{u=1}^{x}\sum_{m=s}^{x} b(m) = \frac{1}{x-1}\sum_{u=1}^{x}\left[1 - \sum_{m=1}^{u-1} b(m)\right]$$

$$= \frac{1}{x-1}\sum_{u=1}^{x}\left[1 - b(u) - \sum_{m=1}^{u} b(m)\right]$$

$$= \frac{x}{x-1} - \frac{1}{x-1} - \frac{1}{x-1}\sum_{u=1}^{x}\sum_{m=1}^{u} b(m)$$

$$= 1 - \frac{1}{x-1}\sum_{u=1}^{x}\sum_{m=1}^{u} b(m) \geq 1 - \frac{1}{x-1}\sum_{u=1}^{x}\sum_{m=1}^{u} b'(m) = c\ n(b') \tag{5.4}$$

Also,

$$\text{for any two } C,\ C' \in 2^{G} \text{ with } C \subset C',\ t_G^{(b,\varphi)}(C) \leq t_G^{(b,\varphi)}(C\blacksquare') \tag{5.5}$$

$$\text{for any two } b,\ b' \in B^{\langle x \rangle} \text{ with } b < b',\ t_G^{(b,\varphi)}(C) \leq t_G^{(b',\varphi)}(C); \tag{5.6}$$

for any two φ, $\left(\varphi' \in 2^{F}\right)^{G}$ with $\Theta < \Theta'$ (here "<" is $\varphi(j) < \varphi'(j)$ for all $j \in G$),

$$t_G^{(b,\varphi)}(C) \leq t_G^{(b,\varphi')}(C) \tag{5.7}$$

Readers may refer the paper [9] for proof of the equations (5.1) to (5.7).

5.3 PROPOSED MODEL

Figure 5.1 shows the proposed smart device algorithm that employs a Parameterized Fuzzy Measures Decision Making Model (PFMDMM) Based on Preference Leveled Evaluation Functions. The model is fed the input channels listed in Table 5.1 in order to determine the optimum channel for the node. This incorporates a range-based RSSI (Signal Strength Indicator). The node's quality signals are then obtained using PFMDMM and properly formatted for feeding into the ANN for classification. With the use of BPNN, the signal weights for the BPNN training stage are determined by an arbitrary value according to the values in Table 5.1, and they are then tuned for maximum performance during the iterative learning process. To determine if the system that was obtained accurately classifies the signal to the best preferred pattern and other signal portions, the BPNN is tested against a variety of test samples of signals during the testing phase.

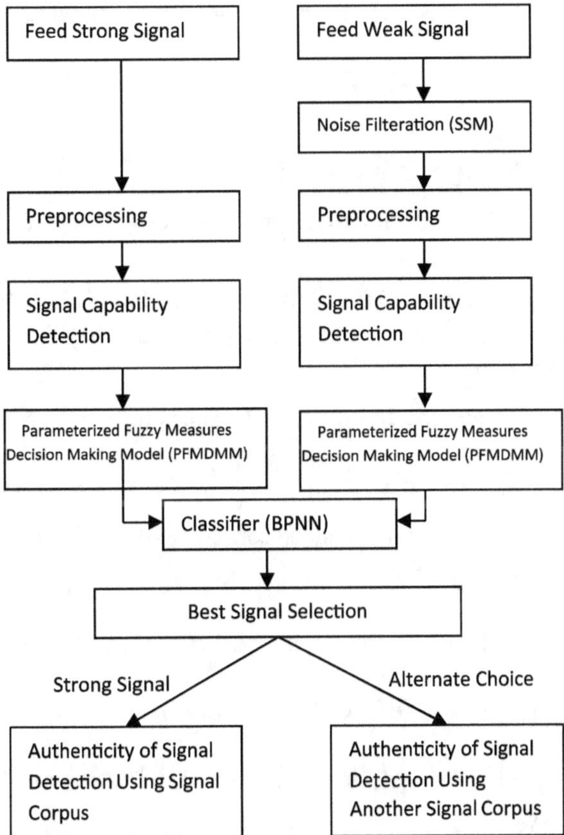

Figure 5.1 Block diagram of proposed PFMDMM based on preference leveled evaluation functions for best signal detection in smart device.

Algorithm 5.1: PFMDMM Based on Preference Leveled Evaluation Functions for Best Signal Detection in Smart Device

Input: Input channels of 2400 MHz range signals.

Output: Categorizes the signal to best preferred signal and other signal parts.

Start

1. Input channels of 2400 MHz range signals.
2. A range-based RSSI (Signal Strength Indicator) for signal range detection.
3. PFMDMM to recognize best signals for the node; arranged in correct format to feed into the ANN for classification.
4. *Training Stage:* Weights of the Feed-Forward Neural Network were given by some arbitrary value as per Table 5.1 and it is then tuned for optimal during the iterative learning procedure with the help of back propagation algorithm.
5. *Testing Stage:* The neural network is tested against a variety of test samples of signals, to ensure whether the acquired system correctly categorizes the signal to best preferred signal and other signal parts.
6. Categorizes the signal to best preferred signal and other signal parts.

Stop

5.3.1 Signal Range Detection

The steps for range-based RSSI based signal range detection given in [14].

5.3.2 Parameterized Fuzzy Measures Decision Making Model Clustering Method (PFMDMM) for Signal Detection

With φ: G → [0, 1]4, assume the normalized weights function $b \in B^{\langle 3 \rangle}$ is assigned to be w = (0.2, 0.5, 0.7).

Then calculate $\lambda_b = (0.2, 0.4, 1)$ and $\hat{\lambda}_b = (0.9,\ 0.7,\ 0)$, with $\mathrm{cn}(b) = \mathrm{cn}(\lambda_b) = \mathrm{con}(\hat{\lambda}_b) = 0.76$ showing a fairly added conservative measure in our preference. Then, using (5.2) we have for t_G:

$$t_G^{(b,\varphi)}(\{1\}) = t_F(\{f \in F | (\varphi(1))(f) \geq \hat{\lambda}_b(1)\}) = t_F(\{4\}) = 0.3,$$

$$t_G^{(b,\varphi)}(\{2\}) = t_F(\{f \in F | (\varphi(2))(f) \geq \hat{\lambda}_b(1)\}) = t_F(\{4\}) = 0.5,$$

$$t_G^{(b,\varphi)}(\{3\}) = t_F(\{f \in F | (\varphi(3))(f) \geq \hat{\lambda}_b(1)\}) = t_F(\{3\}) = 0.5,$$

$$t_G^{(b,\varphi)}(\{1,2\}) = t_F(\{f \in F | (\varphi(1) \vee \varphi(2))(f) \geq \hat{\lambda}_b(2)\}) = t_F(\{1,2,3,4\}) = 1,$$

$$t_G^{(b,\varphi)}(\{1,3\}) = t_F(\{f \in F | (\varphi(1) \vee \varphi(3))(f) \geq \hat{\lambda}_b(2)\}) = t_F(\{1,3,4\}) = 0.9,$$

$$t_G^{(b,\varphi)}\left(\{2,3\}\right) = t_F(\{f \in F | \left(\varphi(2) \vee \varphi(3)\right)(f) \geq \hat{\lambda}_b(2)\}) = t_F\left(\{1,3,4\}\right) = 0.9,$$

$$t_G^{(b,\varphi)}\left(\{1,2,3\}\right) = t_F(\{f \in F | \left(\varphi(1) \vee \varphi(2) \vee \varphi(3)\right)(f) \geq \hat{\lambda}_b(3)\}) = t_F\left(\{1,2,3,4\}\right) = 1.$$

5.3.3 Deep Learning Model for Best Signal Selection

To develop a deep learning model for wireless best signal selection, we can use a Convolutional Neural Network (CNN) with appropriate modifications to suit the task of signal strength prediction [18]. The goal of the model is to take as input the information about available wireless signals and predict the best signal to connect to for optimal performance [19]. Here's an outline of the steps to build such a deep learning model:

If we wish to incorporate the idea of deep learning into the design of smart devices, we can establish five parameters or variables at the input side: Inp = W, L, h, r, f = "Patch width," "Patch length," "Thickness of the dielectric substrate," "Dielectric constant," and "Operating frequency." If we provide our deep neural network this input (Inp) data, we should anticipate optimum values for these five parameters as an output. Keep in mind that a deep learning network can only function successfully if we offer enough training data, since it will learn based on training data that are of extremely high quality. The large-scale (1000 hours) corpus of read English speech (v. 2009–05–08) dataset [20] was utilized in the proposed model's deployment to train the BPNN with anticipated/best permeates for any access point. Back propagation neural networks (BPNNs) were employed in the proposed model to recommend the optimal signal for the access node.

5.4 EXPERIMENTAL RESULTS

With the aid of the PFMDMM algorithm created in MATLAB, the signal vectors in the study are divided into the best preferred signal and other signal components. Signals in the 2.4 GHz range are fed into the model's input channels in order to determine the optimum channel for the node. The received signal strength indicator (RSSI) approach for signal range detection is used. Then, the best signals for the node are chosen and put in the appropriate manner to feed into the Artificial Neural Network for classification using the Parameterized Fuzzy Measures Decision Making Model (PFMDMM) clustering approach. The signal strengths for BPNN's training stage were determined by an arbitrary value according to the values in Table 5.1 and were then optimized during the iterative learning process with the aid of the back propagation algorithm. To check whether the learned system accurately classifies the signal to the best desired signal alongside other signal portions, the neural network is tested against a variety of test samples of signals. To obtain effective and quick signal identification of architecture, the PFMDMM is applied. According to the research, the system is typically 94% efficient for signal identification.

5.4.1 Data Source

The LibriSpeech dataset primarily consists of audio recordings of audiobooks along with their corresponding transcriptions. However, when using the dataset for automatic speech recognition (ASR) research and development, it's common to extract various features from the audio data to facilitate training and processing by ASR models. Here are the typical steps involved in preparing features from the LibriSpeech audio recordings:

1. Audio Preprocessing: The raw audio recordings are usually in WAV format. Before extracting features, you might need to perform preprocessing steps such as resampling to a common sample rate (e.g., 16 kHz), and possibly converting to mono if the audio is in stereo.
2. Feature Extraction: The most common features extracted from audio data for ASR are spectrogram-based features. These features provide a way to represent the audio information in a format that's suitable for input to neural networks. Some common feature extraction techniques include:
3. Mel Frequency Cepstral Coefficients (MFCCs): MFCCs are a set of coefficients that capture the mel-frequency characteristics of the audio signal. They are widely used in ASR due to their effectiveness in representing speech.
4. Filterbank Energies: Filterbank energies are another representation that captures the energy content in different frequency bands. They are often used alongside MFCCs.
5. Spectrograms: Spectrograms are visual representations of the audio signal's frequency content over time. They can be used directly or transformed into other features like log mel spectrograms.
6. Feature Normalization: Once features are extracted, it's common to perform normalization to ensure consistent ranges and improve model convergence during training. This might involve mean and variance normalization or other techniques.
7. Data Splitting: After feature extraction and normalization, the dataset is usually split into training, validation, and testing sets. This ensures that the ASR model can be trained, validated, and evaluated effectively.
8. Model Training: With the features and data splits prepared, you can train ASR models using various architectures such as recurrent neural networks (RNNs), convolutional neural networks (CNNs), and attention-based models.

It's important to note that feature extraction and processing are crucial steps in ASR pipeline, and the specific features and methods used might vary depending on the research and model architecture. If you're interested in working with the LibriSpeech dataset, make sure to refer to the documentation and guidelines provided by the dataset creators and the ASR framework you're using.

Table 5.1 MFCC Features of Data of cu/Device Dataset for Normal Speech (v. 2009–05–08)

26	12.93514	12.78101	12.07424	7.64229	12.93514
27	7.55225	4.16052	1.61571	2.12973	7.55225
28	1.16899	2.67745	2.7283	1.93349	1.16899
29	2.05475	1.94528	4.33331	0.08896	2.05475
30	8.37376	6.05577	7.8039	9.49206	8.37376
31	1.49524	1.11541	0.88058	2.48971	1.49524
32	3.32171	1.50352	3.74667	1.33721	3.332171
33	0.23734	4.24566	2.51809	5.3929	0.23734
34	3.7705	5.9152	0.18575	0.07774	3.7705
35	0.68559	0.25562	0.93503	1.93664	0.68559
36	5.76186	7.02887	2.88118	1.2207	5.76186

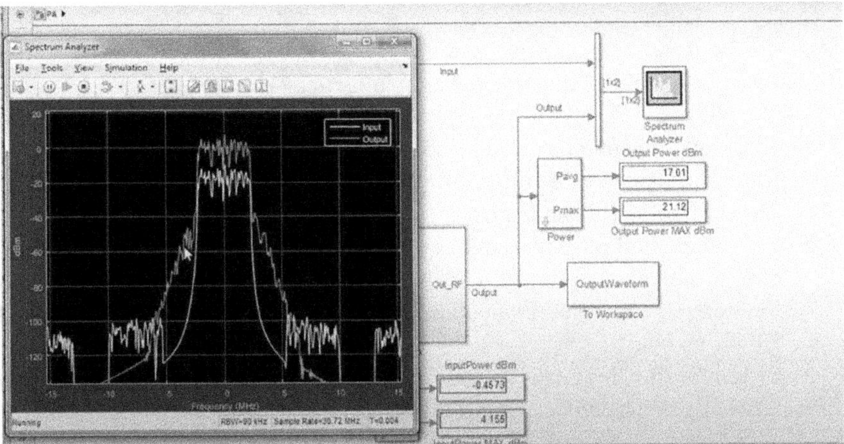

Figure 5.2 **PFMDMM based signal detestation simulation application.**

The LibriSpeech dataset can be obtained from the official website or repository where it is hosted. As of my last update in September 2021, the dataset was available for free download from the source [21]. This research has taken MFCC features of data of cu/device dataset for normal speech. Also a depiction of MFCC features of data of cu/device dataset for normal speech is provided in Table 5.1.

5.4.2 Illustrative Example

Figure 5.2 shows the screenshot of the PFMDMM based signal detestation. The proposed signal recommended by the network for a node is the one shown in blue.

Experiments with Laboratory Data: Experiments are carried out on the lab data to confirm the effectiveness of BPNN models that use the PFMDMM technology for detection of signals. Producer accuracy, user accuracy, and overall accuracy were determined to be 91.50%, 95.32%, and 94.99% accordingly for accurate signal detection metal.

5.5 CONCLUSION

This work presents a novel BPNN model that uses the PFMDMM signal classification system to determine the optimal signal for a node. The trials' findings demonstrated that the parameterized fuzzy measures decision-making model for best signal recognition in smart device exhibits encouraging performances in terms of best signal proposal for a WLAN node. This model is built on preference leveled assessment functions.

To the best of our knowledge, this is the first study to employ the preference-leveled evaluation functions-based Parameterized Fuzzy Measures Decision Making Model for signal categorization.

The main outcomes of this research are, in particular:

- The proposed approach is capable of providing satellite-based picture clustering decisions made within constrained time periods, which aids in tackling the geological challenges.
- For the method of feature extraction, this study suggests a novel application of the Parameterized Fuzzy Measures Decision Making Model Based on Preference Leveled Evaluation Functions for signal classification based BPNN architecture.
- In relation to the test, the provided cu/device dataset (v. 2009–05–08) has been taken into account to assess the suggested methodology. This dataset includes strong signal measurements that were gathered to build a parametric model for directional 802.11 ad-hoc devices operating at (2.4 GHz).
- Particularly, by utilizing the decision-making skills of Parameterized Fuzzy Measures Decision Making Model Based on Preference Leveled Evaluation Functions, this study defined a novel automated signal recommendation system.

REFERENCES

1. Yang, J., Wang, H., Lv, Z., & Wang, H. Design of miniaturized dual-band microstrip device for WLAN application. Sensors, 16(7), 2016, 983.
2. Karthick, M. Design of 2.4 GHz patch devicee for WLAN applications. In 2015 IEEE Seventh National Conference on Computing, Communication and Information Systems (NCCCIS) (pp. 1–4). 2015, IEEE.

3. Singh, S. P., & Singh, B. Non-authoritative naming resolution of DNS. In National Workshop on Network Security & Management, pp. 247–252, 2011.
4. Singh, S. P., & Passi, A. Real time communication. International Journal of Recent Development in Engineering and Technology (IJRDET), 2(3), 2014, 141–144.
5. Singh, S. P., & Kaur, G. Wireless communication. International Journal of Advanced Research in Computer Science and Software Engineering (IJARCSSE), 4(3), 2014, 957–961.
6. Singh, S. P., & Marwaha, R. Wi-Fi security. International Journal of Scientific Research in Computer Science Applications and Management Studies (IJSRCSAMS), 3(2), 2014.
7. Prasad, L., Ramesh, B., Kumar, K. S. R., & Vinay, K. P. Design and implementation of multiband microstrip patch device for wireless applications. Advanced Electromagnetics, 7(3), 2018, 104–107.
8. Latha, S., & Anand, P. R. Circular polarized microstrip patch array device for wireless LAN applications. In 2016 International Conference on Wireless Communications, Signal Processing and Networking (WiSPNET) (pp. 1197–1200). 2016, IEEE.
9. Elkorany, A. S., Mousa, A. N., Ahmad, S., Saleeb, D. A., Ghaffar, A., Soruri, M.,... Limiti, E. Implementation of a miniaturized planar tri-band microstrip patch device for wireless sensors in mobile applications. Sensors, 22(2), 2022, 667.
10. Prakasam, V., & Sandeep, P. Dual edge-fed left hand and right hand circularly polarized rectangular micro-strip patch device for wireless communication applications. IRO Journal on Sustainable Wireless Systems, 2(3), 2020, 107–117.
11. Firos, A., Prakash, N., Gorthi, R., Soni, M., Kumar, S., & Balaraju, V. Fault detection in power transmission lines using ai model. In 2023 IEEE International Conference on Integrated Circuits and Communication Systems (ICICACS), Raichur, India, 2023, pp. 1–6. doi: 10.1109/ICICACS57338.2023.10100005.
12. Naik, K. K. Asymmetric CPW-fed SRR patch device for WLAN/WiMAX applications. AEU-International Journal of Electronics and Communications, 93, 2018, 103–108.
13. Patel, R., & Upadhyaya, T. K. Compact planar dual band device for WLAN application. Progress in Electromagnetics Research Letters, 70, 2017, 89–97.
14. Dzogovic, B., Santos, B., Noll, J., Feng, B., & Van Do, T. Enabling smart home with 5G network slicing. In 2019 IEEE 4th International Conference on Computer and Communication Systems (ICCCS) (pp. 543–548). 2019, IEEE.
15. Islam, M. M., Funabiki, N., Sudibyo, R. W., Munene, K. I., & Kao, W. C. A dynamic access-point transmission power minimization method using PI feedback control in elastic WLAN system for IoT applications. Internet of Things, 8, 2019, 100089.
16. Tightiz, L., Yang, H., & Piran, M. J. A survey on enhanced smart micro-grid management system with modern wireless technology contribution. Energies, 13(9), 2020, 2258.
17. Tramarin, F., Mok, A. K., & Han, S. Real-time and reliable industrial control over wireless lans: Algorithms, protocols, and future directions. Proceedings of the IEEE, 107(6), 2019, 1027–1052.
18. Srisuji, T., & Nandagopal, C. Analysis on microstrip patch devices for wireless communication. In 2015 2nd International Conference on Electronics and Communication Systems (ICECS) (pp. 538–541). 2015, IEEE.

19. Vahora, A., & Pandya, K. Implementation of cylindrical dielectric resonator device array for Wi-Fi/wireless LAN/satellite applications. Progress in Electromagnetics Research M, 90, 2020, 157–166.
20. "LibriSpeech ASR corpus." www.openslr.org/12, November 2013.
21. Mahmood, A., Javaid, N., & Razzaq, S. A review of wireless communications for smart grid. Renewable and Sustainable Energy Reviews, 41, 2015, 248–260.

Chapter 6

A State-of-the-Art 360° Run-Down of Cloud, Edge, Dew, and Fog Computing

Vijaya Kumbhar, Ashvini Shende, Parag Tamhankar, Yudhishthir Raut, and Anirudh Mangore

6.1 INTRODUCTION

Before the emergence of cloud computing, organizations mainly relied on traditional client-server architectures, where applications and data were hosted on centralized servers and accessed by clients through a network connection. This approach had limitations in terms of scalability, flexibility, and cost-effectiveness, especially as data volumes and user demands grew.

Other technologies that preceded cloud computing include:

1. ***Mainframe Computing***: Mainframe computers were large, centralized systems that were widely used in the 1960s and 1970s. They were designed to handle large volumes of data processing and were typically used by large organizations such as banks and government agencies [1].
2. ***Grid Computing***: Grid computing appeared in the late 1990s and early 2000s [84]. It enabled organizations to share computing resources across different networks, allowing for more efficient use of processing power and storage [2].
3. ***Utility Computing***: Utility computing enables organizations to consume computing resources like energy and water [83]. Large organizations with high computing needs employed it [47, 3].
4. ***Application Service Providers (ASPs)***: ASPs were companies that provided access to software applications over the internet [29]. They were popular in the late 1990s and early 2000s, especially among small and medium-sized businesses that could not afford to develop and maintain their own software applications [4][104].
5. ***Virtualization***: Virtualization technology enabled organizations to create virtual instances of operating systems and applications on one real server. It allowed for greater flexibility and scalability, as multiple virtual servers could be created and managed on a single physical machine [5][102].

DOI: 10.1201/9781003480181-6

These technologies laid the groundwork for the emergence of cloud computing, which combined distributed computing, virtualization, and utility computing to create a scalable and flexible computing model that has transformed the way that organizations consume and manage computing resources [13].

6.2 CLOUD COMPUTING

Cloud computing is a model for delivering computing services over the internet on demand, allowing users to access a shared pool of computing resources that can be rapidly provisioned and released with minimal management effort or service provider interaction [12].

Here are some ***definitions*** of cloud computing:

- The ***National Institute of Standards and Technology (NIST)*** describes cloud computing as "a model for enabling ubiquitous, convenient, on-demand network access to a shared pool of configurable computing resources (e.g., networks, servers, storage, applications, and services) that can be rapidly provisioned and released with minimal management effort or service provider interaction" [6][44].
- ***Amazon Web Services (AWS)*** states cloud computing as "the on-demand delivery of IT resources and applications via the internet with pay-as-you-go pricing" [7][98].
- ***Microsoft*** describes cloud computing as "the delivery of computing services—including servers, storage, databases, networking, software, analytics, and intelligence—over the Internet ("the cloud") to offer faster innovation, flexible resources, and economies of scale" [8][122].

6.2.1 Components of Cloud Computing

"Cloud computing is a technology that provides on-demand access to a shared pool of computing resources over the internet" [35]. The components of it are divided into three main groups: infrastructure, platform, and software. Here are some examples of each component:

1. ***Infrastructure as a Service (IaaS)***: This component provides access to virtualized computing resources, such as servers, storage, and networking, over the internet. Examples of IaaS providers include Amazon Web Services (AWS), Microsoft Azure, and Google Cloud Platform (GCP) [6][32].
2. ***Platform as a Service (PaaS)***: This component provides a platform for developing, deploying, and managing applications over the internet. PaaS providers offer preconfigured computing environments, such as programming languages, development tools, and databases. Examples

of PaaS providers include Heroku, Google App Engine, and Microsoft Azure [9][32].

3. ***Software as a Service (SaaS)***: This component provides access to cloud-based software applications over the internet, such as email, customer relationship management (CRM), and accounting software. SaaS providers handle all the infrastructure and platform components, allowing users to access the software from any device with an internet connection. Examples of SaaS providers include Salesforce, Microsoft Office 365, and Zoom [5][10].
4. ***Cloud Storage***: This component provides scalable and flexible storage services over the internet, enabling users to store and access data from anywhere. Examples of cloud storage providers include Amazon S3, Microsoft Azure Storage, and Google Cloud Storage [11][55].
5. ***Cloud Security***: This component provides security measures to protect cloud-based resources from unauthorized access, data breaches, and other security threats. Cloud security solutions include encryption, access control, identity management, and threat detection. Examples of cloud security providers include AWS Security, Azure Security, and Google Cloud Security [12].

Hence, cloud computing consists of several components that work together to provide a flexible, scalable, and cost-effective way to access computing resources over the internet. As technology continues to evolve, we can expect to see more advancements and innovations in the field of cloud computing [98][47].

6.2.2 Architectural Model of Cloud Computing

The architectural model of cloud computing typically consists of several key components, including:

1. ***Cloud Service Models***: Cloud computing offers different service models, including Infrastructure as a Service (IaaS), Platform as a Service (PaaS), and Software as a Service (SaaS). These models provide varying levels of control and responsibility to users, from managing infrastructure resources to using pre-built applications [111].
2. ***Cloud Deployment Models***: Cloud computing can be deployed in different ways, such as Public Cloud, Private Cloud, Hybrid Cloud, and Community Cloud. Each deployment model caters to specific needs regarding data privacy, scalability, and resource sharing [111].
3. ***Cloud Infrastructure***: The cloud infrastructure includes physical servers, storage devices, networking equipment, and data centers that form the backbone of cloud services [112][113].
4. ***Virtualization***: Virtualization is a crucial technology in cloud computing, allowing multiple virtual machines (VMs) or containers to run on a single physical server, optimizing resource utilization [113].

5. ***Cloud Management and Orchestration***: Cloud management tools help manage cloud resources, monitor performance, and automate provisioning and scaling. Orchestration tools handle complex workflows and automated tasks across cloud environments [114].
6. ***Security and Compliance***: Cloud security measures are essential to protect data, applications, and infrastructure from unauthorized access and attacks. Compliance frameworks ensure that cloud providers meet industry-specific regulations and standards [115].

6.2.3 Timeline of Cloud Computing

Here is a timeline of the major developments in cloud computing:

Table 6.1 Evolution of Cloud Computing

Ref.	*Year*	*Service Provider*	*Type of Service*	*Product/ Service Name*	*Particulars*
[93]	1960s	—	Conceptual Development	—	The concept of utility computing is introduced by John McCarthy, who proposes that computing resources should be sold like a utility, similar to electricity or water
[105] [116]	1999	Salesforce	Software as a Service (SaaS)	Salesforce. com	Salesforce.com becomes one of the first companies to deliver enterprise applications over the internet, marking the beginning of Software as a Service (SaaS)
[66]	2002	Amazon	Infrastructure services	Amazon Web Services	Amazon Web Services (AWS) is launched, initially as a platform for internal use but later becoming a leading provider of cloud infrastructure services
[75] [86]	2006	Google	Software as a Service (SaaS)	Google Apps (Gmail, Google Docs, and Google Calendar)	Google launches Google Apps, a suite of web-based productivity tools, including Gmail, Google Docs, and Google Calendar
[58]	2008	NASA and Rackspace	Public And Private Clouds	OpenStack	NASA and Rackspace team up to launch OpenStack, an open-source cloud computing platform for public and private clouds

Ref.	*Year*	*Service Provider*	*Type of Service*	*Product/ Service Name*	*Particulars*
[24] [59]	2009	Microsoft	Infrastructure as a Service (laas) and Platform as a Service (PaaS)	Azure	Microsoft launches its Azure cloud computing platform, providing Infrastructure as a Service (IaaS) and Platform as a Service (PaaS) capabilities
[38] [39]	2010	Oracle	Infrastructure as a Service (IaaS), Platform as a Service (PaaS), and Software as a Service (SaaS)	Oracle Cloud Platform	Oracle introduces its cloud offerings, including Infrastructure as a Service (IaaS), Platform as a Service (PaaS), and Software as a Service (SaaS)
[88]	2011	IBM	IaaS, PaaS, and SaaS	Smart Cloud portfolio	IBM launches its Smart Cloud portfolio of cloud services, including IaaS, PaaS, and SaaS (IBM, n.d.).
[54]	2012	Google	Infrastructure as a Service (IaaS)	Google Compute Engine	Google launches Google Compute Engine, a service for running virtual machines on Google's infrastructure
[9]	2013	National Security Agency (NSA)	Cloud Security and Privacy	Spark	The National Security Agency (NSA) revelations spark concerns about cloud security and privacy, leading to increased interest in hybrid and private clouds
[106]	2015	Cloud Security Alliance	Cloud Adoption	—	The adoption of cloud computing continues to grow, with more than 90% of businesses using cloud services in some capacity, according to a survey by the Cloud Security Alliance
[99]	2021	—	—	—	The COVID-19 pandemic accelerates the shift to cloud computing, with organizations relying on cloud services to support remote work and digital transformation initiatives

Overall, cloud computing has evolved rapidly over the past several decades, with new technologies and services continually being developed and adopted by businesses and individuals around the world [63].

6.2.4 Working of Cloud Computing

Cloud computing is a model of delivering computing services over the internet [5]. It involves using remote servers to store, manage, and process data instead of local servers or personal computers [46]. Cloud computing has several benefits, including scalability, accessibility, flexibility, and cost-effectiveness.

Here's a breakdown of how cloud computing works [13–17]:

1. ***Infrastructure***: Cloud computing service providers offer computing resources such as servers, storage, databases, and networking to customers. These resources are located in data centers around the world [16].
2. ***Service Models***: Cloud computing offers three primary service models:
   a. ***Software as a Service (SaaS)***: This model provides customers with ready-to-use software applications that run on the provider's infrastructure [14].
   b. ***Platform as a Service (PaaS)***: This model provides customers with a platform to build, deploy, and manage their own applications on the provider's infrastructure [42].
   c. ***Infrastructure as a Service (IaaS)***: This model provides customers with access to virtualized computing resources, including servers, storage, and networking, which they can use to build and manage their own IT environments [14].
3. ***Deployment Models***: Cloud computing also offers four deployment models [100]:
   a. ***Public Cloud***: This model offers computing resources over the internet to anyone who wants to use them.
   b. ***Private Cloud***: This model provides computing resources to a single organization and is not accessible to the public [27].
   c. ***Hybrid Cloud***: This model combines public and private clouds to allow organizations to take advantage of the benefits of both [51].
   d. ***Community Cloud***: This model provides computing resources to a group of organizations with similar needs.
4. ***Access***: Customers can access their computing resources using a web browser or a software application that communicates with the cloud computing service provider's application programming interface (API).
5. ***Payment***: Cloud computing service providers charge customers based on their usage of computing resources, such as the number of servers used, the amount of storage used, and the amount of data transferred.

6.2.5 Topmost Service Providers of Cloud Computing

Here are some of the top cloud computing service providers [20]:

- ***Amazon Web Services (AWS)***: AWS is a cloud computing platform that provides a wide range of services, including computing, storage, and databases, to individuals, businesses, and governments [20]. It is considered the largest cloud computing platform in the world, with millions of customers across the globe [13][24].

- ***Microsoft Azure***: Azure is a cloud computing service provided by Microsoft that offers a wide range of services, including computing, storage, and databases, to help businesses build [23], deploy, and manage applications on a global network of Microsoft-managed data centers [14][34].
- ***Google Cloud Platform***: Google Cloud is a cloud computing platform provided by Google [34] that provides a wide range of services, including computing, storage, and databases, to help businesses build and deploy applications on Google's infrastructure [15][49].
- ***IBM Cloud***: IBM Cloud is a cloud computing service provided by IBM [61] that offers a range of services, including computing, storage, and databases, to help businesses build, deploy, and manage applications on IBM's infrastructure [16][52].
- ***Oracle Cloud***: Oracle Cloud is a cloud computing service provided by Oracle that offers a wide range of services, including computing, storage, and databases [23], to help businesses build, deploy, and manage applications on Oracle's infrastructure [18][24].

6.2.6 Benefits of Cloud Computing

Cloud computing is a model of providing computing resources over the internet, including servers, storage, databases, networking, software, analytics, and intelligence, to help organizations innovate and operate more efficiently at scale [40].

Some of the benefits of cloud computing are [19–22]:

- ***Scalability***: Cloud computing allows organizations to scale their computing resources up or down as needed, without having to invest in expensive infrastructure. This helps organizations to better manage their costs and respond to changing business demands quickly [33].
- ***Flexibility***: Cloud computing provides organizations with the flexibility to access their applications and data from anywhere, on any device, with an internet connection. This enables organizations to work remotely, collaborate with others, and respond to customer needs more effectively [5].
- ***Cost savings***: Cloud computing eliminates the need for organizations to invest in expensive hardware and software upfront [41], which can help them save on capital expenses. Additionally, organizations can pay only for the computing resources they use, which can help them save on operational expenses [15].
- ***Security***: Cloud computing providers offer robust security measures to protect data from theft, unauthorized access, and data loss [71]. Additionally, cloud providers offer disaster recovery and backup services to ensure business continuity in the event of a disaster.

- ***Innovation***: Cloud computing providers offer a range of services, such as artificial intelligence, machine learning, and big data analytics, which organizations can use to drive innovation and gain insights from their data [27].

6.2.7 Limitations of Cloud Computing

Cloud computing has brought about significant advancements in the field of information technology by providing a wide range of services such as storage, computation, and networking to users through the Internet [4][35]. However, it also has some limitations that organizations and individuals need to be aware of.

Here are some of the limitations of cloud computing [97]:

1. ***Security Risks***: Cloud computing poses significant security risks, particularly concerning data privacy, control, and protection. The data that is transmitted and stored in the cloud can be accessed by unauthorized users, leading to data breaches, which can compromise sensitive information [78]. Moreover, cloud providers may share user data with third-party service providers or governments, raising concerns about the confidentiality and privacy of data [4][23][82].
2. ***Dependence on Internet Connectivity***: Cloud computing services are entirely dependent on an internet connection, which can be a limitation for organizations or individuals with limited or unreliable internet connectivity. This dependence on the internet can lead to delays in accessing critical information, particularly in areas with weak or no internet connectivity [24].
3. ***Limited Control and Customization***: Cloud computing providers offer pre-configured services that may not meet the specific requirements of an organization. Additionally, the cloud provider retains control over the infrastructure, making it challenging for users to customize or modify the services according to their needs [25].
4. ***Potential for Downtime***: Cloud computing services are subject to downtime, which can have a significant impact on businesses that rely on these services [18]. Downtime can occur due to several factors such as hardware or software failures, power outages, and natural disasters [26].
5. ***Data Transfer Costs***: Cloud computing services require significant data transfer between the client and the server, which can lead to additional costs for the user. The costs can increase significantly if the data transfer is frequent or if large amounts of data need to be transferred [6][26].

6.2.8 Applications of Cloud Computing

Cloud computing has revolutionized the way businesses and individuals store, access, and process data [5].

Here are some of the popular applications of cloud computing in various industries [107]:

1. ***Data Storage and Backup***: Cloud computing provides businesses and individuals with a cost-effective way to store and back up their data. Cloud storage services such as Google Drive, Dropbox, and Microsoft OneDrive are widely used for their ease of use, scalability, and accessibility [27].
2. ***Software Development and Testing***: Cloud computing allows software developers to access virtual environments, tools, and platforms for developing and testing their applications. This approach provides developers with flexibility and agility in managing their development environments and reduces the time and costs associated with setting up infrastructure [28].
3. ***Big Data Analytics***: Cloud computing provides businesses with the ability to store and analyze large amounts of data quickly and efficiently [4]. Cloud-based big data analytics platforms such as Amazon Web Services (AWS) and Google Cloud Platform (GCP) are widely used to process large data sets and derive insights from them [29][65].
4. ***E-commerce and Retail***: Cloud computing provides e-commerce and retail businesses with the ability to scale their infrastructure quickly and efficiently based on demand. Cloud-based e-commerce platforms such as Shopify and BigCommerce allow businesses to set up their online stores quickly and efficiently [30].
5. ***Healthcare***: Cloud computing provides healthcare providers with the ability to store, access, and share patient data securely and efficiently [5]. Cloud-based healthcare platforms such as Athenahealth and Cerner Corporation are widely used to manage patient data and provide healthcare services [31].

6.2.9 Future Scope of Cloud Computing

Looking into the future, cloud computing is expected to continue evolving and expanding, offering new opportunities and advancements in the technology landscape.

Here are some potential future scopes of cloud computing [133][134[135][136]:

1. ***Edge Cloud Computing***: The rise of edge computing is expected to complement traditional cloud services by bringing computing resources closer to end-users and devices. Edge cloud computing enables low-latency processing, real-time data analysis, and reduced network congestion. It allows for efficient handling of data at the edge of the network, making it well-suited for applications like IoT, smart cities, and autonomous vehicles.
2. ***Hybrid Cloud Solutions***: Organizations are likely to adopt hybrid cloud solutions that combine public cloud, private cloud, and on-premises infrastructure. Hybrid cloud models offer flexibility, data sovereignty, and the ability to leverage the benefits of both public and private cloud environments while meeting specific business needs.

3. ***Serverless Computing***: Serverless computing, also known as Function as a Service (FaaS), allows developers to run code without managing the underlying infrastructure. This model is expected to gain more popularity as it simplifies development and deployment, enables cost optimization, and scales automatically based on demand.
4. ***Quantum Computing in the Cloud***: As quantum computing technologies advance, cloud service providers may offer quantum computing services to researchers and businesses. Quantum cloud computing has the potential to tackle complex problems that are currently infeasible with classical computing.
5. ***Multi-Cloud and Interoperability***: Organizations may increasingly adopt multi-cloud strategies to avoid vendor lock-in and enhance redundancy. Interoperability between different cloud providers and platforms will become crucial to enable seamless data and workload portability across cloud environments.
6. ***Enhanced Security and Privacy***: With the growing importance of data security and privacy, cloud providers will continue to invest in advanced security measures, encryption technologies, and compliance frameworks to protect sensitive data and meet regulatory requirements.
7. ***AI-Enabled Cloud Services***: Cloud providers are likely to integrate artificial intelligence (AI) and machine learning (ML) capabilities into their services, enabling intelligent data analysis, automation, and personalized user experiences.
8. ***Green Cloud Computing***: Environmental concerns may drive the adoption of green cloud computing practices, focusing on energy-efficient data centers, renewable energy sources, and sustainable computing.

6.3 DEW COMPUTING

Dew computing is a paradigm for computing that extends the concept of cloud computing to the edge of the network [29]. It enables computing and storage resources to be distributed in a decentralized way, closer to the devices and users that need them [10].

Here are some ***definitions*** of dew computing:

- According to Mahdi Ben Alaya and Ching-Hsien Hsu, "dew computing is a paradigm that extends cloud computing to the edge of the network [3], providing services and resources to end-users closer to their devices" [32].
- Xiangjian He and Zibin Zheng define dew computing as "a new paradigm for computing that shifts the focus from the central cloud to the edge of the network, providing pervasive, reliable, and real-time computing and storage services for users and devices" [33].

- Tao Zhang and Xiaoming Fu define dew computing as "a new computing paradigm that focuses on the edge of the network to provide distributed computing and storage resources, with the goal of reducing the latency, bandwidth, and energy consumption of cloud computing" [34].

So, dew computing is a paradigm that aims to provide computing and storage resources closer to the edge of the network, enabling faster and more efficient processing of data and applications [1][73].

6.3.1 Components of Dew Computing

Dew computing, also known as edge computing, is a distributed computing paradigm that extends cloud computing to the edge of the network [3], bringing computation and data storage closer to where it is needed [6].

Here are the main components of dew computing:

1. ***Edge Devices***: These are the devices that are used to collect and process data at the edge of the network, such as smartphones, sensors, and IoT devices [57][95]. These devices are typically resource-constrained and have limited processing power, memory, and storage capacity [22][35][36].
2. ***Edge Nodes***: These are the computing nodes that are deployed at the edge of the network to provide computing and storage resources. Edge nodes can be physical devices, such as routers, gateways, and micro servers, or virtual instances that run on cloud infrastructure [1][35][36].
3. ***Fog Computing***: This is a layer of computing infrastructure that is located between the edge devices and the cloud data centers [67]. Fog computing provides a platform for processing and analyzing data in real time, improving response times and reducing network congestion [79]. Fog computing is often used to run data analytics and machine learning algorithms on edge data [35][36][87].
4. ***Cloud Data Centers***: These are the central data centers that provide computing and storage resources for cloud computing. Cloud data centers are used to store and process large amounts of data that cannot be processed at the edge due to resource constraints [18][35][36].
5. ***Communication Protocols***: These are the protocols used to communicate between edge devices, edge nodes, fog computing infrastructure, and cloud data centers. Communication protocols are designed to be lightweight, low-latency, and energy-efficient, to support communication between resource-constrained devices [35][36].

6.3.2 Architectural Model of Dew Computing

Dew computing is a new paradigm in computing that involves the use of decentralized resources and distributed infrastructure for data processing and storage [11].

It is based on the concept of dew, which is a metaphorical term used to describe the moisture that forms on the surface of objects in the early morning hours [77].

An architectural model of dew computing typically consists of four layers: device layer, fog layer, edge layer, and cloud layer [8]. Each layer has a specific function in the overall architecture of dew computing [31].

1. ***Device Layer***: This layer consists of various devices that are used for data collection and processing, such as sensors, mobile phones, and IoT devices. These devices are often low-powered and have limited storage and computational capabilities [31][37–39].
2. ***Fog Layer***: The fog layer is the first layer of processing in dew computing. It consists of a network of edge devices that are closer to the data source than the cloud. The fog layer can perform basic data filtering and aggregation tasks, reducing the amount of data that needs to be sent to the cloud for further processing [37–39].
3. ***Edge Layer***: The edge layer is the second layer of processing in dew computing. It consists of more powerful devices that are closer to the data source than the cloud. The edge layer can perform more complex data processing tasks, such as machine learning and predictive analytics [37–39].
4. ***Cloud Layer***: The cloud layer is the final layer of processing in dew computing. It consists of a centralized data center that can perform advanced data processing and storage tasks. The cloud layer can also be used for long-term storage and backup [37–39].

6.3.3 Timeline of Dew Computing

Dew computing is a relatively new paradigm in computing that is focused on bringing computing resources closer to the edge of the network to reduce latency and improve performance. Here is a timeline of the development of dew computing, along with relevant references:

Table 6.2 Evolution of Dew Computing

Ref.	*Year*	*Development*	*Details*
[40]	2015	Initial Concept	The concept of dew computing was first proposed by Prof. Fernando G.S.L. da Silva
[41]	2016	Further Research	The concept of dew computing was further developed and explored
[42]	2017	Dew Computing Symposium	The first Dew Computing Symposium was held in Boston, USA, bringing together researchers and practitioners to discuss the latest developments in dew computing
[43]	2018	Dew Computing Applications	Researchers began exploring applications of dew computing in areas such as healthcare, smart cities, and Internet of Things (IoT) devices

Ref.	Year	Development	Details
[44]	2019	Dew Computing in Industry	Companies such as Huawei began exploring the use of dew computing in industrial applications such as manufacturing and energy
[45]	2020	Dew Computing Architecture	Researchers proposed a dew computing architecture that leverages edge computing, fog computing, and cloud computing to enable efficient and scalable data processing

The concept of dew computing is still in its early stages, but it has shown promise in improving the performance and efficiency of computing systems. As research and development continue, we can expect to see more applications and advancements in the field of dew computing.

6.3.4 Working of Dew Computing

Dew computing, also known as fog-to-cloud computing, is an extension of the fog computing paradigm that further distributes computing resources to the extreme edge of the network, closer to the data source. In dew computing, small-scale computing devices, such as microcontrollers or single-board computers, are deployed at the very edge of the network to process data locally before sending relevant information to the fog or cloud for further analysis and storage. This approach aims to reduce latency, bandwidth consumption, and dependency on centralized cloud resources [121–124].

The working of dew computing involves the following steps [121–124]:

1. ***Data Collection***: Dew computing begins with the collection of data from sensors, IoT devices, or other edge devices at the extreme edge of the network.
2. ***Local Processing***: The collected data is processed locally on the dew devices, where lightweight computing resources, such as microcontrollers or low-power processors, are utilized for data filtering, aggregation, and simple analytics.
3. ***Data Filtering and Selection***: Dew devices perform data filtering to select the most relevant information or events that need to be sent to the higher layers of the network for further processing.
4. ***Communication with Fog/Cloud***: The filtered data is then transmitted to the fog or cloud computing resources for more in-depth analysis, complex computations, and long-term storage.
5. ***Feedback Loop***: The dew computing devices may also receive instructions or updates from the fog or cloud to adapt their local processing and filtering strategies based on changing requirements or network conditions.

By leveraging dew computing, organizations can achieve improved performance, reduced latency, and better utilization of network resources, especially in scenarios with limited connectivity or strict latency requirements.

6.3.5 Benefits of Dew Computing

Dew computing is a relatively new paradigm in computing that is focused on bringing computing resources closer to the edge of the network to reduce latency and improve performance. Here are some benefits of dew computing, along with relevant references:

1. ***Reduced Latency***: Dew computing can reduce the latency of data processing and communication by moving computing resources closer to where the data is generated or consumed. This can improve the performance of applications that require real-time data processing [40].
2. ***Improved Security***: By keeping data and processing closer to the edge of the network, dew computing can improve security by reducing the amount of data that needs to be transmitted across the network [46].
3. ***Cost-Effective***: Dew computing can be a cost-effective alternative to traditional cloud computing, as it eliminates the need for expensive data centers and reduces the amount of data that needs to be transmitted over the network [41].
4. ***Scalability***: Dew computing can be easily scaled up or down by adding or removing computing resources at the edge of the network. This can improve the efficiency and agility of computing systems [44].
5. ***Better Resource Utilization***: Dew computing can improve the utilization of computing resources by distributing processing and storage capabilities closer to where they are needed, which can reduce the load on centralized systems and improve overall efficiency [45].

Dew computing offers several benefits that can improve the performance, security, and cost-effectiveness of computing systems. As research and development continue, we can expect to see more applications and advancements in the field of dew computing.

6.3.6 Limitations of Dew Computing

While dew computing offers several benefits, it also has some limitations that need to be considered. Here are some limitations of dew computing, along with relevant references:

1. ***Limited Resources***: Computing resources at the edge of the network are often limited in terms of processing power, memory, and storage capacity. This can limit the types of applications that can be supported and the amount of data that can be processed [40].

2. ***Security Risks***: By distributing computing resources, dew computing can introduce new security risks, such as unauthorized access and data breaches. The lack of centralized control can make it difficult to monitor and secure computing resources at the edge of the network [47].
3. ***Complexity***: Dew computing introduces additional complexity to computing systems, as computing resources are distributed across the network. This can make it more difficult to manage and maintain the system, especially for non-expert users [41].
4. ***Dependence on Network Connectivity***: Dew computing relies on network connectivity to transmit data and access computing resources. This means that performance and availability can be impacted by network congestion, latency, and reliability issues [44].
5. ***Compatibility Issues***: Dew computing may not be compatible with existing computing systems and applications, which can make it difficult to integrate with existing infrastructure [45].

Dew computing has some limitations that need to be addressed to realize its full potential. As the field continues to evolve, researchers and practitioners are working to overcome these limitations and develop more efficient, secure, and user-friendly dew computing systems.

6.3.7 Applications of Dew Computing

Dew computing is a new paradigm in computing that focuses on bringing computing resources closer to the edge of the network. Here are some applications of dew computing, along with relevant references:

1. ***Internet of Things (IoT)***: Dew computing can be used to process and analyze data from IoT devices, such as sensors and cameras, in real time. By bringing computing resources closer to where the data is generated, dew computing can reduce latency and improve the performance of IoT applications [46].
2. ***Smart Cities***: Dew computing can be used to support smart city applications, such as traffic management, public safety, and environmental monitoring. By processing and analyzing data at the edge of the network, dew computing can enable real-time decision-making and improve the efficiency of city operations [47].
3. ***Edge Computing***: Dew computing can be seen as a form of edge computing, where computing resources are located at the edge of the network to support real-time data processing and decision-making. By reducing the amount of data that needs to be transmitted to centralized data centers, dew computing can improve the efficiency and scalability of edge computing systems [44].
4. ***Healthcare***: Dew computing can be used to support healthcare applications, such as remote patient monitoring and telemedicine. By processing and

analyzing data at the edge of the network, dew computing can enable real-time monitoring and decision-making, and improve the quality of care [48].

5. ***Agriculture***: Dew computing can be used to support precision agriculture applications, such as soil monitoring and crop management. By processing and analyzing data at the edge of the network, dew computing can enable real-time decision-making and improve the efficiency of farming operations [49].

Dew computing has several applications in various domains, from IoT to healthcare, that can benefit from real-time data processing and decision-making. As research and development continue, we can expect to see more innovative applications and advancements in the field of dew computing.

6.3.8 Future Scope of Dew Computing

Dew (Distributed Environment for the Web) computing is an emerging paradigm that holds significant potential for the future of computing. While its adoption is still in the early stages, the future scope of Dew computing is promising and can have several implications for various industries.

Here are some potential future directions and applications of Dew computing [137–139]:

1. ***Edge Computing Advancements***: Dew computing can complement and advance edge computing capabilities by further extending computing resources closer to end-users and IoT devices. This could lead to reduced latency, improved real-time data processing, and enhanced user experiences.
2. ***Decentralized Applications (DApps)***: Dew computing's decentralized nature aligns well with the principles of blockchain technology. It can enable the development of decentralized applications (DApps) that are more resilient, secure, and scalable.
3. ***Internet of Things (IoT) Enablement***: Dew computing can support the rapid growth of IoT devices by offering edge computing resources for data processing and analytics, leading to efficient IoT deployments.
4. ***Enhanced Content Delivery***: Dew computing can optimize content delivery by caching and serving content from distributed nodes closer to users, reducing the load on central servers and improving content delivery speed.
5. ***Federated Learning and AI***: Dew computing can facilitate federated learning approaches, where machine learning models are trained on data from distributed sources without centralized data aggregation, ensuring data privacy and security.
6. ***Edge-to-Cloud Orchestration***: The future may witness more sophisticated orchestration mechanisms between edge computing and cloud computing, where Dew computing plays a crucial role in bridging the gap between the two.

It is important to note that Dew computing is still a developing concept, and its future scope will largely depend on technological advancements, standardization, and industry adoption. As the field of computing evolves, the potential applications and benefits of Dew computing are likely to expand.

6.4 FOG COMPUTING

Fog computing, also known as ***edge fog*** computing or ***fog*** networking, is a distributed computing architecture that extends the cloud computing model to the edge of the network. In fog computing, resources and applications are distributed in a hierarchical manner between the cloud and edge devices, with the goal of improving performance, reducing latency, and enabling real-time data processing and analysis.

Mist computing, sometimes called fog computing, is a distributed computing paradigm that extends cloud computing to the edge of the network, closer to where data is generated and consumed.

Here are some ***definitions*** of fog computing:

- "***Fog computing*** is a paradigm that extends cloud computing and services to the edge of the network. Like cloud computing, fog provides data, compute, storage, and application services to end-users. However, fog computing is more distributed and closer to end-users, enabling real-time applications and services" [50].
- "***Fog computing*** is a decentralized computing infrastructure in which data, compute, storage, and applications are distributed in the most logical, efficient place between the data source and the cloud" [51].
- "***Fog computing*** is a system-level horizontal architecture that distributes resources and services of computing, storage, control, and networking anywhere along the continuum from Cloud to Things" [52].

6.4.1 Components of Fog Computing

Fog computing is an extension of cloud computing that brings computing resources closer to the edge of the network, enabling faster data processing and reducing latency. It involves a distributed computing architecture that utilizes local edge devices, such as routers, gateways, and edge servers, to perform data processing and storage tasks. Here are the key components of fog computing:

- ***Edge Devices***: These are the devices located at the edge of the network, such as sensors, cameras, smartphones, and IoT devices. They collect and generate data from the physical world and act as data sources for fog computing applications [102].

- ***Edge Servers***: Edge servers are computing nodes located at the edge of the network that process and analyze data from edge devices locally. They perform real-time data processing and filtering, reducing the need to send all data to the centralized cloud for analysis [103].
- ***Fog Nodes/Gateways***: Fog nodes or gateways act as intermediaries between edge devices and the centralized cloud. They aggregate and preprocess data from multiple edge devices and send only relevant data to the cloud, optimizing bandwidth usage [104].
- ***Fog Controllers***: Fog controllers are responsible for managing the resources and tasks distributed across the fog nodes and edge servers. They ensure efficient data routing, load balancing, and task scheduling within the fog computing infrastructure [105].
- ***Cloud Infrastructure***: While fog computing aims to bring data processing closer to the edge, it still relies on the centralized cloud for certain tasks, such as long-term storage, complex analytics, and resource-intensive computations [105].

6.4.2 Architectural Model of Fog Computing

The architectural model of fog computing, also known as edge computing, involves the distribution of computing resources and services closer to the edge of the network, closer to the data source. This approach aims to reduce latency, improve data processing efficiency, and enable real-time decision-making.

The key components of the fog computing architecture include:

- ***Edge Devices***: These are the IoT devices, sensors, and actuators that generate data at the edge of the network. They serve as the entry point for data into the fog computing infrastructure [116].
- ***Fog Nodes/Gateways***: Fog nodes or gateways are intermediary devices located at the edge of the network. They collect and preprocess data from edge devices before sending it to the cloud or data center for further processing [117].
- ***Fog Computing Infrastructure***: This layer includes the computational resources, storage, and networking capabilities deployed at the edge of the network to support fog computing services [118].
- ***Fog Services and Applications***: These are the software applications and services running on the fog nodes to perform data analytics, real-time processing, and decision-making at the edge [119].
- ***Cloud/Data Center***: While fog computing brings processing closer to the edge, cloud services and data centers are still part of the architecture. Complex and resource-intensive tasks may be offloaded to the cloud for more in-depth analysis and storage [120].

6.4.3 Timeline of Fog Computing

Fog computing is a relatively new paradigm in computing that involves the use of distributed resources for data processing and storage. Here is a timeline of key events in the development of fog computing [53–57]:

Table 6.3 Evolution of Fog Computing

Year	*Contribution By*	*Development*	*Details*
2008	Cisco	Concept Introduction	Introduced the concept of "fog computing" in a white paper that describes a distributed computing architecture that extends cloud computing to the edge of the network
2012	Researchers at Carnegie Mellon University	Cyber-physical fog computing	Introduced the concept of "cyber-physical fog computing" in a paper that describes a framework for integrating sensing, actuation, and computation in distributed systems
2014	OpenFog	Consortium Foundation for promotion of fog computing	The OpenFog Consortium is founded by Cisco, Intel, Microsoft, and other companies to promote the development of fog computing technologies and standards
2015	IEEE	Publishes a paper on "fog computing and its role in the internet of things"	The IEEE publishes a paper on "fog computing and its role in the internet of things" that describes the benefits of fog computing in reducing latency, improving security, and enabling real-time decision-making
2017	Fog World Congress	Launched as a conference dedicated to fog computing	Launched as a conference dedicated to fog computing, bringing together researchers, industry leaders, and policymakers to discuss the latest developments in fog computing
2018	National Institute of Standards and Technology (NIST)	Developed the framework for fog computing	Published a report on fog computing that provides a framework for understanding fog computing and its applications
2019	OpenFog Consortium and Industrial Internet Consortium	OpenFog Consortium merges with the Industrial Internet Consortium to form the new Industrial Internet Consortium	Promoted the development and adoption of fog computing technologies

6.4.4 Working of Fog Computing

Fog computing is a decentralized computing paradigm that extends cloud computing capabilities to the edge of the network, closer to the data source and end-users. It works by distributing computing resources and services to multiple fog nodes or edge devices, allowing data processing, storage, and analytics to occur at the network's edge, reducing latency and bandwidth consumption [125][126][127][128].

The working of fog computing involves the following steps [125][126][127][128]:

1. ***Data Collection***: Fog computing begins with the collection of data from various sensors, IoT devices, or end-user devices at the edge of the network.
2. ***Local Processing***: The collected data is processed locally on the fog nodes or edge devices. Fog nodes have more computing power than traditional edge devices, enabling them to perform real-time data processing, filtering, and aggregation.
3. ***Data Analysis and Decision-Making***: Fog nodes can analyze the processed data and make real-time decisions based on local conditions and predefined rules. This enables faster response times and reduces the need to send all data to the central cloud for analysis.
4. ***Communication with Cloud***: If necessary, the processed data or relevant information is sent to the central cloud for further analysis, storage, and long-term data processing. The cloud acts as a central repository for data that requires more extensive computations or historical analysis.
5. ***Feedback Loop***: The cloud can send feedback or instructions back to the fog nodes for dynamic adjustments in processing or decision-making based on changing requirements or network conditions.

By implementing fog computing, organizations can achieve lower latency, improved performance, reduced data traffic, and enhanced scalability, making it well-suited for real-time applications and IoT deployments.

6.4.5 Benefits of Fog Computing

Fog computing is a distributed computing paradigm that brings computing resources closer to the edge of the network, enabling faster processing of data and reducing the latency associated with cloud-based solutions. Here are some of the key benefits of fog computing [58–60]:

1. ***Reduced Latency***: One of the primary benefits of fog computing is reduced latency. By bringing computing resources closer to the edge of the network, fog computing can process data in real time, enabling faster decision-making and response times.
2. ***Improved Security***: Fog computing can improve the security of data by keeping sensitive information closer to its source and reducing the amount

of data that needs to be transmitted over the network. Fog computing can also provide additional security layers by using local firewalls, intrusion detection systems, and other security measures.

3. ***Increased Efficiency***: Fog computing can improve the efficiency of data processing by distributing the workload across multiple devices and reducing the need for centralized data centers. This can result in lower energy consumption, reduced costs, and improved scalability.
4. ***Enhanced Reliability***: Fog computing can increase the reliability of data processing by providing redundant computing resources that can take over in case of a failure. This can ensure that critical applications remain operational even in the event of a hardware or software failure.
5. ***Support for Mobile Devices***: Fog computing can support the growing number of mobile devices that are connected to the network by providing a localized computing infrastructure that can handle the increased processing demands.

6.4.6 Limitations of Fog Computing

While fog computing offers many benefits over traditional cloud computing, there are also some limitations and challenges that need to be addressed. Here are some of the key limitations of fog computing [61–64]:

1. ***Limited Scalability***: Fog computing is limited in terms of scalability due to its distributed nature. While fog computing can handle small to medium-sized workloads, it may struggle to handle large-scale workloads.
2. ***Complex Management***: Fog computing can be complex to manage due to its distributed nature. This can lead to challenges in deploying and maintaining fog computing infrastructures.
3. ***Higher Costs***: Fog computing can be more expensive than traditional cloud computing due to the need for additional hardware and software resources at the edge of the network.
4. ***Lack of Standards***: There is currently a lack of standards in the fog computing space, which can lead to interoperability issues and limit the adoption of fog computing solutions.
5. ***Security Concerns***: While fog computing can enhance the security of data by keeping sensitive information closer to its source, it also presents new security challenges. For example, fog computing infrastructures may be more vulnerable to cyberattacks due to their distributed nature.

6.4.7 Applications of Fog Computing

Fog computing is a versatile technology that has a wide range of applications across various industries. Here are some of the key applications of fog computing [61,65–68]:

1. ***Smart Cities***: Fog computing can be used in smart cities to manage traffic flows, monitor air quality, and provide real-time updates on public transportation.
2. ***Healthcare***: In healthcare, fog computing can be used to monitor patient health, provide real-time medical data to doctors, and support remote patient monitoring.
3. ***Industrial Internet of Things (IIoT)***: Fog computing can be used in IIoT applications to monitor and control machinery, optimize production processes, and improve worker safety.
4. ***Smart Grids***: In the energy sector, fog computing can be used to optimize energy consumption, monitor grid stability, and improve the efficiency of energy distribution.
5. ***Autonomous Vehicles***: Fog computing can be used in autonomous vehicles to process sensor data in real time, enabling faster decision-making and improving safety.
6. ***Retail***: In retail, fog computing can be used to personalize shopping experiences, optimize inventory management, and improve customer service.

6.4.8 Future Scope of Fog Computing

Fog computing is an evolving paradigm that has gained traction due to its potential to address the limitations of cloud computing in edge environments. The future scope of fog computing is promising, and it is expected to have several implications for various industries.

Here are some potential future directions and applications of fog computing [140–147]:

1. ***Internet of Things (IoT) and Industry 4.0***: Fog computing is well-suited for IoT deployments and Industry 4.0 applications, where a massive number of IoT devices generate data at the edge. Fog computing can provide real-time data processing, analytics, and decision-making capabilities, enabling efficient and intelligent IoT systems.
2. ***5G and Mobile Edge Computing (MEC)***: With the rollout of 5G networks, fog computing is expected to play a crucial role in supporting low-latency and high-bandwidth applications. It can work in tandem with Mobile Edge Computing (MEC) to provide ultra-responsive and high-performance services to mobile users.
3. ***Autonomous Vehicles and Smart Transportation***: Fog computing can enhance the capabilities of autonomous vehicles and smart transportation systems by enabling real-time data processing, vehicle-to-vehicle communication, and situational awareness at the edge.
4. ***Smart Cities and Urban Infrastructure***: Fog computing can be instrumental in building smart cities, where edge devices and sensors collect data to optimize urban infrastructure, manage traffic, and improve public services.

5. ***Edge AI and Machine Learning***: Fog computing can facilitate the deployment of AI and machine learning models at the edge, enabling intelligent decision-making without relying heavily on centralized cloud resources.
6. ***Healthcare and Telemedicine***: In the healthcare sector, fog computing can enable remote patient monitoring, real-time data analysis, and support telemedicine services, providing better healthcare access and response.
7. ***Augmented Reality (AR) and Virtual Reality (VR)***: Fog computing can enhance AR and VR experiences by reducing latency and offloading computational tasks to edge nodes, ensuring immersive and real-time interactions.

It is important to note that fog computing is an evolving field, and its future scope will depend on technological advancements, standardization efforts, and industry adoption. As the demand for edge computing capabilities continues to grow, fog computing is expected to play a crucial role in enabling edge intelligence and supporting a wide range of applications.

6.5 EDGE COMPUTING

Edge computing is a distributed computing paradigm that brings computing and data storage closer to the location where it is needed to improve response time, reduce latency, and conserve bandwidth. Here are some definitions of edge computing:

- "***Edge computing*** is a distributed computing paradigm that enables data to be processed closer to where it is generated in order to reduce latency and improve application performance" [69].
- "***Edge computing*** is a method of optimizing cloud computing systems by performing data processing at the edge of the network, near the source of the data" [70].
- "***Edge computing*** refers to the practice of processing data near the edge of the network, where the data is being generated, instead of in centralized data-processing warehouses" [71].
- "***Edge computing*** is a decentralized approach to computing that enables the computation and data storage to be carried out near the sources of data" [72].
- "***Edge computing*** is a model in which computation is done on distributed devices (the edge) rather than on a central server or in the cloud" [73].

Edge computing is becoming increasingly important as the Internet of Things (IoT) and other technologies generate massive amounts of data that need to be processed quickly and efficiently.

6.5.1 Components of Edge Computing

Edge computing is a decentralized computing paradigm that brings data processing closer to the source of data generation, reducing latency and bandwidth usage.

Here are the key components of edge computing:

- ***Edge Devices***: These are the devices located at the edge of the network, such as IoT devices, sensors, smartphones, and edge servers. They collect and generate data from the physical world and act as data sources for edge computing applications [106][107].
- ***Edge Servers***: Edge servers are computing nodes located at the edge of the network, closer to the edge devices. They process and analyze data locally, enabling real-time decision-making and reducing the need to send all data to centralized data centers [106][107].
- ***Edge Gateways***: Edge gateways act as intermediaries between edge devices and the centralized cloud or data center. They aggregate and preprocess data from multiple edge devices before sending it to the cloud, optimizing data transmission and reducing latency [108].
- ***Edge Analytics***: Edge analytics involves performing data analytics and processing tasks at the edge of the network. This allows for real-time insights and immediate responses to critical events, without relying on distant data centers [109].
- ***Edge Management and Orchestration***: Edge management systems are responsible for managing and orchestrating the resources and tasks distributed across the edge nodes and devices. They ensure efficient resource allocation, load balancing, and task scheduling within the edge computing infrastructure [110].

6.5.2 Architectural Model of Edge Computing

The architectural model of edge computing involves decentralized computing resources located closer to the data source, reducing latency and enabling real-time data processing.

The key components of the edge computing architecture are as follows:

- ***Edge Devices***: These are the IoT devices, sensors, and endpoints that generate and collect data at the edge of the network.
- ***Edge Nodes/Gateways***: Edge nodes or gateways are intermediate devices located at the edge of the network, responsible for collecting, preprocessing, and filtering data from edge devices before transmitting it to the central data center or cloud.
- ***Edge Computing Infrastructure***: This layer includes the edge servers, edge data centers, and networking equipment deployed at the edge to support edge computing services and applications.

- ***Edge Services and Applications***: These are the software applications and services running on the edge nodes to perform data analytics, real-time processing, and decision-making at the edge.
- ***Cloud/Data Center***: The cloud or data center is still part of the architecture, serving as the centralized computing and storage resource for complex data processing, historical data analysis, and long-term storage.

6.5.3 Timeline of Edge Computing

Here is a timeline of the key developments in edge computing:

Table 6.4 Evolution of Edge Computing

Ref.	*Year*	*Contribution*	*By*
[74]	2001		
[75]	2006		
	2008		
	2012		
[76]	2014	Introduced Azure IoT Edge, a service that allows customers to deploy cloud services to the edge of the network	Microsoft
[77]	2015	Consortium was established to promote the development and adoption of edge computing technologies in industrial applications	Industrial Internet Consortium (IIC)
[78]	2018	EdgeX Foundry, an open-source framework for edge computing, was launched	Linux Foundation
[79]	2019	Launched StarlingX, an open-source software platform for edge computing	OpenStack Foundation
[80]	2020	Launched the Edge Computing Consortium to promote collaboration and research in edge computing	World Economic Forum

6.5.4 Working of Edge Computing

Edge computing is a distributed computing paradigm that brings computing resources and data storage closer to the location where data is generated, such as IoT devices or sensors, rather than relying on a centralized cloud server.

The working of edge computing involves the following steps [129][130][131][132]:

1. ***Data Collection***: Edge computing starts with the collection of data from various sensors, IoT devices, or end-user devices at the edge of the network.

2. ***Local Processing***: The collected data is processed locally on the edge devices or edge servers. These edge devices have computing capabilities that allow them to perform real-time data processing, analysis, and filtering.
3. ***Data Storage***: Edge devices can also store relevant data locally, reducing the need to send all data to the central cloud for storage. This local storage ensures data availability and can be used for historical analysis.
4. ***Decision-Making***: Edge devices can make real-time decisions based on the locally processed data, enabling faster response times and reducing the dependency on the central cloud for decision-making.
5. ***Communication with Cloud***: If required, the processed data or important information can be sent to the central cloud for further analysis, storage, and long-term data processing. The cloud serves as a central repository for data that requires more extensive computations or historical analysis.

Edge computing enhances the performance, reduces latency, and optimizes data traffic, making it suitable for applications that require real-time processing, such as industrial automation, autonomous vehicles, and video analytics.

6.5.5 Benefits of Edge Computing

Edge computing is a powerful technology that provides several benefits to users and organizations. Here are some of the key benefits of edge computing [81–86]:

1. ***Low Latency***: Edge computing reduces network latency by processing data locally at the edge of the network, enabling faster response times for time-sensitive applications.
2. ***Improved Security***: Edge computing can improve security by processing sensitive data locally, reducing the risk of data breaches and ensuring compliance with data protection regulations.
3. ***Increased Reliability***: Edge computing can improve reliability by reducing the dependence on a centralized cloud infrastructure and enabling distributed processing across multiple edge devices.
4. ***Reduced Bandwidth Usage***: Edge computing can reduce bandwidth usage by processing data locally and only transmitting the necessary data to the cloud.
5. ***Scalability***: Edge computing can support scalable and flexible deployments by enabling the use of lightweight edge devices that can be easily added or removed from the network.
6. ***Cost-Effectiveness***: Edge computing can reduce costs by reducing the need for expensive cloud infrastructure and enabling the use of low-cost edge devices.

6.5.6 Limitations of Edge Computing

While edge computing offers several benefits, there are also some limitations and challenges associated with this technology. Here are some of the key limitations of edge computing [81–87]:

1. ***Limited Processing Power***: Edge devices typically have limited processing power, memory, and storage capacity, which can limit the types and complexity of applications that can be run on them.
2. ***Network Connectivity***: Edge devices rely on network connectivity to communicate with other devices and the cloud, and network reliability and latency can affect the performance of edge computing applications.
3. ***Security Concerns***: Edge devices are often deployed in unsecured environments, making them vulnerable to security threats such as hacking and malware attacks.
4. ***Management and Maintenance***: Managing and maintaining a large number of edge devices can be challenging, especially when they are deployed in remote or hard-to-reach locations.
5. ***Integration with Legacy Systems***: Integrating edge computing with legacy systems can be difficult, especially when there is a lack of standardization and compatibility between different systems.

6.5.7 Applications of Edge Computing

Edge computing has a wide range of applications across different industries and domains. Here are some of the key applications of edge computing [81][82][88–91]:

1. ***Industrial Automation***: Edge computing can be used for real-time monitoring, control, and optimization of industrial processes, enabling faster response times and reducing downtime and maintenance costs.
2. ***Healthcare***: Edge computing can be used for remote patient monitoring, real-time analysis of medical data, and personalized healthcare services, improving the quality and efficiency of healthcare delivery.
3. ***Smart Cities***: Edge computing can be used for real-time monitoring and management of city infrastructure, such as traffic lights, waste management systems, and energy grids, improving the sustainability and livability of cities[99].
4. ***Autonomous Vehicles***: Edge computing can be used for real-time processing and analysis of sensor data from autonomous vehicles, enabling faster and more accurate decision-making and improving safety and reliability.
5. ***Gaming and Entertainment***: Edge computing can be used for low-latency gaming and streaming services, improving the user experience, and enabling new forms of immersive entertainment.
6. ***Retail***: Edge computing can be used for real-time inventory management, personalized marketing, and customer analytics, improving the efficiency and effectiveness of retail operations.

6.5.8 Future Scope of Edge Computing

Edge computing is a rapidly evolving technology with significant potential for future development and innovation. Here are some of the key areas of future scope for edge computing [81][82][91][92][93][94]:

1. ***5G Networks***: The rollout of 5G networks is expected to enable more advanced edge computing applications, with higher bandwidth and lower latency, allowing for real-time processing and analysis of large volumes of data.
2. ***Artificial Intelligence***: Edge computing can be combined with artificial intelligence (AI) technologies to enable real-time decision-making and autonomous operations, with applications in areas such as autonomous vehicles, robotics, and smart homes.
3. ***Augmented Reality and Virtual Reality***: Edge computing can enable low-latency, high-performance augmented and virtual reality experiences, with applications in areas such as gaming, training, and remote collaboration.
4. ***Blockchain***: Edge computing can be used to support distributed ledger technologies such as blockchain, enabling decentralized, secure, and transparent data processing and analysis.
5. ***Energy Efficiency***: Edge computing can be used to reduce the energy consumption of data centers by offloading processing tasks to edge devices, reducing the need for data transmission and storage.
6. ***Healthcare***: Edge computing can be used to support personalized healthcare services, with real-time monitoring, analysis, and feedback based on individual patient data.

6.6 CLOUD, DEW, EDGE, AND FOG COMPUTING: AN IN-DEPTH NOTIONAL ESTIMATION

6.6.1 Evolution of Cloud, Dew, Edge, and Fog Computing

The four cutting edge technologies in cloud computing domain are evolved by the times listed broadly in following table [95]:

Table 6.5 Timeline of cloud oriented technologies

Sr. No.	*Technology*	*Year*	*Development*
1	Cloud Computing	2000	Salesforce.com becomes the first cloud-based Software as a Service (SaaS) provider.
		2006	Amazon launches Amazon Web Services (AWS), popularizing cloud Infrastructure as a Service (IaaS).
		2008	Google releases Google App Engine, a Platform as a Service (PaaS) offering.
		2010	Microsoft launches Azure, its cloud platform offering.

Sr. No.	Technology	Year	Development
2	Dew Computing	2015	The term "Dew Computing" is first introduced by the Chinese Academy of Sciences.
		2016	The Dew Computing Research Group is formed at the University of Bridgeport.
		2018	The first Dew Computing Conference is held in Vancouver, Canada.
3	Edge Computing	2003	Researchers at the University of California, Berkeley, publish a paper on "The Case for VM-based Cloudlets in Mobile Computing."
		2007	IBM researchers propose "Smart Cloud" architecture for mobile devices.
		2012	Cisco introduces the concept of "Fog Computing," which later becomes synonymous with edge computing.
		2015	The OpenFog Consortium is formed to develop standards for fog computing.
4	Fog Computing	2012	Cisco introduces the concept of "Fog Computing," a distributed computing architecture for the Internet of Things (IoT).
		2015	The OpenFog Consortium is formed to develop standards for fog computing.
		2018	The Industrial Internet Consortium releases a reference architecture for fog computing.

6.6.2 Issues and Prevailing Solutions on Cloud, Dew, Fog, and Edge Computing

Some possible issues and solutions related to Cloud, Dew, Fog, and Edge Computing are listed in following table [96]:

Table 6.6 Issues and current Solutions of cloud technologies

Sr. No.	Issue	Solution
1	Resource management can be a challenge in cloud computing due to its centralized nature, while resource availability can be limited in dew computing due to its reliance on local resources.	Hybrid cloud-dew models can be used to combine the benefits of both paradigms and optimize resource management.
2	Latency can be a significant concern in cloud computing due to data transmission between remote data centers, while edge computing may face bandwidth constraints and hardware limitations.	Fog computing, which provides a middle ground between cloud and edge computing, can help address latency issues by placing computing resources closer to the end-users.

(Continued)

Table 6.6 (Continued)

Sr. No.	*Issue*	*Solution*
3	Security and privacy can be a concern in all four paradigms, but fog and edge computing face additional challenges due to their distributed nature and limited resources.	Encryption, access control, and intrusion detection systems can be used to enhance security and privacy in fog and edge computing.
4	Scalability can be a challenge in cloud computing due to the centralized infrastructure, while edge computing may not be able to handle large-scale data processing.	Fog computing can provide a scalable and flexible computing model by combining the strengths of cloud and edge computing.
5	Adaptability can be a challenge in cloud computing due to the rigid infrastructure, while dew computing may not be able to adapt to changing workloads or network conditions.	Edge and fog computing can provide more adaptable and dynamic computing models by utilizing local resources and enabling real-time decision-making.
6	Interoperability and standardization can be a challenge in all four paradigms, as there are multiple platforms and technologies used in each paradigm.	Industry-wide standards and open-source platforms can be developed to promote interoperability and facilitate cross-platform compatibility.
7	Environmental impact can be a concern in cloud computing due to the energy consumption of data centers, while edge computing may also have high energy costs due to the need for multiple devices and infrastructure.	Energy-efficient hardware and software, renewable energy sources, and sustainable computing practices can be adopted to reduce the environmental impact of Cloud, Dew, Fog, and Edge Computing.

6.6.3 Cloud, Dew, Edge, and Fog Computing: A State-of-Art Difference

Here is a 360° evaluation of Cloud, Dew, Fog, and Edge Computing based on various parameters:

Table 6.7 Cloud, Dew, Edge, and Fog Computing: Key Distinctions

Parameter	*Cloud Computing*	*Dew Computing*	*Edge Computing*	*Fog Computing*
Definition	A model for delivering computing resources as a service over the internet. It involves the use of remote servers to store, manage, and process data.	A new computing paradigm that extends cloud computing to the edge of the network. It enables resource sharing and collaboration among nearby devices and services.	A distributed computing model that processes data and services at or near the source of data generation. It enables faster response times, reduced bandwidth usage, and improved security.	A decentralized computing architecture that brings computation and data storage closer to the end-users and devices. It enables low-latency and real-time processing of data.

Parameter	*Cloud Computing*	*Dew Computing*	*Edge Computing*	*Fog Computing*
Infrastructure	Centralized infrastructure with many servers and data centers	Decentralized infrastructure with a small number of nearby devices and services	Distributed infrastructure with computing resources at or near the source of data generation	Distributed infrastructure with intermediate nodes between cloud and edge
Data Processing	High processing power and storage capacity for large-scale data processing [101]	Limited processing power and storage capacity for small-scale data processing [101]	Limited processing power and storage capacity for localized data processing [101]	Intermediate processing power and storage capacity for real-time data processing [101]
Latency	High latency due to data transmission between remote data centers	Low latency due to nearby resources, but limited availability	Low latency due to processing data at or near the source of data generation	Low latency due to distributed architecture and nearby resources
Security	Centralized security mechanisms with dedicated security personnel	Limited security mechanisms and reliance on local security measures	Limited security mechanisms and reliance on local security measures	Intermediate security mechanisms with distributed security protocols
Scalability	High scalability due to the centralized infrastructure	Limited scalability due to the decentralized infrastructure	Limited scalability due to the distributed architecture and limited resources	Intermediate scalability due to the distributed architecture and nearby resources
Energy Efficiency	High energy consumption due to data centers and large-scale infrastructure	Moderate energy consumption due to the small-scale infrastructure	Low energy consumption due to the localized infrastructure	Moderate energy consumption due to the distributed architecture and intermediate resources

6.6.4 Research Gap Investigation of Cloud, Dew, Fog, and Edge Computing

The field of Cloud, Dew, Fog, and Edge Computing has been extensively researched and developed in recent years, with many studies focusing on various aspects of these computing paradigms. However, there are still several areas where research gaps exist. Here are some potential research gaps in Cloud, Dew, Fog, and Edge Computing:

- ***Standardization***: Despite the significant advancements in Cloud, Dew, Fog, and Edge Computing, there is a lack of standardization in terms of architectures, protocols, and interfaces. This can hinder interoperability and portability across different platforms and devices.

- ***Resource Management***: Efficient resource management is crucial for the success of Cloud, Dew, Fog, and Edge Computing. However, there is a lack of comprehensive resource management frameworks that can balance resource utilization, performance, and energy efficiency in a dynamic and heterogeneous environment.
- ***Security and Privacy***: As Cloud, Dew, Fog, and Edge Computing involve distributed and heterogeneous systems, ensuring security and privacy is a challenging task. There is a need for more robust and scalable security and privacy mechanisms that can protect against various types of attacks and threats.
- ***Quality of Service***: Cloud, Dew, Fog, and Edge Computing aim to provide high-quality services to end-users. However, there is a lack of effective QoS frameworks that can ensure consistent performance and reliability across different devices, networks, and applications.
- ***Application Development***: Developing applications for Cloud, Dew, Fog, and Edge Computing requires specialized skills and knowledge. There is a need for more user-friendly and accessible application development platforms that can simplify the development and deployment of complex distributed applications.

To address these research gaps, future studies can focus on developing novel architectures, frameworks, algorithms, and tools that can enhance the performance, scalability, security, and usability of Cloud, Dew, Fog, and Edge Computing.

6.6.5 Future Challenges of Cloud, Dew, Fog, and Edge Computing

Cloud, Dew, Fog, and Edge Computing have revolutionized the way we think about computing and have enabled the development of new applications and services that were not possible before. However, there are several challenges that need to be addressed to further enhance the capabilities and potential of these computing paradigms. Here are some of the future challenges of Cloud, Dew, Fog, and Edge Computing:

- ***Scalability***:
  As the number of devices and data sources connected to Cloud, Dew, Fog, and Edge Computing continues to grow, there is a need for more scalable and flexible architectures and frameworks that can handle large-scale and dynamic workloads.
- ***Interoperability***:
  As Cloud, Dew, Fog, and Edge Computing involve a range of different devices, systems, and networks, interoperability remains a key challenge.

There is a need for standardized interfaces and protocols that can facilitate seamless integration and communication across different platforms.

- ***Energy Efficiency***:
  Cloud, Dew, Fog, and Edge Computing require significant amounts of energy to operate, and as the number of devices and systems increases, this energy consumption is expected to grow. There is a need for more energy-efficient architectures and algorithms that can reduce energy consumption while maintaining high levels of performance and reliability.
- ***Privacy and Security***:
  As Cloud, Dew, Fog, and Edge Computing involve distributed and heterogeneous systems, ensuring privacy and security remains a challenge. There is a need for more robust and scalable security and privacy mechanisms that can protect against various types of attacks and threats.
- ***Real-Time Processing***:
  As more applications and services rely on real-time data processing, there is a need for faster and more efficient data processing techniques that can handle large volumes of data in real time.

To address these future challenges, researchers and practitioners can work on developing new architectures, frameworks, algorithms, and tools that can improve the scalability, interoperability, energy efficiency, privacy and security, and real-time processing capabilities of Cloud, Dew, Fog, and Edge Computing.

6.7 CONCLUSION

This chapter provides a comprehensive comparison of the four computing paradigms. The comparison is based on several criteria such as resource availability, latency, security, privacy, scalability, and adaptability. The chapter highlights the key features of each paradigm, potential applications, and limitations. The chapter aims to help researchers and practitioners make informed decisions when selecting the most appropriate computing model for their needs.

REFERENCES

[1] Mainframe computing: IBM Archives. (n.d.). Mainframe computing. Retrieved from www.ibm.com/ibm/history/ibm100/us/en/icons/mainframe/

[2] Grid computing: Foster, I., Kesselman, C., & Tuecke, S. (2001). The anatomy of the grid: Enabling scalable virtual organizations. International Journal of High Performance Computing Applications, 15(3), 200–222. doi: 10.1177/109434200101500302

[3] Utility computing: Armbrust, M., Fox, A., Griffith, R., Joseph, A. D., Katz, R., Konwinski, A.,... Zaharia, M. (2010). A view of cloud computing. Communications of the ACM, 53(4), 50–58. doi: 10.1145/1721654.1721672

[4] Application service providers (ASPs): Saini, S., & Yadav, R. K. (2014). Evolution of cloud computing and its challenges. International Journal of Computer Applications, 98(9), 14–19. doi: 10.5120/17170-3082

[5] Virtualization: VMware. (n.d.). The history of virtualization. Retrieved from www.vmware.com/content/dam/digitalmarketing/vmware/en/pdf/whitepaper/products/virtualization/VMware-White-Paper-The-History-of-Virtualization.pdf

[6] Mell, P., & Grance, T. (2011). The NIST definition of cloud computing. National Institute of Standards and Technology.

[7] Mathew, S., & Varia, J. (2014). Overview of amazon web services. Amazon Whitepapers, 105, 1–22.

[8] Parchure, A. (2020). Benefits of using cloud computing in education field. International Journal of Computer Applications, 176(1), 10–15.

[9] Gartner. (2021). Magic quadrant for enterprise low-code application platforms. Retrieved from www.gartner.com/doc/reprints?id=1-244FAO6L&ct=210524&st=sb

[10] Chong, F., Carraro, G., & Wolter, J. (2010). SaaS and cloud computing: Key features and differentiators. Microsoft Corporation, White Paper, 1–16.

[11] Leavitt, N. (2010). Is cloud storage ready for prime time? Computer, 43(4), 15–17.

[12] Alliance, C. S. (2018). Security guidance for critical areas of focus in cloud computing v4. 0. Cloud Security Alliance.

[13] Amazon Web Services. (n.d.). What is cloud computing? Retrieved from https://aws.amazon.com/what-is-cloud-computing/

[14] Microsoft Azure. (n.d.). What is cloud computing? Retrieved from https://azure.microsoft.com/en-us/overview/what-is-cloud-computing/

[15] Google Cloud. (n.d.). What is cloud computing? Retrieved from https://cloud.google.com/what-is-cloud-computing

[16] IBM Cloud. (n.d.). What is cloud computing? Retrieved from www.ibm.com/cloud/learn/what-is-cloud-computing

[17] Red Hat. (2022, March 23). What is cloud computing? Retrieved from www.redhat.com/en/topics/cloud-computing/what-is-cloud-computing

[18] Oracle Cloud. (n.d.). Oracle cloud. Retrieved from www.oracle.com/cloud/

[19] Amazon Web Services. (n.d.). Benefits of cloud computing. Retrieved from https://aws.amazon.com/what-is-cloud-computing/

[20] Microsoft Azure. (n.d.). Why Azure? Retrieved from https://azure.microsoft.com/en-us/overview/why-azure/

[21] Google Cloud. (n.d.). Why Google cloud? Retrieved from https://cloud.google.com/why-google-cloud

[22] Salesforce. (n.d.). Cloud computing benefits. Retrieved from www.salesforce.com/cloud-computing/benefits/

[23] Khan, S., & Khan, S. U. (2016). Security and privacy issues in cloud computing: A review. Journal of Information Security, 7(2), 73–91.

[24] Gupte, M. (2013). Cloud computing advantages and disadvantages. International Journal of Engineering Research and Applications, 3(6), 2029–2034.

[25] Buyya, R., Yeo, C. S., Venugopal, S., Broberg, J., & Brandic, I. (2009). Cloud computing and emerging IT platforms: Vision, hype, and reality for delivering computing as the 5th utility. Future Generation Computer Systems, 25(6), 599–616.

[26] Garg, S. K., Versteeg, S., & Buyya, R. (2013). A framework for ranking of cloud computing services. Future Generation Computer Systems, 29(4), 1012–1023.

[27] Subramanian, R., Anbazhagan, M., & Premalatha, L. (2019). Cloud-based data storage and backup services: A review. International Journal of Innovative Technology and Exploring Engineering, 8(8S3), 195–198.

[28] Chao, K. M., Wang, C. H., & Wang, K. H. (2016). A cloud computing-based framework for software development and testing. Journal of Information Science and Engineering, 32(5), 1095–1111.

[29] Jin, H., Yan, W., Gao, B., & Xu, X. (2019). Big data analytics in cloud computing: A review. IEEE Access, 7, 47380–47395.

[30] Teregowda, P., & Goudar, R. H. (2017). Cloud computing applications in retail and e-commerce. International Journal of Innovative Research in Computer and Communication Engineering, 5(2), 2613–2623.

[31] Shaikh, S. A., & Arora, R. (2019). Cloud computing in healthcare: A review. International Journal of Computer Applications, 182(47), 1–6.

[32] Ben Alaya, M., & Hsu, C.-H. (2019). Dew computing: A paradigm to bridge cloud computing and edge computing. IEEE Cloud Computing, 6(5), 16–23. doi: 10.1109/MCC.2019.2916155

[33] He, X., & Zheng, Z. (2019). Dew computing: Perspectives, concepts and technologies. Springer. doi: 10.1007/978-3-030-16946-6

[34] Zhang, T., & Fu, X. (2018). Dew computing: A review, reflection, and vision for future research. Proceedings of the IEEE, 106(1), 11–24. doi: 10.1109/JPROC.2017.2766229

[35] Satyanarayanan, M. (2017). The emergence of edge computing. Computer, 50(1), 30–39.

[36] Bonomi, F., Milito, R., Zhu, J., & Addepalli, S. (2012). Fog computing and its role in the Internet of Things. In Proceedings of the First Edition of the MCC Workshop on Mobile Cloud Computing (pp. 13–16). ACM.

[37] Xiao, Z., & Zhang, Q. (2017). Dew computing: A new paradigm for computing after cloud and edge computing. IEEE Access, 5, 20427–20436.

[38] Wang, Y., Xu, Z., & Wang, Y. (2020). Dew computing: Concept, model and applications. Future Generation Computer Systems, 107, 645–653.

[39] Zhang, Q., Zhang, L., & Chen, Y. (2020). Dew computing-based intelligent maintenance system for industrial Internet of Things. IEEE Transactions on Industrial Informatics, 16(6), 4146–4154.

[40] da Silva, F. G. S. L. (2015). Dew computing: Towards user-centric computing. Journal of Parallel and Distributed Computing, 86, 14–29.

[41] da Silva, F. G. S. L., Delicato, F. C., Figueiredo, R. J., Pires, P. F., & Batista, T. (2016). Dew computing and its applications in practical systems. Journal of Systems Architecture, 66, 1–13.

[42] Dew Computing Symposium. (2017). Retrieved from https://sites.google.com/view/dewcom-2017/home

[43] Wang, C., Zhang, Y., & Tian, H. (2018). Dew computing-based healthcare monitoring system. Journal of Medical Systems, 42(6), 113.

[44] Li, L., Li, J., & Wang, J. (2019). Dew computing-based industrial big data processing platform. Journal of Parallel and Distributed Computing, 126, 15–27.

[45] Hu, X., Guo, J., & Yang, Y. (2020). Dew computing architecture: An edge computing approach for distributed data processing in the Internet of Things. IEEE Access, 8, 35278–35288.

[46] Gubbi, J., Buyya, R., Marusic, S., & Palaniswami, M. (2013). Internet of Things (IoT): A vision, architectural elements, and future directions. Future Generation Computer Systems, 29(7), 1645–1660.
[47] Ahmed, A., Ahmed, E., Yaqoob, I., Gani, A., Imran, M., Guizani, M., & Noor, R. M. (2017). Security issues in fog computing: A comprehensive review. Journal of Network and Computer Applications, 88, 10–28.
[48] Kao, Y. T., Tsai, C. F., & Tsai, W. Y. (2020). Dew computing in healthcare: A review. Journal of Medical Systems, 44(6), 110.
[49] Zanella, A., Bui, N., Castellani, A., Vangelista, L., & Zorzi, M. (2014). Internet of Things for smart cities. IEEE Internet of Things Journal, 1(1), 22–32.
[50] Bonomi, F., Milito, R., Zhu, J., & Addepalli, S. (2012). Fog computing and its role in the Internet of Things. In Proceedings of the First Edition of the MCC Workshop on Mobile Cloud Computing (pp. 13–16). doi: 10.1145/2342509.2342513
[51] OpenFog Consortium. (n.d.). What is fog computing? Retrieved from www.openfogconsortium.org/about-us/what-is-fog-computing/
[52] IEEE. (n.d.). What is fog computing? Retrieved from https://iot.ieee.org/images/files/pdf/IEEE_Future_Directions_Fog_Computing.pdf
[53] Bonomi, F., Milito, R., Natarajan, P., & Zhu, J. (2012). Fog computing: A platform for Internet of Things and analytics. In M. Giacobbe, C. Catrini, & P. Bellavista (Eds.), Big Data and Internet of Things: A Roadmap for Smart Environments (pp. 169–186). Springer.
[54] Yi, S., Li, C., & Li, Q. (2015). A survey of fog computing: Concepts, applications and issues. In Proceedings of the IEEE Conference on Communications and Network Security (pp. 104–111). IEEE.
[55] National Institute of Standards and Technology. (2018). Fog computing conceptual model (NIST Special Publication 500–325).
[56] Basir, R., Qaisar, S., Ali, M., Aldwairi, M., Ashraf, M. I., Mahmood, A., & Gidlund, M. (2019). Fog computing enabling industrial internet of things: State-of-the-art and research challenges. Sensors, 19(21), 4807.
[57] Fog World Congress. (2022). About the fog world congress. Retrieved from www.fogworldcongress.com/about-us/
[58] Kaur, A., & Singh, A. (2021). Fog computing: Architecture, applications, and security issues. International Journal of Network Security, 23(2), 245–252.
[59] OpenFog Consortium. (2017). OpenFog Reference Architecture for Fog Computing.
[60] Pahl, C., & Banaei-Kashani, F. (2016). Fog computing in industry 4.0 environment: A survey. International Journal of Computer Applications, 146(11), 44–53.
[61] Zhang, Q., & Zhang, L. (2018). Fog computing: Concepts, architectures, and applications. Springer.
[62] Al-Fuqaha, A., Guizani, M., Mohammadi, M., Aledhari, M., & Ayyash, M. (2015). Internet of Things: A survey on enabling technologies, protocols, and applications. IEEE Communications Surveys & Tutorials, 17(4), 2347–2376.
[63] Satyanarayanan, M., Bahl, P., & Caceres, R. (2009). The case for vm-based cloudlets in mobile computing. IEEE Pervasive Computing, 8(4), 14–23.
[64] Dastjerdi, A. V., Gupta, H., Calheiros, R. N., Ghosh, S. K., & Buyya, R. (2016). Fog computing: Principles, architectures, and applications. Internet of Things, 1, 81–99.

[65] Shi, W., Cao, J., Zhang, Q., Li, Y., & Xu, L. (2016). Edge computing: Vision and challenges. IEEE Internet of Things Journal, 3(5), 637–646.
[66] Patel, N., & Patel, M. (2018). Fog computing: Survey of trends, architectures, requirements, and research directions. Journal of Ambient Intelligence and Humanized Computing, 9(5), 1747–1777.
[67] Wang, C., Liu, Y., Zomaya, A. Y., & Zhou, B. B. (2019). Fog computing: Focusing on mobile users at the edge. Mobile Networks and Applications, 24(2), 392–408.
[68] Cucinotta, T., & De Donato, W. (2016). A survey on fog computing: Concepts, applications and issues. Journal of Network and Computer Applications, 98, 27–42.
[69] Cisco. (n.d.). Edge computing. Retrieved from www.cisco.com/c/en/us/solutions/cloud/what-is-edge-computing.html
[70] TechTarget. (2021). What is edge computing? Definition, use cases and more. Retrieved from https://internetofthingsagenda.techtarget.com/definition/edge-computing
[71] Investopedia. (2021). Edge computing. Retrieved from www.investopedia.com/terms/e/edge-computing.asp
[72] IEEE. (2018). Edge computing. Retrieved from https://ieeexplore.ieee.org/abstract/document/8371563
[73] MIT Technology Review. (2018). Edge computing is going to make the internet much faster. Retrieved from www.technologyreview.com/2018/06/04/141958/edge-computing-is-going-to-make-the-internet-much-faster/
[74] Satyanarayanan, M. (2017). The emergence of edge computing. Computer, 50(1), 30–39.
[75] Shi, W., Cao, J., Zhang, Q., Li, Y., & Xu, L. (2016). Edge computing: Vision and challenges. IEEE Internet of Things Journal, 3(5), 637–646.
[76] Li, Y., & Yang, Y. (2019). Edge computing in the industrial Internet of Things. IEEE Transactions on Industrial Informatics, 15(6), 3578–3586.
[77] Microsoft. (2017). Cloud intelligence deployed locally on IoT edge devices.
[78] Linux Foundation. (2018). EdgeX Foundry Documentation.
[79] OpenStack Foundation. (2019). StarlingX.
[80] World Economic Forum. (2020). Edge Computing Consortium.
[81] Satyanarayanan, M. (2017). The emergence of edge computing. Computer, 50(1), 30–39.
[82] Shi, W., Cao, J., Zhang, Q., Li, Y., & Xu, L. (2016). Edge computing: Vision and challenges. IEEE Internet of Things Journal, 3(5), 637–646.
[83] Bonomi, F., Milito, R., Zhu, J., & Addepalli, S. (2012). Fog computing and its role in the Internet of Things. In Proceedings of the First Edition of the MCC Workshop on Mobile Cloud Computing (pp. 13–16).
[84] Fernández-Caramés, T. M., & Fraga-Lamas, P. (2018). A review on the use of blockchain for the Internet of Things. IEEE Access, 6, 32979–33001.
[85] Zhang, Y., Zhang, J., Chen, M., & Hu, S. (2018). A survey on edge computing for the Internet of Things. IEEE Access, 6, 6900–6919.
[86] Bhattacharya, S., Chakraborty, S., & Saha, H. N. (2019). Edge computing: A survey on architecture, applications, and research directions. Journal of Network and Computer Applications, 135, 1–18.
[87] Kulkarni, S., & Chaudhari, K. (2021). Edge computing: Review of research and future directions. Sustainable Computing: Informatics and Systems, 31, 100459.

[88] Yi, S., Li, C., Li, Q., & Song, H. (2015). Fog computing: Architecture and issues. In Proceedings of the 2015 Workshop on Mobile Big Data (pp. 37–42).

[89] Huang, Y., Zheng, X., Liu, Y., Chen, Y., & Zeng, Z. (2019). Applications of edge computing in the Internet of Things: A survey. IEEE Access, 7, 63780–63800.

[90] Abbas, H., Zhang, L., & Chen, X. (2019). A survey on edge computing technologies for IoT: Security and privacy challenges. Journal of Network and Computer Applications, 135, 27–42.

[91] Kamble, S., & Gupta, R. (2021). Edge computing applications in various domains: A comprehensive review. Future Computing and Informatics Journal, 6(1), 1–13.

[92] Li, Y., Zhang, J., Chen, M., & Hu, S. (2019). Edge computing: Opportunities and challenges. Electronics, 8(11), 1270.

[93] Gia, T. N., & Lee, G. M. (2019). A survey on the edge computing for the Internet of Things. Journal of Information Processing Systems, 15(1), 1–22.

[94] Banerjee, S., Roy, S., & Sengupta, S. (2020). Edge computing: A review of recent advances and future trends. Journal of Network and Computer Applications, 154, 102491.

[95] Bonomi, F., Milito, R., Zhu, J., & Addepalli, S. (2012). Fog computing and its role in the Internet of Things. In Proceedings of the First Edition of the MCC Workshop on Mobile Cloud Computing (pp. 13–16). ACM.

[96] Aazam, M., & Huh, E. N. (2014). Cloud, dew, fog and edge computing: A proposal for a taxonomy of data-centric computing paradigms. arXiv preprint arXiv:1412.0684.

[97] NASA & Rackspace. (2010). OpenStack: Cloud software shifts into high gear. Retrieved from www.nasa.gov/pdf/501579main_Nebula2010_Web.pdf

[98] Sissodia, R., Rauthan, M. S., & Barthwal, V. (2023). Chapter 8 Survey 9 and research issues on fog computing. IGI Global.

[99] Hossain, M. D., Sultana, T., Akhter, S., Hossain, M. I., et al. (2023). The role of microservice approach in edge computing: Opportunities, challenges, and research directions. ICT Express.

[100] Edge computing. (2019). Springer Science and Business Media LLC.

[101] Hossain, K., Rahman, M., & Roy, S. (2019). IoT data compression and optimization techniques in cloud storage. International Journal of Cloud Applications and Computing.

[102] Bonomi, F., Milito, R., Zhu, J., & Addepalli, S. (2012). Fog computing and its role in the Internet of Things. In Proceedings of the First Edition of the MCC Workshop on Mobile Cloud Computing (pp. 13–16).

[103] Yi, S., Li, C., & Li, Q. (2015). A survey of fog computing: Concepts, applications and issues. In Proceedings of the 2015 Workshop on Mobile Big Data (pp. 37–42).

[104] Baccarelli, E., Mazzenga, F., & Biagi, M. (2016). Internet of Things-based smart homes: Systems, architectures, and proposed technologies. IEEE Internet of Things Journal, 3(5), 1–12.

[105] Kumar, M., Khurana, S., Singh, P., & Jangra, A. (2021). Fog computing and its applications: A review. International Journal of Computer Applications, 181(41), 17–20.

[106] Shi, W., Cao, J., Zhang, Q., Li, Y., & Xu, L. (2016). Edge computing: Vision and challenges. IEEE Internet of Things Journal, 3(5), 637–646.

[107] Satyanarayanan, M. (2017). The emergence of edge computing. Computer, 50(1), 30–39.
[108] Shi, W., Wu, J., Cao, J., Zhang, Q., & Liu, Y. (2016). Edge computing: A new frontier for distributed computing. Proceedings of the IEEE, 104(5), 687–700.
[109] Kaur, A., Verma, A., & Mallick, P. K. (2020). Edge computing: A survey and analysis of state-of-the-art. Future Generation Computer Systems, 106, 692–713.
[110] Gubbi, J., Buyya, R., Marusic, S., & Palaniswami, M. (2013). Internet of Things (IoT): A vision, architectural elements, and future directions. Future Generation Computer Systems, 29(7), 1645–1660.
[111] Mell, P., & Grance, T. (2011). The NIST definition of cloud computing. National Institute of Standards and Technology.
[112] Armbrust, M., Fox, A., Griffith, R., Joseph, A. D., Katz, R., Konwinski, A.,... Zaharia, M. (2010). A view of cloud computing. Communications of the ACM, 53(4), 50–58.
[113] Vaquero, L. M., Rodero-Merino, L., Caceres, J., & Lindner, M. (2009). A break in the clouds: Towards a cloud definition. ACM SIGCOMM Computer Communication Review, 39(1), 50–55.
[114] Gartner. (2019). Cloud computing. Gartner IT Glossary.
[115] Amazon Web Services. (2023). Cloud computing. Retrieved from https://aws.amazon.com/what-is-cloud-computing/
[116] Bonomi, F., Milito, R., Zhu, J., & Addepalli, S. (2012). Fog computing and its role in the Internet of Things. In Proceedings of the First Edition of the MCC Workshop on Mobile Cloud Computing (pp. 13–16).
[117] Mukherjee, S., Datta, S., & Dutta, S. (2017). Fog computing: A taxonomy, survey and future directions. In 2017 Second International Conference on Fog and Mobile Edge Computing (FMEC) (pp. 1–6).
[118] Shi, W., Cao, J., Zhang, Q., Li, Y., & Xu, L. (2016). Edge computing: Vision and challenges. IEEE Internet of Things Journal, 3(5), 637–646.
[119] Gartner. (2019). Edge Computing. Gartner IT Glossary.
[120] Cisco. (2015). Fog computing and the Internet of Things: Extend the cloud to where the things are. Cisco White Paper.
[121] Yao, L., Wang, W., Chen, Y., & Wu, H. (2018). Dew computing: Principles, paradigms, and applications. Journal of Parallel and Distributed Computing, 117, 149–165.
[122] Cao, J., Li, K., Xie, L., Lin, X., & Wu, D. (2019). Dew computing: A new paradigm for computing at the edge of the network. IEEE Internet of Things Journal, 6(1), 160–170.
[123] Xu, Y., Huang, D., Liu, W., Wang, Y., & Li, Y. (2018). Dew computing: Definition, framework and architecture. In 2018 IEEE International Conference on Smart Cloud (SmartCloud) (pp. 76–83).
[124] Yao, L., Wu, H., Wang, W., Wang, T., & Hu, L. (2020). Collaborative dew computing on wireless IoT for real-time image processing. IEEE Transactions on Industrial Informatics, 16(10), 6378–6387.
[125] Bonomi, F., Milito, R., Natarajan, P., & Zhu, J. (2012). Fog computing: A platform for Internet of Things and analytics. In Big Data and Internet of Things: A Roadmap for Smart Environments (pp. 169–186). Springer.

[126] Zhang, Y., Wen, Y., & Kumar, N. (2015). A survey on fog computing: Architecture, key technologies, applications and open issues. Journal of Network and Computer Applications, 98, 27–42.
[127] Yi, S., Qin, Z., Li, Q., & Vasilakos, A. V. (2015). Security and privacy of fog computing: A review. IEEE Internet of Things Journal, 3(6), 791–800.
[128] Li, W., Lu, Y., Sheng, Q. Z., Xiang, Y., & Li, X. (2018). Fog computing for sustainable smart cities: A survey. ACM Computing Surveys (CSUR), 51(3), 1–37.
[129] Satyanarayanan, M., Bahl, P., Caceres, R., & Davies, N. (2009). The case for VM-based cloudlets in mobile computing. IEEE Pervasive Computing, 8(4), 14–23.
[130] Shi, W., Cao, J., Zhang, Q., Li, Y., & Xu, L. (2016). Edge computing: Vision and challenges. IEEE Internet of Things Journal, 3(5), 637–646.
[131] Shi, W., Cao, J., Zhang, Q., Li, Y., & Xu, L. (2016). Edge computing: Vision and challenges. IEEE Internet of Things Journal, 3(5), 637–646.
[132] Khan, A. M., Salah, K., Alshehri, M., & Madani, S. A. (2019). Edge computing: A survey. Future Generation Computer Systems, 97, 219–235.
[133] Gartner. (2020). Top Strategic Technology Trends for 2021. www.gartner.com/en/newsroom/press-releases/2020-10-19-gartner-identifies-the-top-strategic-technology-trends-for-2021
[134] Deloitte. (2020). Tech Trends 2020. https://www2.deloitte.com/global/en/pages/technology-media-and-telecommunications/articles/tech-trends.html
[135] Forbes. (2021). Cloud Computing: What to Expect in 2021 and Beyond. www.forbes.com/sites/louiscolumbus/2021/01/03/cloud-computing-what-to-expect-in-2021-and-beyond/?sh=75ea184b5fe3
[136] IBM. (2021). 7 Cloud Computing Trends to Watch in 2021. www.ibm.com/cloud/blog/7-cloud-computing-trends-to-watch-in-2021
[137] Li, W., Zuo, C., Yang, L. T., & Min, G. (2019). Distributed environment for the web: From cloud computing to dew computing. Future Generation Computer Systems, 92, 791–803.
[138] Li, W., Zuo, C., Min, G., & Yang, L. T. (2020). Dew computing: A decentralized computing paradigm. IEEE Cloud Computing, 7(2), 36–45.
[139] Khan, M. A., & Madani, S. A. (2018). A survey of dew computing: Architected and applications. International Journal of Grid and High-Performance Computing, 10(1), 49–69.
[140] Bonomi, F., Milito, R., Natarajan, P., & Zhu, J. (2012). Fog computing: A platform for Internet of Things and analytics. In Big Data and Internet of Things: A Roadmap for Smart Environments (pp. 169–186). Springer.
[141] Yi, S., Qin, Z., Li, Q., & Vasilakos, A. V. (2015). Security and privacy of fog computing: A survey. Proceedings of the IEEE, 104(8), 1449–1472.
[142] Kaur, S., Singh, M., Garg, S., & Chana, I. (2018). Fog computing: A taxonomy, survey and future directions. Computer Networks, 139, 120–151.
[143] Singh, S. P., Shrivastav, S., Singla, N., & Kaur, H. (2020). Cloud computing: Security challenges and issues. Journal of Natural Remedies, 21(2), 101–106.
[144] Kour, R., & Singh, S. P. (2016). An improved data classification technique for data security in cloud computing. In Communication and Computing Systems (pp. 959–964). Taylor & Francis Group.

[145] Gupta, U., Pantola, D., Bhardwaj, A., & Singh, S. P. (2022). Next-generation networks enabled technologies. In Next Generation Communication Networks for Industrial Internet of Things Systems (p. 191). CRC Press.

[146] Shrivastava, R. K., Singh, S. P., Banerjee, A., & Kaur G. (2022). Heartbeat classification using sequential method. In Proceedings of Data Analytics and Management: ICDAM (pp. 293–300). Springer.

[147] Zhang, Y., Zhang, T., & Qin, B. (2020). A survey on fog computing: Architecture, key technologies, applications and open issues. Journal of Network and Computer Applications, 168, 102727.

Chapter 7

Millimetre Wave V2X Communications in 5G for Achieving Reliability in Vehicle Drive

Jayanta Kumar Ray, Rogina Sultana, Rabindranath Bera, Sanjib Sil, Quazi Mohmmad Alfred, and Imtiaj Ahmed

7.1 INTRODUCTION

In recent years, there have been several technological advancements that have led the automotive industry to bring out new types of conveyances like the connected and autonomous car. These utilize an extensive range of features to identify barriers, handle driving related videos and images, and provide momentary command of a car. Numerous onboard sensors must be utilized, such as automotive radars, visual cameras, infrared and LiDAR devices, ultrasonic sensors, and motion sensors, to accomplish this task. Recently, there has been an increasing interest in situations where vehicles' communications connect with other systems such as infrastructures, proximate vehicles, and humans. This concept is known as vehicle to everything (V2X), and it has gained momentum in the field. In reality, the high quantity of vehicles present in urban areas, which have a significant number of sensors, may present a difficulty for network machinists because of the instance load which needs to be dealt with in a dependable and timely manner. Taking into account the restricted operating capabilities of numerous vehicle sensors, transmitting unprocessed sensor data with minimal delay becomes increasingly challenging when scaling to higher speeds [1]. Thankfully, the newly developing 5G cellular systems which utilize millimetre wave radios functioning at incredibly high frequencies and delivering gigabit-per-second throughputs provide a promising solution to the upcoming network capacity issue, even with high data rate sensors [2]. Previous research in the field concentrated on network connectivity and traffic control load in existing cellular networks, taking into account a number of critical technical specifications, including access delay, access likelihood, service-quality guarantees, energy consumption, and network utilization. Millimetre wave is a good frequency range to use for very fast V2V communications due to lots of availability of continuous spectrum resources [3]. For instance, the IEEE 802.11bd specification ensures interoperability with IEEE 802.11p and IEEE 802.11ad while adopting modern V2X applications [4]. It will have a peak data rate of more than 6.75 Gbps. The 3GPP (3rd-Generation Partnership Project) is also defining New Radio V2X (NR-V2X), which has a peak data rate of 20 Gbps [5, 6]. In the last several years, the area of linked automobiles has advanced

 DOI: 10.1201/9781003480181-7

quickly. The integration of cognitive computing methods, vehicular networks, as well as automotive hardware and software technologies has made these advancements feasible. Experts in both the public and private sectors create and enhance vehicle designs that will focus on intercar communication and sensor capabilities. Introducing self-driving cars has many benefits, but it might also cause challenges with maintaining road safety and proper working technology.

New technologies will help by minimizing traffic accidents to make roads safer, save lives, improve traffic flow optimization, protect sustainable environment, and meet the needs of drivers; this will have a big impact on people and society throughout its entirety [7]. It is important to remember that a smart city cannot exist without dependable linked cars [8, 9]. Government sectors have boosted their spending in recent years to enhance traffic control and make improvements to vehicular communications. Car makers are putting many sensors in cars to help people drive safely as more smart transportation is more extensively used. These sensors will produce data at a rapid pace. They come in a variety of sorts, from simple ultrasonic sensors for measuring distance to sophisticated radar-based sensors. Radar- and camera-based sensor technologies provide a vehicle with the data it needs to make wise judgements and enhance communication [10, 11].

7.2 IMPORTANCE OF DEDICATED SHORT-RANGE COMMUNICATION (DSRC)

Dedicated Short-Range Communication is the modern usual for tying together automobiles. DSRC can be used to create preliminary vehicle to vehicle (V2V), vehicle to infrastructure (V2I), and perhaps vehicle to everything (V2X) systems of communications. By 2017, the National Highway Traffic Safety Agency will make it mandatory for all new cars to have DSRC capabilities [7].

In this study, we provide justification for using millimetre wave (mmWave) spectrum for fully connected cars in addition to DSRC or 4G V2X. Consumers may currently purchase mmWave as IEEE 802.11ad, and fifth generation (5G) communications networks are most likely to support it. Gigabit-per-second data rates are made achievable by using mmWave, opening the door to connection to high capacity communication channels and enabling the transmission of raw sensor data between operational infrastructure and automobiles. In actuality, mmWave is not brand-new to the car industry. The mmWave spectrum is already used by modern vehicle radars [2]. mmWave has been researched for use in automotive communications for more than ten years [8]. Even the International Organization for Standardization has developed a standard for vehicle communications at mmWave [7]. In order to support basic safety and non-safety V2X uses, IEEE 802.11 p [12] is a unified communication technology of PHY & MAC layers for Wireless Access in Vehicular Environment (WAVE) [13]. This technology forms the foundation for Dedicated Short-Range Communication (DSRC) in the United States, ITS-G5 in

Europe, which drives in the 5.9 GHz ITS band, and ITS Connect in Japan, which drives in the 760 MHz ITS band [14].

The DSRC, which allows V2X communications, is now a widely used communication protocol which supports vehicular networks. Long Term Evolution-Advanced (LTE-A) was selected by the authors in [6] for use in automotive networks. Compared to DSRC, LTE can support data rates of up to 100 Mbps and has a transmission delay that is less than 100 ms. Hence, the vast volumes of data that connected cars will produce cannot be sent via either DSRC or LTE-A [7].

7.3 INTERNET OF VEHICLE (IoV)

Several academic and industrial researchers have lately been interested in using the expansive spectrum of the underused mmWave frequency range [7]. The needs of mobile user traffic can be met by this unique and promising communication technology. To support IoV applications, it is feasible to have DSRC and LTE cooperate effectively [15].

One of the newest IoT application areas is the Internet of Vehicles (IoV). Due to the quick development of computing and communication technologies, IoV can claim to be a widely used technology. As the Internet of Things is added to vehicle technology, Vehicle Adhoc Networks (VANETs) are being emerged as Internet of Vehicles (IoV). Earlier the coverage area of VANET [16] was small and constrained by mobile networks and number of connected vehicles. It was also not suited for large cities where traffic jams, buildings, bad driver behaviours, and complex road networks were present. In IoV, there are two important components i.e. vehicular networking and vehicular intelligence. The vehicular networking comprised of vehicle-interconnection, vehicle-telematics and mobile internet. The integration of artificial intelligence with human (driver) ability is summarily called vehicular intelligence in IoV. It is an integrated network environment having the interconnections among multiple users, vehicles, things and networks. Hence the IoV is a distinct genre of IoT. It deals with the diverse parameters of transportation such as traffic, vehicle's speed, infrastructures in the city. To enable, It is essential to integrate automobiles into the Internet of Vehicles (IoV) network, which is supported by 5G and other cutting-edge wireless technologies. Wireless access solutions need to be effective in order for IoV to be managed and believable. The five types of vehicular communications in IoV are vehicle to vehicle (V2V), vehicle to infrastructure (V2I), vehicle to roadside unit (V2R), vehicle to personal device (V2P), and vehicle to sensor (V2S) Figure 7.1) [17]. The IoV is said to have a brain called Cloud.

One Both drivers and passengers can access the internet using Internet of Vehicles (IoV) [18, 19]. In an IoV setting, automobiles have sensors that generate a lot of data that is sent towards the cloud [20]. High speed mmWave connectivity for IoV must be enabled in order for such a communication to be successful. V2V and vehicle to infrastructure communication may both be done using mmWave

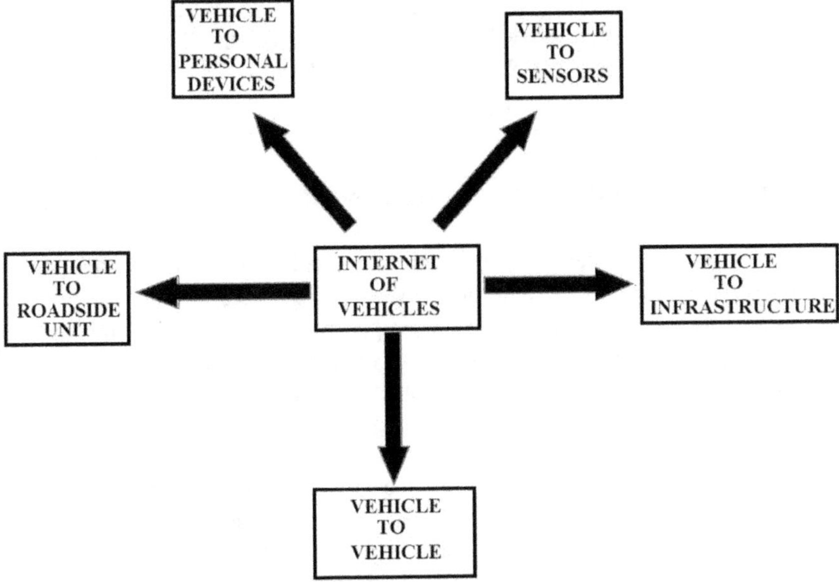

Figure 7.1 Types of Vehicular Communications in IoV.

technology (V2I). Although millimetre wave V2I links are utilized for sending transport data to the cloud, mmWave links allow vehicles to exchange data from their sensors with other adjacent vehicles. This can be accomplished by using a variety of channels, including the 5G bands of 28 GHz & 38 GHz, to enable mmWave communications for IoV. Moreover, 24 GHz & 76 GHz frequency bands can be employed for vehicle positioning [21].

7.4 ENHANCEMENT OF LTE V2X AND TOWARDS mmWAVE V2X

Almost 200 nations across the world presently use LTE, which is currently being widely adopted [22]. The 3GPP's standardization work for LTE-based V2X officially commence with Release 14. These 3GPP projects aim to improve LTE systems and make it possible for cars to exchange data with other cars, people, and infrastructure in order to regulate traffic flow, provide traffic warnings, and improve road safety [3]. Due to its widely used network, Users Equipment (UE), secured network, exceptionally high spectrum efficiency, extensive coverage, high mobility, high dependability, low latency, longer battery life, and other features, LTE has the potential to enable a variety of V2X applications. The 3GPP includes a list of numerous use cases and the accompanying criteria [23]. Its development involves a variety of communications channels in addition to DL and UL improvements involving base stations and UE. For instance, Multimedia Broadcast Multicast Services (MBMS) is just a wide cast/multicast service from many cells to numerous UEs, whereas Single-Cell Point-to-Multipoint (SC-PTM) is a television service from one single cell to several UEs [3, 24].

7.5 USE OF mmWAVE FOR MASSIVE V2X SCENARIOS

By 2020, 200 mMTC devices per car are predicted to be supported by today's wireless systems thanks to their faster transmission rates and shorter transfer delays as well as the significant capacity improvements promised by the upcoming 5G mmWave standard [25]. It is obvious that delivering enormous amounts of sensory data results in greater traffic loads that are incompatible with the present microwave-based network infrastructures (particularly through bandwidth-hungry radars and cameras). To this aim, enhanced multiple-input multiple-output (MIMO) functioning is made possible by the usage of mmWave bands, which also offer bigger quantities of radio spectrum and permit higher order modulation. As a result, this improves spectral efficiency and makes it possible to reach greater transmission rates (i.e., Gbit/s), which is an improvement over the performance of typical systems that operate at carrier frequencies below 6 GHz [1]. In V2X, mmWave gives good support for enhancing automated driving. Because of its capacity to transmit data at a fast pace, IEEE 802.11p and LTE V2X are well complemented by mmWave V2X. Due to capacity restrictions, LTE V2X and IEEE 802.11p may only provide data rates of up to 28.8 Mbps and 27 Mbps, respectively. Yet, because of the large bandwidth offered by mmWave, mmWave V2X can sustain a transmission rate that can reach to 1 Gbps or more. Hence, mmWave V2X is advantageous in enabling sophisticated V2X use cases that demand larger data speeds, such as camera/Li DAR sensors and data exchange.

7.6 mmWAVE V2X FOR AUTOMATED DRIVING

The only practical technology for secure automatic driving is mmWave. Based on the established data rate, the maximum permitted speed of the vehicles is decided to offer reliable automated vehicles. Also, it discusses particular difficulties with mmWave for V2X, such as beam alignment and coverage improvement. Future automobiles will come with a variety of vehicular sensors and ever-improving computer power, allowing them to sense their environment. When used in conjunction with other vehicular sensors, line-of-sight (LOS) is not required in order for V2X to identify other cars or at-risk road users. This capacity considerably improves how intelligent vehicles are seen, and it makes it possible for surrounding vehicles to acquire pertinent information for applications that deal with safety choices [14]. Additional advanced V2X use cases are described in the 3GPP article TR 22.886, "Study on Enhancing 3GPP Support for 5G V2X Services" [26], including Vehicle Platooning, Extended Sensors, Advanced Driving, and Remote Driving. In order to enable these next-generation V2X applications, the transmission of BSMs and CAMs will need to be augmented by the transmission of larger messages that include processing sensor data, raw sensor, vehicle purpose information, cooperation, confirmation of a planned manoeuvre, etc. The demands on data throughput, dependability, latency, communication range, and speed are significantly higher for these "enhanced V2X" (eV2X) application cases [14].

7.7 DESIGN OF NEW RADIO (5G CELLULAR COMMUNICATIONS)

The natural/inherent ability of a 5G cellular system to effectively utilize different frequencies is one of its fundamental characteristics and technological advantages. The NR system, which evolved from the carrier aggregation architecture first used in LTE, offers multiple alternatives for managing radio-resources and spectrum with the goal of supporting diverse device and application types and facilitating communication in various Frequency bands. Future NR systems are designed to provide communication at up to 100 GHz frequency ranges [27]. The NR system architecture in Release 15's initial iteration only specifies cellular radio-interface (also known as Uu radio) and supports specific frequency ranges (FR): FR1 [28] and FR2 [29]. Whereas FR2 covers the lower mmWave spectrum from 24 GHz to 52.6 GHz, FR1 comprises frequencies across 450 MHz to 6 GHz. Future versions are anticipated to improve on the support for more frequency bands, and 3GPP Release 16 will specify a sidelink air interface architecture for V2X use cases. Active antenna arrays can be used in mmWave communication with a robust architecture that combines analog and digital beamforming. The NR radio-layer procedures contain support for multi-beam operation after providing the bare minimum of user mobility and access for DL and UL data transmission [30]. With regard to the mmWave V2X system design, Release 15 marks the completion of the initial release of the NR proposed system enabling cellular communication, which is capable of providing a limited number of vehicle-to-vehicle solutions well over Uu radio interface by facilitating communication in both frequency ranges, including mmWave (FR2). The NR Uu radio interface design still has to be improved in order to support multicast/broadcast in DL broadcasts, a valuable feature for many V2X applications [14].

7.8 5G—WIRELESS BACKBONE OF IoV

5G is the latest technology in which connection among wireless technologies, networks, and applications takes place in a simultaneous chain. The following stage of mobile communication standards after 4G, it is the most recent generation of wireless mobile communication technologies. 5G provides dynamic speed having

Figure 7.2 5G radio access.

much more efficiency. As a result, massive development of wireless technologies has been evolved.

Developments in LTE, LTE-Advanced, and LTE-Pro technologies architect the foundation of 5G technology (Figure 7.2). Standardization of 5G technology makes it possible after the introduction of new air interface called 5G NR [10]. It had been released on December 2017, applicable for NSA (None Stand Alone) 5G NR cases. Here, existing LTE radio and core networks are used. The use of 5G NR provides the facilities of high data rates in NSA mode. As a result, 5G technology can be applicable using existing 4G setup. In September 2018, the 5G NR SA (Stand Alone) [11] was completed. It specifies full user and capability of a control plane using 5G new core network architecture. There is no connection between the core network and the radio access network.

The service offered by NR is to establish connection between the user's mobile device and the base station. Also, it offers room for potential technological advancement in the future. The enhanced Mobile Broadband (eMBB) and Massive Machine-Type Communication (mMTC), and Ultra-Reliable as well as Low Latency Communication (uRLLC) service categories were established by the International Telecommunication Union (ITU) in accordance with future requirements [31]. These are the three families of usage scenarios which require a completely different network service and pose requirements that are different. On the other side, these are the 5G IMT application cases and essential specifications for 2020.

- Enhanced Mobile BroadBand (eMBB)—The services having the characterization of high data rates which includes new enterprise services, application along with the exploding consumption of multimedia, collaborative working and social communications such as AR/VR (Augmented Reality/Virtual Reality) and videos having different forms and formats are focussed. 20 GB/s of downlink peak data rate and 10 GB/s of uplink peak data rate. Low throughput, improved spectral efficiency, and increased coverage characterize the latency.
- Massive Machine-Type Communication (mMTC)—The services that require for connection density, such as those typical for smart city and home, smart factories and smart agriculture use cases are focussed. It has an

extremely high connection density, energy optimization, minimal complexity, and extensive coverage.

- URLLC (Ultra-Reliable and Low Latency Communications)—Focus is placed on latency-sensitive services like autonomous driving, remote monitoring, drone control, etc. The emphasis is on latency-sensitive services, such as monitoring system, drone control, autonomous driving, etc. As a result, lags in browsing, videos, controlling of drones, and robots can be avoided. As it involves machines, the connection must be extremely reliable and latency-free. Its usage is for public safety, autonomous vehicles, and telemedicine. It also includes the critical missions having high availability and location precision.

7.9 TIME SENSITIVE NETWORKING (TSN) FOR LOW LATENCY

Time Sensitive Networking is a collection of standards developed according to the principles of audio visual bridging (AVB) (TSN). It is going to develop vehicles with high bandwidth, interoperability, and affordability. TSN is simply the IEEE 802.1Q recognized standard technology that provides deterministic communications over standard Ethernet. For real-time applications, it can be administered centrally, deliver on its promise, and minimize jitter. The goal of TSN development is to make deterministic communication over regular Ethernet possible. Time is the most important aspect of TSN because it is a time-oriented technology. TSN provides the transfer of data over a predefined and fixed period of time from one site to another. The central network controller (CNC), bridges, end devices, TSN flow, and centralized user configuration (CNC) are the five components that make up TSN systems [32]. The existing AVB (Audio/Video Bridging) has been renamed as the TSN. TSN is a modification and enrichment of AVB [17]. AVB/TSN works by making sure that each audio or video stream has its own space on the network, without disrupting other data transmission. Bandwidth, security, interoperability, low cost, latency, and synchronization are the primary characteristics of TSN. Professional audio and video, automobile control, industrial settings, and other domains are the typical applications of TSN. The development of a TSN standard file by IEEE 802.1 had been made in order to provide full real-time communication. It can be broken down into three main parts: time synchronization, scheduling, and traffic shaping; stream management; and fault tolerance. TSN offers the assured data transmission within the promised window of time. The assured time frame denotes extremely little data loss, constrained low latency, and low delay variance. For time-sensitive or mission-critical data flow, a variety of applications require Quality of Service (QoS). The fifth-generation (5G) cellular network includes time synchronization. The radio network components are the ones that the Precision Time Protocol uses to time-synchronize. Low latency, time synchronization, resource management, and reliability are the four elements that make up URLLC. The four elements of TSN, on the other hand, are reliability, resource management, time synchronization, and traffic shaping [18]. The TSN's traffic shaping is responsible for the reduced latency of URLLC. The features of TSN have a good match with 5G URLLC. The 5G URLLC and TSN are combined and integrated. As a result, there is established deterministic connection between the ends.

7.10 COMP FOR HIGH RELIABILITY

The Coordinated Multi-Point approach is a potential method to attain great dependability (CoMP). One important method for reducing the intercell interference in 4G mobile communication is the broadcast and receipt of CoMP [33]. The CoMP system is a large-scale multiple-input, multiple-output (MIMO), distributed antenna group (DAG) system. CoMP strategies can be divided into four categories: JT (joint transmission), coordinated beamforming (CS/CB), coordination scheduling, and transmission point selection. The accuracy and latency of the feedback have an impact on how well the CoMP system performs. Options for increased dependability, capacity, and geographic variety are available in the CoMP. There are three CoMP options: coherent joint transmission with interference nulling (CJT-MU), coherent joint transmission without interference nulling (CJT-SU), and non-coherent joint transmission (NCJT) [34]. The User Equipment (UE) may see all the possibilities. Due to interference nulling, CJT-MU manages scheduling for a large number of users on the same time and frequency resources. CJT-SU controls the schedule of a single user over a specified range of time frequency resources. Instead of using sources for precoding in NCJT, static precoding matrices like SDCID (short delay cyclic interruption diversity) or precoding cycle are employed. Similar to CJTSU, due to the lack of interference, NCJT schedules a single user on a particular set of time-frequency resources. Similar to the CJT-MU for diversity gain, the signal-to-interference-to-interference plus noise ratio (SNR) gains of CJT-MU and CJT-SU are greater than those of NCJT.

7.11 V2X COMMUNICATIONS

The term "intelligent transportation systems" (ITS) [35] refers to a collection of technology advancements made to increase the efficiency and security of land transportation. Over the past few years, there have been significant developments in the area of vehicular communication systems. Vehicle automation technology is getting increasing attention. Nowadays, research on vehicle to everything (V2X) communications is becoming more and more well-liked in the field of mobile wireless connectivity. The aim of connected vehicle technologies is the tackling of some of the biggest challenges in the Intelligent Transport System (ITS) which includes safety, mobility, and environment. The usage of technologies by ITS include the communication of vehicle with other vehicle, roadside infrastructure and pedestrian. They are sometimes referred to as V2X communications. Accident rates might be reduced thanks to V2X communications. We are in a race to develop safe driving vehicles, and it will take a combination of technologies to get us there. Many collaborations and technologies are anticipated to be used in tandem to make driving with autonomous cars safer, greener, and more effective. By the transmission of position, speed, and direction, V2X offers a higher level of predictability and determinism. When the specifications under first conditions are finished in September 2016, the introduction of C-V2X will enable direct communications operating in the ITS 5.9 GHz band without the need for networks' support, making it perfect for V2V, V2I, and V2P communications. Without a Subscriber Identification Module (SIM), a cellular subscription, or network support, an automobile can be connected to other vehicles and roadside infrastructure [36]. The radio layer used in DSRC and ITS-G5 is IEEE 802.11p, which prepared the way for essential safety services including forward collision warning. With enhanced dependability, longer range, low latency, and non-line-of-sight (NLOS) capabilities, C-V2X Release 14 provides improved safety use cases at faster vehicle speeds and challenging road conditions.

The key technology is cellular vehicle to everything (CVE) that allows automobiles to interact with nearly everything around them (C-V2X). It opens the way for autonomous driving and contributes to increasing traffic safety by offering 360° non-line of sight (NLOS) monitoring and a better degree of predictability. Vehicle to vehicle (V2V), vehicle to infrastructure (V2I), vehicle to pedestrian (V2P), and vehicle to cloud (V2C) are all included in the term of "C-V2X" (Figure 7.3). Qualcomm C-V2X is a crucial component of the Qualcomm Snapdragon Automotive Platform.

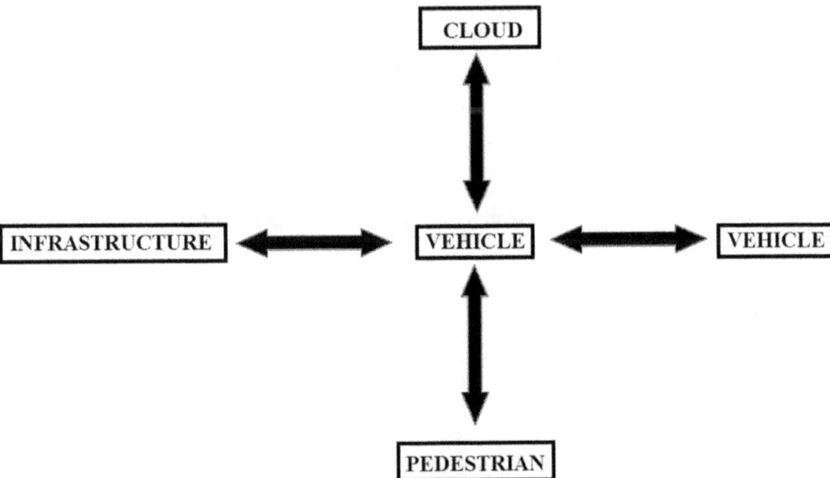

Figure 7.3 C-V2X technology.

Cellular vehicle to everything (CVE) (C-V2X) is the generic name for the core technology that enables automobiles to connect with nearly everything around them. By providing non-line of sight (NLOS) 360° monitoring and increased predictability, it helps to increase traffic safety, and it paves the path for autonomous driving. The term "C-V2X" includes vehicle to vehicle (V2V), vehicle to infrastructure (V2I), vehicle to pedestrian (V2P), and vehicle to cloud (V2C) (Figure 7.3). Qualcomm C-V2X is an essential component of a Qualcomm Snapdragon Automotive Platform. In each OFDM subcarrier, C-V2X devices may broadcast using 16 Quadrature Amplitude Modulation (QAM) or Quadrature Phase Shift Keying (QPSK) techniques. The C-V2X users are responsible for transmitting control data and reference signals. C-V2X has included 5G NR technologies that offer higher bandwidth, wideband carrier support, extremely low latency, and excellent dependability in Release 15 [36]. They are intended for use in conjunction with driver awareness, shared-driver behaviour, collision avoidance, and fully autonomous driving. Accurate location and range, fast throughput, real-time updates, and other characteristics are among the 5G NR features. A significant problem for C-V2X Version 15 is direct communications using the same ITS 5.9 GHz frequency. Advanced services as well as fundamental and improved safety services are made possible by C-V2X. By enabling the expansion and broad ecosystem of the C-V2X system, automotive and telecom companies are paving the way for safer autonomous driving. Qualcomm is accountable for playing a crucial part in the creation of C-V2X. The IEEE 802.11p and C-V2X solutions provided by vehicle to everything (V2X) communications, which enable automated driving, is one of the technical advancements Qualcomm is pushing in the connected automobile and autonomous driving. Figure 7.4 shows direct V2X communication.

After publishing the specifications enabling cellular V2X (C-V2X) in Release 14 and 15, the Third Generation Partnership Project's (3GPP) objective is to develop the 5G architecture for more opportunities in Release 16 in order to meet the most stringent requirements for V2X performance (Figure 7.5). The top automakers and telecoms firms who enable interoperable end-to-end 5G supported V2X connections formed the 5G Automotive Association (5GAA) in September 2016 [37–41].

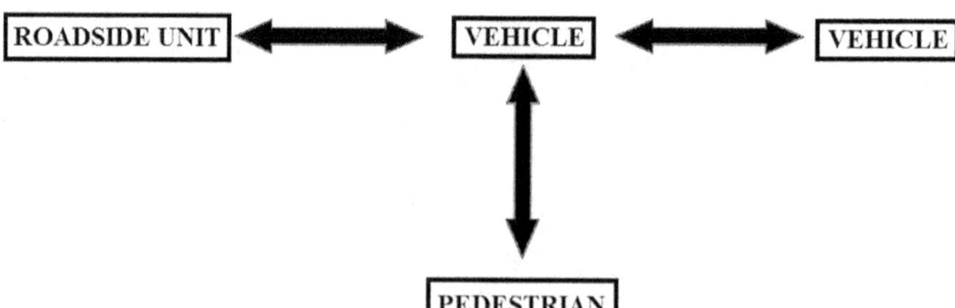

Figure 7.4 Direct V2X technology.

The Long Term Evolution (LTE) initiative of the 3GPP, which includes C-V2X, is helping to extend the role of cellular networks while also steadily and quickly expanding the function of the 3GPP. The first level was defined in Release 14 and finished in June 2017. As a result, Release 15 includes a detailed roadmap for future improvements. The additional capabilities anticipated by 5G have been defined in Release 16.

The important achievement in 3GPP Release 14 [38] is C-V2X open interaction. It happens as a result of an industry-specific partnership between the automotive and communications sectors. The continued development of Cellular C-V2X is a crucial step towards the next stage in cellular technology, 5G NR. Figure 7.6 depicts the development of the functions offered by present and next 3GPP versions of mobile technology. Backward compatibility has been shown to be a natural and effective notion in 3GPP deployments and networks. They are crucial to enabling fundamental use case functioning between various vehicle generations. Version 16 offers significant improvements in bandwidth, latency, and dependability that match the demands of large data rates. Examples include better safety, coordinated and cooperatively autonomous driving, high-quality entertainment, and changes to maps and software. Several release dates for prospective standard releases as well as historical time frames are shown in Figure 7.6.

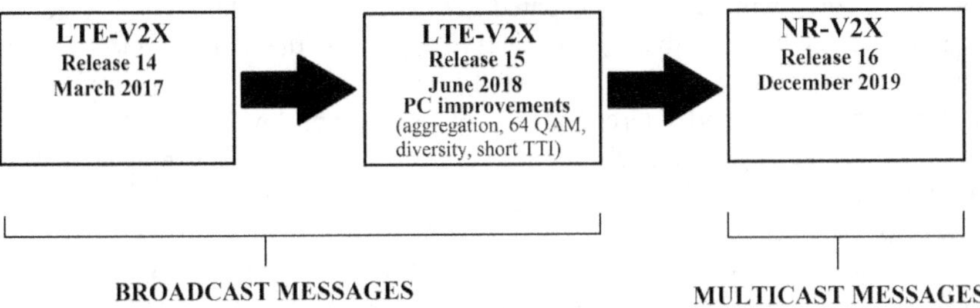

Figure 7.5 3GPP time plan from LTE-V2X to NR-V2X.

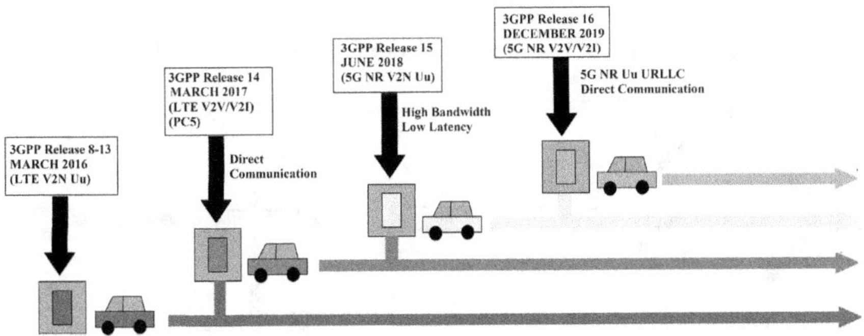

Figure 7.6 C-V2X showing different releases.

The capabilities of New Radio (5G NR) have now been presented in 3GPP Release 15. In terms of network communication only at Uu interface, it is similar to 4G LTE's successor. For V2N network communications, Release 15 uses the faster data speeds and shorter latencies of 5G NR. The first 5G NR chipsets should be available in 2019; these chipsets should also be used in automobiles within that same year. In 2021, such cars will be put into service. Version 15 mentions a few small improvements to the PC5 radio for direct communications, such as broadcast diversity and high order modulation (64 QAM). In 3GPP Release 16, "Ultra Reliable Low Latency Communications" (URLLC) is the term used to describe the enhanced 5G NR capabilities described in terms of short-range direct connection with enormous bandwidth and low latency. For 5G-V2X, the exchange of control messages for platoons of really near driving vehicles (just a few metres gap), the exchange of vehicle trajectories to prevent collisions, and the sharing of sensor data that makes the link (e.g., video data) are all essential (cooperative decision-making). Teleoperated driving, software updates, high definition sensor sharing, real-time situational awareness, high definition maps, autonomous vehicle cooperation in emergency situations, etc. are a few examples of application scenarios. These use cases, which can be applied for the first time, enable enhanced key performance indicators (KPIs). Currently in the definition phase, Release 16 of the 3GPP should be functionally frozen by the end of 2019. Version 16 deployments are anticipated to begin in 2023. Although 5G NR and LTE versions have very distinct physical radio layers, the chipsets and supporting communication stacks will support a variety of radio technologies, ensuring seamless operation and interoperability at the service level. The objective of 5GAA is to speed up the global rollout of cellular vehicle to everything (C-V2X). A completely integrated digital transportation system with 5G will be realized as a consequence.

The transmission modes of C-V2X are mentioned in Figure 7.7 which specify the direct (sidelink) and network (uplink/downlink).

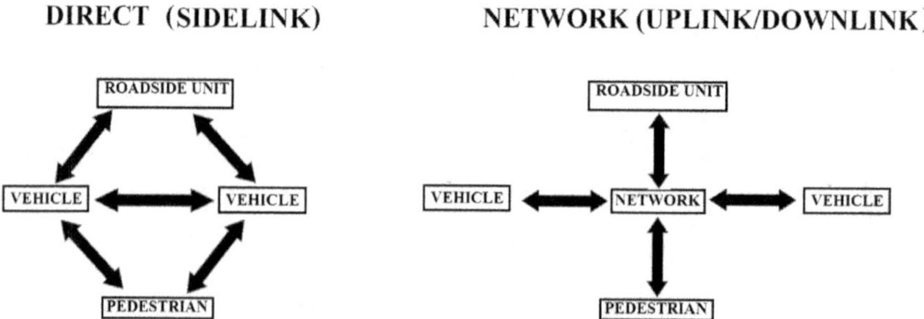

Figure 7.7 Transmission modes of C-V2X.

7.12 CHALLENGES FOR mmWAVE IN V2X COMMUNICATIONS

The development of mmWave V2X communication systems faces many challenges. In this section, we focus on three main problems: the lack of accurate mmWave vehicular channel models, the required penetration rate of mmWave V2X capable cars, and the absence of simple and quick mmWave beam alignment solutions for vehicular communications. At frequencies under 6 GHz, wave propagation between cars has been investigated, and several channel models have indeed been put forth. In order to expand these models to mmWave bands, modifications are required to the graph-based models, geometry- and non-geometry-based stochastic models, and beam propagation approaches. The additional parameters must be inferred from the measurement data [2]. At 60 GHz, measurements of propagation and angle-of-arrival have also been made [2]. To describe the original signal across antenna arrays in vehicle settings, further mmWave band data are required. Measurements must be made and included within mmWave V2X channels of communication to account for the impacts of antenna placement and the interference brought on by surrounding objects like buildings, cars, and pedestrians. In order to fully use V2X (particularly V2V) communications in the early stages of mmWave V2X communication system adoption, a problem would be the relatively low rate of V2X compatible cars. With the deployment of narrow mmWave beams and sophisticated MAC protocols, high penetration rates that may result in significant interference under situations of very congested traffic could be avoided. It is possible to build integrated mmWave communication and radar systems and increase the penetration depth of mmWave V2X-capable vehicles during the early deployment stage because automotive radars now employ mmWave spectrum. Additionally, combined technology will reduce expenses, consume less electricity, and free up space inside cars. On the basis of the IEEE 802.11ad technology, a preliminary research was done to develop a hybrid mmWave radar and communication system for vehicle situations (operating in the 60 GHz unlicensed band). It was shown that range forecast accuracy could be approximately 0.1 metres and velocity estimate precision could be within 0.1 metres per second by using the IEEE 802.11ad preamble as a radar signal using conventional Wi-Fi receiver methods. For current mmWave automobile radars operating in the 76–81 GHz frequency range, it would potentially be possible to add communications infrastructure having mmWave beam alignment. In mmWave communications, the transmitter and receiver's beam alignment is essential because mmWave systems frequently employ sharp transmit and receive beams. Following correct beam alignment, typical communication methods, such as accurate channel estimate and data transmission, may be carried out with enough link margin. Testing every conceivable sending and receiving beam in turn is the brute force method of beam alignment, which has a significant overhead. The overhead associated with beam alignment in Wi-Fi and cellular networks has received a lot of attention. Nevertheless, given the increased mobility of vehicles, mmWave V2X communications may necessitate the use of even more sophisticated beam alignment algorithms. Beam alignment strategies for mmWave V2I & V2V communications may differ. Beam realignment brought on by obstruction should be correctly addressed in mmWave V2I & V2V instances.

7.13 SENSING FOR NEXT GENERATION VEHICLES

The three primary automobile sensors—radars, cameras, and LIDARs—for both current-generation and prospective cars are examined in this section, along with their advantages, disadvantages, and data output. Automobile radars are already in use, most recently within 76 and 81 GHz frequencies, in the mmWave spectrum. Adaptive cruise control uses long-range radars. Blind spot identification, lane change assistance, as well as cross traffic alerts are supported by medium range radars. Short-range radars are useful for pre-crash applications and parking. Radars use active sensing to gather data on the presence, location, and speed of nearby vehicles by emitting a particular waveform, often a modulated signal continuous waveform. In contrast to communication systems, which are normally standardized, different manufacturers use different waveforms and antenna configurations. The European Telecommunications Standards Institute (ETSI) has not outlined general waveforms or signal processing, although it has established specific waveform-related properties. Spectrum, maximum and average allowable power, and out-of-band emission are some of these features. Without making changes to their typical signal processing, radars are not ideal for identifying the different sorts of target objects (i.e., whether the object being searched for is just a vehicle, truck, or motorcycle). In order to reduce noise and provide point-map data particularly for pertinent targets, the data underwent significant post processing. From kbps (for point maps) to hundreds of Mbps (for raw sampled data), radar generates data at different speeds. The visible or infrared spectrums are used by vehicle cameras. They offer some features like side-rearview cameras that can check for blind spots and lane departure, rearview cameras that can prevent backover crashes, interior cameras that can stop the driver from dozing off while driving, and front cameras that can detect speed limit signs and improve night vision (using infrared sensors). In order for automotive cameras to function as (intelligent) safety applications, new computer vision algorithms are needed. The sheer quantity of data produced by automobile cameras is substantial. Narrow laser beams are used by LIDARs to create high-resolution, three-dimensional (3D) digital images with a 360° field of view, where pixels are also related to depth. LIDAR works by scanning an area with a laser and detecting the backscattered signal's time delay. With the right post-processing, in order to build 3D maps and identify moving objects like cars, bicycles, and pedestrians, LIDAR may provide images with a very high resolution.

The chapter flows according to the recent technological advancements of employing 5G in applications for smart vehicles that are described in the article

Table 7.1 Purposes, Drawbacks, and Data Rates of Automotive Radar, Camera, and LIDAR. The Sensor Data Rates Are Obtained from the Specifications of Commercial Products and Conversations with Industrial Partners.

	Purpose	*Drawback*	*Data Rate*
Radar	Target detection, velocity estimation	Hard to distinguish targets	Less than 1 Mbps
Camera	Virtual mirrors for drivers	Need computer vision techniques	a) 100 to 700 Mbps for raw images b) 10 to 90 Mbps for compressed images
Lidar	Target detection and recognition, velocity estimation	High cost	10 to 100 Mbps

1. Numerous research projects for proof of concept in the realization of smart vehicles have also been developed by the SMIT Electronics and Communication Engineering department (ECE) which are discussed in article 2. The RSSI, or received signal strength indicator, is used to assess the C-V2X system's performance. Utilizing the 5G NR waveform and two RF carriers, specifically 3 GHz and 28 GHz, the C-V2x communication system will be implemented. Further, article 2 discussed some preliminary field trials by extending the same experimental set up done at ECE Department, SMIT.

7.14 REALIZATION OF C-V2X SYSTEM AT 28 GHz AND SUB-6 GHz

7.14.1 Antenna Performances of Millimetre Wave at 28 GHz in Comparison to Microwave at Sub-6 GHz

While the performance of a microwave antenna (sub-6 GHz) frequency using a horn antenna varies from 1.5 dBi to 15 dBi and has a beamwidth of 35° to 55° (45° ± 10°), a millimetre wave antenna's gain (28 GHz) frequency using a dish antenna is 38.1 dBi and its HPBW is 2.2°. As a result, their gain differs by 28.1 dBi (or about 30 dBi). Compared to microwave antennas, millimetre wave antennas perform substantially better in terms of directivity.

So

$$(S/N)_{\text{Millimetre}} >> (S/N)_{\text{Microwave}}$$

Finally, it results in increased channel capacity, as shown by

$$C = B\log_2\left(1+\frac{S}{N}\right) \tag{7.1}$$

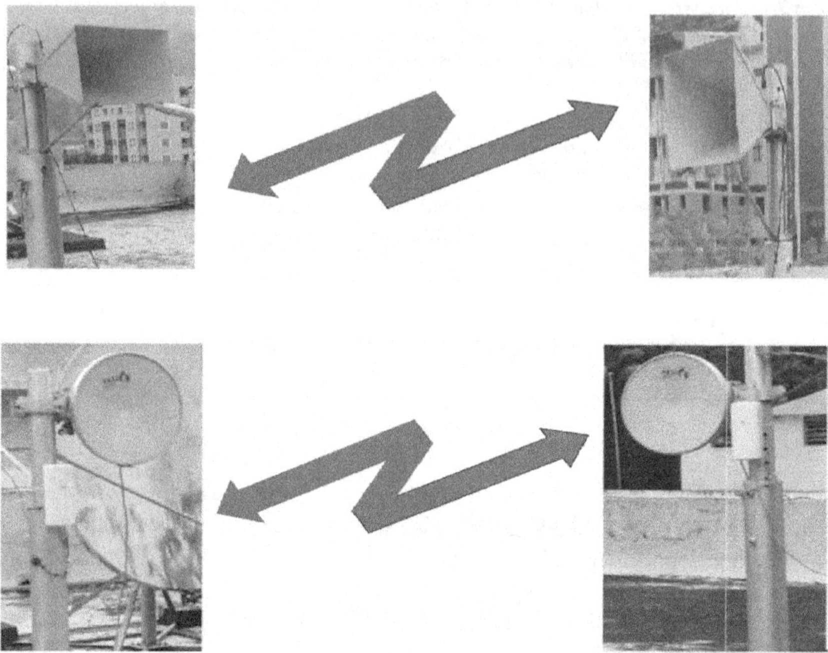

Figure 7.8 (a) Microwave antenna at sub-6 GHz with gain 1.5 dBi to 15 dBi. (b) millimetre wave Antenna at 28 G Hz with gain 38.1 dBi.

7.14.2 Comparison Between Microwave And Millimetre Wave

Table 7.2 RSSI and SNR for 5G NR and 4G LTE

Experimental Platform	*Link Distance (metres)*	*Frequency (GHz)*	*RSSI (dBm)*	*SNR (dB)*	*Remarks*
Hardware	30.48	3 GHz	−97.6	30.6	Millimetre wave C-V2X is far superior than Microwave C-V2X
	30.48	28 GHz	−44	50.7	
	30.36	3 GHz	−97.2	31.3	
	30.36	28 GHz	−43.8	49.4	

where C stands for "channel capacity," B for "bandwidth," and S/N for "received signal to noise ratio."

In their experiment, the authors used both antennas, as depicted visually in Figure 7.8(a) and (b).

Similar spectrum has been acquired from a 28 GHz LOS C-V2X link with a higher RSSI level and better SNR. Both outcomes are tabulated in Table 7.2.

7.15 PRELIMINARY FIELD TRIALS ON ROAD VEHICLES

7.15.1 Communication between Road Side Unit Having 1st 28 GHz Transceiver and Vehicle Having 2nd 28 GHz Transceiver

As shown in Figure 7.9, direct provisioning, direct discovery, name management, and communication are all part of the ProSe function. The service is already in existence. The mobile device is configured with the appropriate parameters for using the direct discovery and sidelink communication functions using the direct provisioning sub-function. The unconstrained direct discovery function processes the mapping of ProSe application credentials and ProSe application codes using the direct discovery name management sub-function.

There are two sections in the ProSe application code: i) the operator's identifier, finished by the Scope field. ii) E bit. The Scope field specifies whether the operator identity is operator-specific, country-specific, or has a global scope. The E bit controls whether the operator's identify who assigned this specific code is present in the ProSe application code.

The ProSe function is further extended from generic D2D communication (as stated earlier) to specific V2V direct communication. The smart vehicle or the Roadside Unit (RSU) will discover about the details of next vehicle in the proximity using wireless direct communication and take actions based on this discovery. This

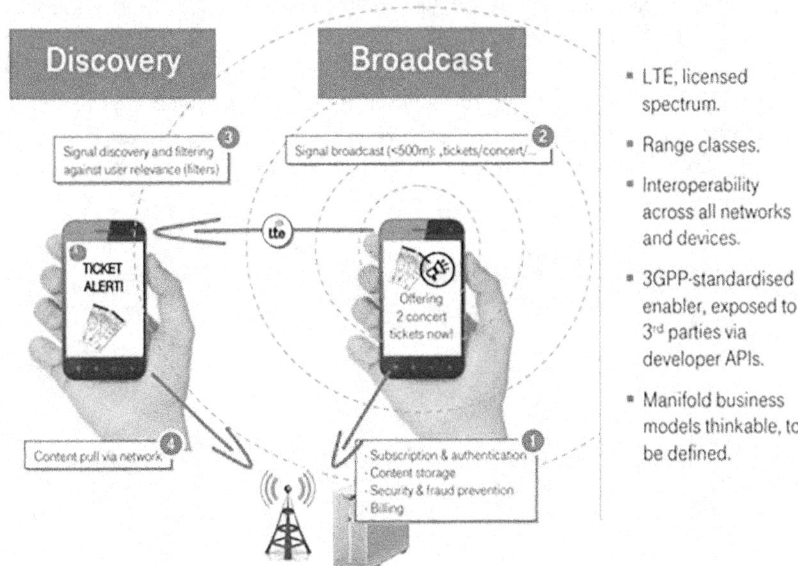

Figure 7.9 LTE Proximity Service (ProSe) started by T-Mobile for device to device discovery and broadcast.

is a totally new approach for vehicular management and is highly useful to achieve "ZERO ROAD COLLISION" in coming days. Effectively, ProSe function is the 3rd generation of RFID identification and localization. ***The authors are motivated with this ProSe based RFID development for V2V safety and collision avoidance and initiated a field trial using their 28 GHz system developed as discussed in article 7.14.***

In the world, there are various cities. In the cities, the number of vehicles is increasing day by day. So the problem of traffic jams increases day by day. In order to avoid traffic jams, construction of over-bridges on the road becomes mandatory. So governments of various countries made a plan to build various over-bridges on congested roads. Those over-bridges are used specifically for small light vehicles and not for large heavy vehicles. So in our experimental setup, we propose to guide the vehicles at the junction as shown in Figure 7.12 (a) and (b).

A typical RFID [38] consists of various components, i.e. RFID tag, RFID reader, and Host system. An antenna and integrated circuitry are typically composed as a tag. The identification code and sometimes additional information like specification of the item or any special care while handling is stored in the integrated circuitry. In case of sensing, the type of vehicle is an important issue. An electronic tagging technology in which radio waves are utilized for the detection and identification of a device or an object having an encoded tag is called Radio Frequency Identification (RFID). In RFID, there involves the storing and retrieving of information or data. RFID contains tag, reader, and backend database. The tag stores one-of-a-kind, recognizable proof data of objects and communication of the tags. As a result, inaccessible recovery of recognizable proof will be permitted. The full duplex communication is to be established between tags and readers. Each of the tags of RFID possesses "read only" and "rewrite" internal memory which depends on the type and application. These signals trigger the tags to answer to the inquiry. As a result, the communication between the most components of the framework is built up and the vicinity vehicle will be found. Figure 7.10 depicts the methodology used for vehicle identification and localization using reader code

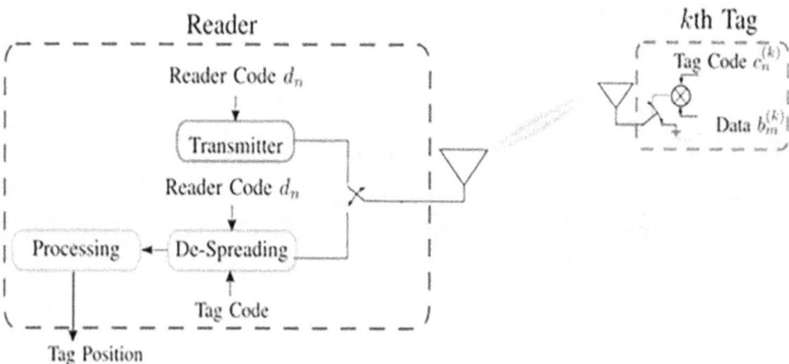

Figure 7.10 Experimental setup at ECE, SMIT lab on user ID verification and localization using 28 GHz mmWave.

dn and Tag code cn. The de-spreading operation needs to be performed at the RFID reader for retrieval of the vehicle information.

7.15.2 Demonstration of RFID Generation 3 with ECE, SMIT RFID System Supporting Vehicular Identification and Localization

28 GHz mmWAVE RFID TAG EMULATION AT ECE, SMIT LAB for precise Road Vehicle Identification and PRECISE Localization with reduced _False_ Alarm_Rate

- TAG is the Subject Vehicle under Identification
- INITIAL VEHICLE DETECTION and Localization using RADAR mode

Technology Used: CW Energy signal radiated from Reader Vehicle will Backscatter from the Vehicle Metallic Body

- FINAL VEHICLE DETECTION and Localization using RFID mode Technology Used:

i) Coded Burst energy from reader will Backscatter From the 28 GHz Radio
ii) The de-spreading operation needs to be performed at the RFID reader for retrieval of the vehicle information connected with the 28 GHz TAG.
   - In this way, using Initial Radar Mode and Final RFID Mode, The Problem of road VEHICLE DETECTION and Localization will be solved very accurately with negligible False Alarm Rate.
   - Figure 7.11 is the experimental setup at ECE, SMIT Lab depicting Emulation environment

RFID READER at 28 GHz
mounted on 1st Vehicle
at ECE, SMIT
Using 28 GHz, Millimeterwave RADIO

RFID TAG at 28 GHz mounted on
2nd Vehicle to be identified and localized
at ECE, SMIT
Using 28 GHz, Millimeterwave RADIO

Figure 7.11 Experimental setup at ECE, SMIT lab.

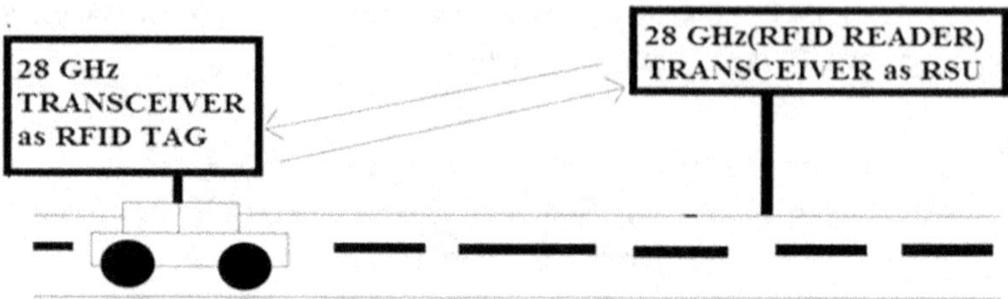

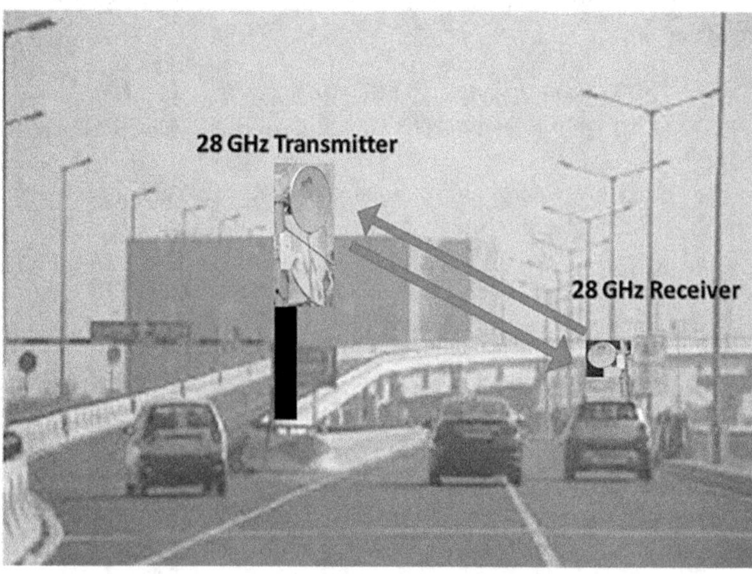

Figure 7.12 (a) RFID tag on vehicle used for device discovery use cases. (b) Prototype installation of 28 GHz communication system with one unit mounted on Roadside UNIT (RSU) and 2nd Unit on Vehicle.

- Figure 7.12 (a) and 9b) is Prototype Installation of 28 GHz communication system with One unit mounted on Roadside UNIT (RSU) and 2nd Unit on Vehicle.

The Field trial on road was completed successfully at the Kolkata metro city as depicted in Figure 7.12 ***(b).***

7.16 SUMMARY AND CONCLUSION

One of today's society's major problems is contingency. This takes the place of the unreliable signal. Currently, Sub-6 GHz microwave technology is used, however the signal dependability is not satisfactory. Millimetre wave will replace microwave in the future. Millimetre wave is next used in an effort to increase the signal's constant quality. When using the Sub-6 GHz 5G NR waveform for C-V2X communication, it has been shown that the millimetre wave at 28 GHz delivers a more dependable signal than the microwave signal. The field trial on road was completed successfully at the Kolkata metro city as depicted in Figure 7.12(b).

REFERENCES

1. A. Orsino, Galilina, "Improving Initial Access Reliability of 5G mmWave Cellular in Massive V2X Communications Scenarios," 2018 IEEE International Conference on Communications (ICC), Kansas City, MO, 2018, pp. 1–7. https://doi.org/10.1109/ICC.2018.8422158.
2. J. Choi, V. Va, N. Gonzalez-Prelcic, R. Daniels, C. R. Bhat, and R. W. Heath, "Millimeter-wave vehicular communication to support massive automotive sensing," IEEE Communications Magazine, vol. 54, no. 12, pp. 160–167, 2016.
3. Y. Yin, T. Yu, K. Maruta, and K. Sakaguchi, "Distributed and scalable radio resource management for mmWave V2V relays towards safe automated driving," Sensors, vol. 22, no. 1, p. 93, 2022. https://doi.org/10.3390/s22010093
4. S. Chen, J. Hu, Y. Shi, Y. Peng, J. Fang, R. Zhao, and L. Zhao, Vehicle-to-everything (v2x) services supported by LTE-based systems and 5G. IEEE Communications Standards Magazine, vol. 1, pp. 70–76, 2017.
5. 3GPP TR38.885. Study on NR Vehicle-to-Everything (V2X). Technical Report (TR) 38.885, 3rd Generation Partnership Project (3GPP), 2019. Available online: https://portal.3gpp.org/desktopmodules/Specifications/SpecificationDetails.aspx?specificationId=3497 (accessed on 27 March 2019).
6. 3GPP TR38.886. V2X Services Based on NR. Technical Report (TR) 38.886, 3rd Generation Partnership Project (3GPP), 2019. Available online: https://portal.3gpp.org/desktopmodules/Specifications/SpecificationDetails.aspx?specificationId=3615 (accessed on 7 July 2020).
7. K. Zrar Ghafoor, et al., "Millimeter-wave communication for Internet of vehicles: Status, challenges, and perspectives," IEEE Internet of Things Journal, vol. 7, no. 9, pp. 8525–8546, 2020. https://doi.org/10.1109/JIOT.2020.2992449.
8. H. Menouar, I. Guvenc, K. Akkaya, A. S. Uluagac, A. Kadri, and A. Tuncer, "UAV-enabled intelligent transportation systems for the smart city: Applications and challenges," IEEE Communications Magazine, vol. 55, no. 3, pp. 22–28, 2017.
9. K. Z. Ghafoor, L. Kong, D. B. Rawat, E. Hosseini and A. S. Sadiq, "Quality of Service Aware Routing Protocol in Software-Defined Internet of Vehicles," in IEEE Internet of Things Journal, vol. 6, no. 2, pp. 2817-2828, April 2019, doi: 10.1109/JIOT.2018.2875482.
10. E. Uhlemann, "Connected-vehicles applications are emerging [connected vehicles]," IEEE Vehicular Technology Magazine, vol. 11, no. 1, pp. 25–96, 2016.
11. G.-P. W. P. on Automotive Vertical Sector. 5g infrastructure public private partnership, tech. rep. [Online]. Available online: https://5g-ppp.eu/wp-content/uploads/2014/02/5G-PPP-White-Paperon-Automotive-Vertical.
12. S. Chen, J. Hu, Y. Shi, Y. Peng, J. Fang, R. Zhao, and L. Zhao, "Vehicle-to-Everything (v2x) Services Supported by LTE-Based Systems and 5G," IEEE Communications Standards Magazine, vol. 1, no. 2, pp. 70–76, 2017.

13. ITU-R; Rep. ITU-R M.2228–1 "Advanced intelligent transport systems radiocommunications," 2015. Available at: https://www.itu.int/pub/R-REP/en
14. K. Sakaguchi, R. Fukatsu, T. Yu, E. Fukuda, K. Mahler, R. Heath, T. Fujii, K. Takahashi, A. Khoryaev, S. Nagata, and T. Shimizu, "Towards mmWave V2X in 5G and beyond to support automated driving," 2020. Available at: https://arxiv.org/abs/2011.09590
15. K. Z. Ghafoor, M. Guizani, L. Kong, H. S. Maghdid, and K. F. Jasim, "Enabling efficient coexistence of dsrc and c-v2x in vehicular networks," IEEE Wireless Communications, 2019.
16. Y. Fangchun, S. Wang, J. Li, Z. Liu, and Q. Sun, "An overview of Internet of vehicles", in IEEE, Wireless Communications over ZigBee for Automotive Inclination Measurement, China Communication (Volume 11, Issue 10), INSPEC Accession Number: 14786602, 2014.
17. O. Kaiwartya, A. H. Abdullah, Y. Cao, A. Altameem, M. Prasad, C.-T. Lin, and X. Liu, "Internet of vehicles: Motivation, layered architecture, network model, challenges, and future aspects," IEEE Access, vol. 4, pp. 5356–5373, 2016. https://doi.org/10.1109/ACCESS.2016.2603219.
18. J. Contreras-Castillo, S. Zeadally, and J. A. Guerrero-Ibanez, "Internet of vehicles: Architecture, protocols, and security," IEEE Internet of Things Journal, vol. 5, no. 5, pp. 3701–3709, 2017.
19. A. S. Sadiq, S. Khan, K. Z. Ghafoor, M. Guizani, and S. Mirjalili, "Transmission power adaption scheme for improving iov awareness exploiting: Evaluation weighted matrix based on piggybacked information," Computer Networks, vol. 137, pp. 147–159, 2018.
20. J. Guerrero-Ibáñez, S. Zeadally, and J. Contreras-Castillo, "Sensor technologies for intelligent transportation systems," Sensors, vol. 18, no. 4, p. 1212, 2018.
21. F. Jameel, Z. Chang, J. Huang, and T. Ristaniemi, "Internet of autonomous vehicles: Architecture, features, and socio-technological challenges," IEEE Wireless Communications, vol. 26, no. 4, pp. 21–29, 2019.
22. GSMA. Available online: www.gsma.com/.
23. 3GPP TR 22.885 v14.0.0, "Study on LTE Support for V2X Services," 2015. Available at: https://portal.3gpp.org/desktopmodules/Specifications/SpecificationDetails.aspx?specificationId=2898
24. 3GPP TS 36.440 v14.0.0, "Evolved Universal Terrestrial Radio Access Network (E-UTRAN); General aspects and principles for interfaces supporting Multimedia Broadcast Multicast Service (MBMS) within E-UTRAN," 2017. Available at: https://portal.3gpp.org/desktopmodules/Specifications/SpecificationDetails.aspx?specificationId=2455
25. N. Lu, N. Cheng, N. Zhang, X. Shen, and J. W. Mark, "Connected vehicles: Solutions and challenges," IEEE Internet of Things Journal, vol. 1, no. 4, pp. 289–299, 2014.
26. 3GPP TR 22.886 V1.0.0: Study on enhancement of 3GPP support for 5G V2X services (Release 15), 3GPP Std., 2016.
27. 5G Americas, "5G spectrum recommendation," 2017. Available online: www.5gamericas.org/files/9114/9324/1786/5GA_5G_Spectrum_Recommendations_2017_FINAL.pdf.
28. 3GPP TS 38.101 "NR; User Equipment (UE) radio trans-mission and reception Part 1 Range 1." Available at: https://portal.3gpp.org/desktopmodules/Specifications/SpecificationDetails.aspx?specificationId=3283

29. 3GPP TS 38.101 "NR; User Equipment (UE) radio trans-mission and reception Part 2 Range 2." Available at: https://portal.3gpp.org/desktopmodules/Specifications/SpecificationDetails.aspx?specificationId=3284
30. J. Liu et al., "Initial Access, Mobility, and User-Centric Multi-Beam Operation in 5G New Radio," in *IEEE Communications Magazine*, vol. 56, no. 3, pp. 35–41, March 2018.
31. S. Marimuthu, "5G NR White Paper," August 2018. Available online: www.researchgate.net/publication/327111521.
32. Cisco Public, "Time-sensitive networking: A technical introduction", White Paper Cisco Public, www.cisco.com/c/dam/en/us/solutions/collateral/industry-solutions/white-paper-c11-738950.pdf
33. www.researchgate.net/publication/265416828_Evolution_of_Limited-Feedback_CoMP_Systems_from_4G_to_5G_CoMP_Features_and_Limited-Feedback_Approaches
34. M. Khoshnevisan, V. Joseph, P. Gupta, F. Meshkati, R. Prakash, and P. Tinnakornsrisuphap, "5G industrial networks with CoMP for URLLC and time sensitive network architecture," IEEE Journal on Selected Areas in Communications, vol. 37, no. 4, pp. 947–959, 2019. https://doi.org/10.1109/JSAC.2019.2898744.
35. Qualcomm, "Accelerating C-V2X toward 5G for autonomous driving [video]", www.qualcomm.com/news/onq/2017/02/24/accelerating-c-v2x-toward-5g-autonomous-driving.
36. Antonella, Claudia, "5G for V2X communications", 5G-Italy-White-eBook, www.5gitaly.eu/2018/wp-content/uploads/2019/01/5G-Italy-White-eBook-5G-for-V2X-Communications.pdf.
37. S. P. Singh and A. Sharma, "Node selection algorithm for routing protocols in VANET," International Journal of Advanced Science and Technology, vol. 96, pp. 43–54, 2016.
38. K. Kaur and S. P. Singh, "Multicasting technique to enhance the performance of DSR routing protocol using various performance metrics under different traffic conditions in mobile ad hoc networks," International Journal of Future Generation Communication and Networking, vol. 9, no. 4, pp. 103–144, 2016.
39. P. Sharma and S. P. Singh, "Mobile ad hoc network," International Journal of Emerging Technology and Advanced Engineering (IJETAE), vol. 4, no. 3, pp. 473–476, 2014.
40. R. Gujral and S. P. Singh, "Security threats," International Journal of Advanced Research in Computer Science and Software Engineering (IJARCSSE), vol. 4, no. 3, pp. 476–479, 2014.
41. GSMA, "Cellular Vehicle to Everything Enabling Intelligent Transport." Available at: https://www.gsma.com/solutions-and-impact/technologies/internet-of-things/gsma_resources/cellular-vehicle-everything-enabling-intelligent-transport/

Chapter 8

Managing and Controlling the COVID-19 Pandemic

The Role of Humanoid Robots

Mamatha A, Sangeetha V, Parkavi. A, and Sreelatha P K

8.1 INTRODUCTION

Since late 2019, the COVID-19 pandemic has swept the globe, presenting significant challenges due to its high infection mortality and contagiousness. Frontline healthcare workers, who frequently interact with patients, face a higher risk of infection. Physical separation has become a widely used disease-prevention strategy. The pandemic poses a serious threat to public health, medical professionals, patients, and healthcare systems worldwide. However, crucial personnel often cannot maintain physical distance, making healthcare facilities essential in combating pandemics [1].

One approach to address this challenge is the use of robots to perform critical tasks while reducing the risk of exposure to COVID-19 for important personnel and their families. Robots simulate human actions in risky settings, minimizing human-to-human contact. Since the virus cannot replicate inside a robot, and robots reduce human-to-human interaction, they offer an effective solution to this problem [2]. As a result of the hardships caused by COVID-19 and lockdowns, many individuals and businesses have turned to robots to overcome the obstacles posed by the pandemic.

The COVID-19 outbreak has highlighted both the potential and limitations of robots as a direct substitute for human work. Due to the increased fatalities among frontline workers, many countries have already implemented various robots to support human personnel. Robotic and autonomous systems offer advantages over traditional human work, such as inherent immunity to viruses and the impossibility of disease-causing germs spreading from human to robot to human.

Humanoids are robots that resemble humans in their anthropometric structure, featuring two hands, two legs, a head, and a backbone. During the pandemic, humanoid service robots have played a vital role in assisting the overburdened global healthcare system. These robots have been utilized in hospitals and other environments, such as quarantines, to ensure cleaning, disinfection, and assistance while minimizing person-to-person interaction. AI-powered humanoid robots can also raise awareness about COVID-19.

DOI: 10.1201/9781003480181-8

By employing humanoid robots, the risk to medical personnel and physicians actively managing the pandemic can be reduced [3]. These robots can perform various tasks, including retrieving a patient's sample for testing, disinfecting hospitals, delivering logistics and nutrition to patients, and collecting clinical information. Their versatile capabilities have proven invaluable in combating the COVID-19 disease and providing crucial support to healthcare professionals during this challenging time.

However, it is essential to recognize that not all situations can accommodate fully autonomous robots. Some of the current robotic solutions still require human supervision, while other autonomous ones can only perform a limited range of basic tasks [4]. Concerns about the cost of robots and their maintenance are particularly pressing for economies in underdeveloped countries, where healthcare personnel may struggle to afford even the most basic materials [5]. Therefore, the future adoption of cutting-edge technologies such as 5G and AI will be crucial to enhance the flexibility and functionality of robots. Additionally, cost-effective production and widespread access to robotic solutions globally should be taken into account.

The COVID-19 pandemic, caused by the novel coronavirus SARS-CoV-2, has had an unprecedented impact on global public health, economies, and societies. Since its emergence in late 2019, countries worldwide have faced significant challenges in containing the spread of the virus, treating patients, and mitigating its impact on healthcare systems [6]. In the quest for innovative and effective solutions, humanoid robots have emerged as a potential tool to aid in managing and controlling the pandemic.

Humanoid robots, designed to resemble and imitate human movements and behaviors, offer unique advantages in pandemic management. Their ability to interact with people, perform tasks autonomously, and navigate complex environments makes them valuable assets in various aspects of pandemic response [7]. This chapter explores the potential roles humanoid robots can play in the ongoing fight against COVID-19.

- Reducing Human-to-Human Interaction: One of the primary modes of COVID-19 transmission is through close human-to-human contact. Humanoid robots equipped with advanced sensors and artificial intelligence can assist in reducing such interactions in high-risk settings. These robots can serve as contactless delivery agents, delivering essential supplies and medications to patients in isolation wards or quarantined areas. By minimizing direct human contact, they help curb the spread of the virus and protect healthcare workers and the general population [8].
- Patient Care and Monitoring: Humanoid robots can be programmed to perform various patient care tasks, providing support to healthcare personnel and reducing their workload. For example, they can monitor patients' vital signs, offer medication reminders, and collect and transmit data to medical professionals remotely. This remote monitoring capability not only limits

the exposure of healthcare workers to infected patients but also enables healthcare providers to manage a higher number of patients effectively.
- Disinfection and Sanitization: Maintaining stringent hygiene and sanitization practices is crucial in controlling the virus's spread. Humanoid robots can be equipped with ultraviolet (UV) light disinfection systems or other sanitizing technologies to decontaminate public spaces, hospitals, and transportation facilities. These robots can access areas that may be difficult for humans to reach and conduct disinfection operations with precision and efficiency.
- Public Awareness and Education: Promoting public awareness and adherence to safety guidelines is vital in managing the pandemic. Humanoid robots can serve as interactive platforms for disseminating information, answering frequently asked questions, and reinforcing safety measures in public spaces like airports, train stations, and shopping malls. Their engaging and approachable nature can help deliver essential messages effectively and consistently.
- Enforcing Social Distancing and Safety Protocols: In settings where maintaining social distancing is challenging, humanoid robots can play a role in enforcing safety protocols. They can patrol public areas, remind individuals to wear masks, and ensure adherence to distancing guidelines. By providing a non-threatening reminder, these robots can contribute to a positive change in behavior and help mitigate the risk of transmission.

The COVID-19 pandemic has spurred an urgency for innovative solutions to manage and control its spread [9]. Humanoid robots offer a promising approach to complement human efforts in pandemic response. From minimizing human-to-human interaction to supporting patient care, disinfection, and public awareness efforts, humanoid robots can significantly contribute to pandemic management.

While these robots are not a standalone solution and should be integrated into a comprehensive strategy, their capabilities in reducing risks, protecting healthcare workers, and promoting public safety make them valuable tools in the ongoing battle against COVID-19 and future pandemics [10]. Continued research, development, and integration of humanoid robots into healthcare systems can foster a safer and more resilient society in the face of infectious diseases.

8.2 LITERATURE STUDY OF MANAGING AND CONTROLLING THE COVID-19 PANDEMIC

8.2.1 The Role of Humanoid Robots

This literature study aims to explore and analyze the existing research and scholarly articles related to the utilization of humanoid robots in managing and controlling the COVID-19 pandemic. The study delves into various aspects of humanoid robot applications, including patient care, disinfection, public awareness, and enforcing safety protocols [11]. By examining the contributions of humanoid robots in pandemic response, this review seeks to highlight their potential as valuable tools in complementing human efforts and enhancing the overall pandemic management strategies.

The COVID-19 pandemic has presented unprecedented challenges to public health systems worldwide. Researchers and experts have been exploring innovative solutions to combat the spread of the virus, and humanoid robots have emerged as a promising technology with various potential applications in pandemic management.

8.2.2 Humanoid Robots in Patient Care and Monitoring

Numerous studies have investigated the use of humanoid robots in patient care and monitoring during the pandemic [12]. These robots have been equipped with advanced sensors and AI capabilities to monitor vital signs, deliver medications, and provide companionship to isolated patients. By reducing human-to-human interaction, humanoid robots help prevent transmission and provide much-needed support to overburdened healthcare systems.

8.2.3 Disinfection and Sanitization with Humanoid Robots

Robots with disinfection capabilities have gained attention as effective tools in combating the spread of the virus in public spaces, hospitals, and transportation hubs [13]. Researchers have explored the use of UV light and other sanitizing technologies on humanoid robots to autonomously and efficiently disinfect surfaces and reduce the risk of contamination.

8.2.4 Humanoid Robots for Public Awareness and Education

Studies have examined the role of humanoid robots in disseminating information and raising public awareness about the importance of following safety guidelines during the pandemic. By engaging with the public in a friendly and interactive manner, these robots help reinforce essential health and safety measures, ultimately contributing to behavioral changes and risk reduction.

8.2.5 Enforcing Safety Protocols with Humanoid Robots

Humanoid robots have been utilized to enforce social distancing and safety protocols in various settings [14]. Researchers have studied the effectiveness of these robots in reminding individuals to wear masks, maintain distance, and follow other preventive measures, particularly in crowded public spaces.

8.2.6 Challenges and Ethical Considerations

While the potential of humanoid robots in pandemic management is promising, studies have also addressed certain challenges and ethical considerations. Some of these challenges include robot deployment costs, technical limitations, and public acceptance [15]. Additionally, researchers have discussed ethical concerns surrounding privacy, data security, and human-robot interaction in healthcare settings.

8.2.7 Integration into Pandemic Management Strategies

Research in this area has also explored how humanoid robots can be integrated into existing pandemic management strategies. Studies have proposed frameworks for efficient robot deployment, coordination with healthcare personnel, and adapting to evolving pandemic situations.

The literature study highlights the significant potential of humanoid robots in managing and controlling the COVID-19 pandemic. From patient care and disinfection to public awareness and safety enforcement, these robots offer versatile and valuable contributions to pandemic response efforts. However, further research and collaboration between the robotics and healthcare communities are necessary to address challenges and maximize the benefits of humanoid robot integration into pandemic management strategies. With continued development and refinement, humanoid robots can play an increasingly important role in future pandemic preparedness and response.

Review of Previous Work on Managing and Controlling the COVID-19 Pandemic: The Role of Humanoid Robots. The literature study on the role of humanoid robots in managing and controlling the COVID-19 pandemic presents a comprehensive and insightful analysis of the potential applications of this technology in pandemic response. The review covers various aspects of humanoid robot deployment, including patient care, disinfection, public awareness, safety protocol enforcement, challenges, ethical considerations, and integration into pandemic management strategies.

8.2.8 Strengths of Using Humanoid Robots

- Comprehensive Coverage: The review provides a comprehensive overview of the different roles humanoid robots can play in pandemic management. It examines their applications in patient care, where robots can reduce human-to-human interaction and assist healthcare personnel in monitoring and delivering medications. Additionally, the review delves into the importance of robot-enabled disinfection in public spaces and hospitals, as well as the significance of humanoid robots in raising public awareness and enforcing safety protocols.
- Balanced Perspective: The review takes a balanced approach, discussing both the potential benefits and the challenges associated with integrating

humanoid robots into pandemic response efforts. By acknowledging the technical limitations, deployment costs, and ethical considerations, the review highlights the need for careful planning and collaboration between stakeholders to maximize the effectiveness of humanoid robot implementation.

- Integration Insights: The review offers valuable insights into how humanoid robots can be integrated into existing pandemic management strategies. By proposing frameworks for efficient robot deployment and coordination with healthcare personnel, the study lays a foundation for future research and practical implementation.

8.2.9 Weaknesses of Using Humanoid Robots

- Limited Focus on Real-World Deployments: While the review discusses the potential applications of humanoid robots in pandemic response, it lacks in-depth analysis and examples of real-world deployments [16]. Including case studies or reports of actual implementations would strengthen the review's practical relevance and provide a more concrete understanding of the challenges faced in deploying humanoid robots during the COVID-19 pandemic.
- Lack of Latest Developments: The review's knowledge cutoff being up to September 2021 might limit the inclusion of recent advancements and breakthroughs in the field of humanoid robotics. Including updates beyond the cutoff date could provide a more current perspective on the state of research and developments related to humanoid robots and COVID-19.
- Need for Future Prospects: The review focuses on the present state of humanoid robots in managing the COVID-19 pandemic. However, adding a section on potential future prospects, emerging technologies, and the implications of humanoid robots in future pandemics or healthcare scenarios would enhance the study's forward-looking impact. The literature study on the role of humanoid robots in managing and controlling the COVID-19 pandemic is a commendable effort to explore the potential applications and benefits of this technology in pandemic response. It covers a wide range of aspects, including patient care, disinfection, public awareness, and safety enforcement. While the study provides valuable insights into the challenges and ethical considerations, it would benefit from including more real-world case studies and updates beyond the knowledge cutoff date to ensure its relevance to the current state of research and developments in this rapidly evolving field. Overall, the review contributes significantly to the understanding of humanoid robots' potential as valuable tools in enhancing pandemic management strategies.

8.3 APPLICATION OF HUMANOID ROBOTS

8.3.1 Humanoid Robots

Humanoid robots have the advantage of being able to interact with humans in more natural and intuitive ways compared to traditional robots [17]. Their potential for social interaction, mobility, and adaptability makes them suitable for a broad range of applications in the future. As technology continues to evolve, humanoid robots are likely to play an increasingly important role in our daily lives.

8.3.2 Disinfecting/Spraying Robots

Disinfecting/spraying robots have proven to be valuable tools in various industries and settings, particularly in situations where human intervention may pose risks or inefficiencies. Some applications of disinfecting/spraying robots include:

- Healthcare Facilities: Disinfecting robots can be used in hospitals, clinics, and other medical settings to sanitize patient rooms, operating theaters, and high-touch surfaces, reducing the risk of healthcare-associated infections.
- Public Transportation: Robots can be deployed in buses, trains, and airplanes to sanitize seating areas, handrails, and other frequently touched surfaces, enhancing passenger safety and confidence.
- Retail Stores and Shopping Malls: Using robots to disinfect shopping areas can help maintain a hygienic environment for customers and staff, especially during disease outbreaks or seasonal illnesses.
- Schools and Educational Institutions: Disinfecting robots can assist in cleaning classrooms, libraries, and communal areas, supporting efforts to create a healthy learning environment for students and educators.
- Hospitality Industry: Hotels and resorts can utilize robots to disinfect guest rooms, lobbies, and common areas, ensuring a clean and safe experience for guests.
- Manufacturing Facilities: Robots can be deployed to disinfect equipment, workstations, and manufacturing environments, promoting a safer workplace for employees and reducing the risk of contamination.
- Food Processing Plants: Disinfecting robots can help maintain sanitation standards in food processing facilities, reducing the risk of foodborne illnesses and ensuring product safety.
- Airports: Robots can be employed to disinfect security checkpoints, waiting areas, and other high-traffic zones within airports, minimizing the spread of germs.
- Sporting Venues and Event Spaces: Disinfecting robots can be used to sanitize seats, restrooms, and concession areas, allowing fans to enjoy events with peace of mind.

Overall, disinfecting/spraying robots offer an effective and efficient means of maintaining a clean and sanitary environment across various industries, contributing to public health and safety. As technology advances, these robots are likely to find even more applications, further improving hygiene standards in different settings.

8.3.3 Hospitality Robots

Hospitality robots are specialized robots designed to provide various services and enhance the guest experience in the hospitality industry. These robots can be deployed in hotels, resorts, restaurants, and other hospitality establishments. Here are some common roles and functions of hospitality robots:

- Concierge Services: Hospitality robots can act as virtual concierges, providing information to guests about hotel amenities, nearby attractions, restaurant recommendations, and local events.
- Room Service: Robots can be used to deliver food, drinks, and other amenities to guest rooms, reducing the need for human staff to make repetitive trips.
- Bellboy and Porter: Robots equipped with storage compartments can carry guests' luggage from the lobby to their rooms, easing the burden on hotel staff and enhancing convenience for guests.
- Cleaning and Maintenance: Some robots are designed for cleaning tasks, such as vacuuming floors, sanitizing surfaces, and ensuring common areas are well-maintained.
- Language Translation: Hospitality robots equipped with natural language processing capabilities can assist non-native speaking guests by providing real-time translation services.
- Entertainment: Robots can engage guests with interactive games, provide personalized entertainment options, or even perform music and dance routines.
- Check-In and Check-Out Assistance: Robots can assist guests with the check-in and check-out processes, streamlining the administrative procedures.
- Security and Surveillance: Some hospitality robots are equipped with cameras and sensors to monitor the premises for security purposes.
- Customer Feedback and Surveys: Robots can interact with guests to gather feedback on their stay, allowing the hotel to improve its services based on real-time data.
- Social Distancing and Safety Measures: During pandemics or health crises, hospitality robots can enforce social distancing measures, deliver hygiene supplies, and assist with contactless check-in/check-out processes.

Hospitality robots are designed to complement human staff and enhance guest experiences by providing efficient and personalized services. They can help reduce operational costs, improve customer satisfaction, and create a unique and memorable experience for visitors. As technology continues to advance, we can expect hospitality robots to play an increasingly important role in the hospitality industry.

8.3.4 Surgical Robots

Surgical robots, also known as robot-assisted surgery or robotic surgical systems, are advanced medical devices that assist surgeons in performing complex surgical procedures with greater precision, dexterity, and control [18]. These robots are not autonomous but are operated by skilled surgeons who use specialized consoles and hand controls to guide the robot's movements. Surgical robots offer several advantages over traditional surgical methods, including:

- Enhanced Precision: Surgical robots provide superior precision, allowing surgeons to perform delicate and intricate procedures with reduced risk of human error.
- Minimally Invasive Surgery: Many surgical robots are designed for minimally invasive procedures, where smaller incisions are made, leading to less pain, reduced scarring, and faster recovery times for patients.
- Greater Dexterity: Surgical robots can mimic human hand movements with a broader range of motion, enabling surgeons to access and operate in hard-to-reach areas more effectively.
- 3D Visualization: Most surgical robots are equipped with high-definition cameras that offer 3D visualization of the surgical site, providing surgeons with a clearer and more detailed view of the area they are operating on.
- Tremor Reduction: The robotic systems are designed to filter out any hand tremors experienced by the surgeon, resulting in more stable and precise movements during surgery.
- Remote Surgery: Some advanced surgical robots have the potential for remote operation, allowing skilled surgeons to perform surgeries on patients located in different geographical locations.
- Reduced Fatigue: The ergonomic design of robotic surgical consoles helps reduce surgeon fatigue during lengthy and complex procedures.
- Training and Skill Development: Surgical robots offer a platform for surgeons to improve their skills and expertise through simulation and practice, leading to improved patient outcomes.

8.3.5 Common Surgical Specialties

Common surgical specialties where robotic systems are used include:

- General Surgery: For procedures like hernia repair and gallbladder surgery.
- Gynecology: For hysterectomy and other gynecological surgeries.
- Urology: For prostatectomy, kidney surgery, and bladder procedures.

- Cardiac Surgery: For complex heart procedures like mitral valve repair and coronary artery bypass grafting.
- Orthopedics: For joint replacement surgeries and spine procedures.

It's important to note that while surgical robots offer numerous benefits, they are not suitable for all types of surgeries and are generally more expensive to implement than traditional surgical methods. Their use is typically reserved for complex and specialized procedures where their advantages justify the cost and resources required. As technology advances, surgical robots are likely to become even more sophisticated and integrated into routine surgical practices, further improving patient outcomes and surgical efficiency.

8.3.6 Management and Treatment of COVID-19 Patients Using Humanoid Robots

Humanoid robots have the potential to play a role in the management and treatment of COVID-19 patients, especially in situations where minimizing direct contact between healthcare workers and patients is beneficial. Here are some ways humanoid robots could be utilized in this context:

- Remote Patient Monitoring: Humanoid robots equipped with cameras and sensors can monitor patients remotely, tracking their vital signs and symptoms. This allows healthcare providers to observe patients from a safe distance and intervene promptly if any concerning changes occur.
- Delivery of Medication and Supplies: Robots can be used to deliver medications, food, and other necessary supplies to COVID-19 patients in isolation, reducing the need for human staff to enter their rooms and minimizing potential exposure.
- Telemedicine Assistance: Humanoid robots can facilitate telemedicine consultations by acting as an intermediary between patients and healthcare providers. The robot can display the provider on a screen and relay information between the patient and the medical professional.
- Disinfection and Sanitization: Robots equipped with UV-C lights or other disinfection technologies can be deployed to sanitize patient rooms and common areas, helping to reduce the risk of transmission.
- Patient Interaction and Companionship: During isolation, COVID-19 patients may experience feelings of loneliness and anxiety. Humanoid robots can provide social interaction and companionship, offering a sense of comfort during their recovery.
- Remote Consultations and Rounds: Healthcare providers can remotely control humanoid robots to conduct virtual patient rounds and consultations, allowing them to communicate with patients without being physically present.

- Health Education: Humanoid robots can be programmed to provide COVID-19 education and guidance, reinforcing safety measures and best practices to prevent the spread of the virus.
- Temperature Screening and Entry Management: Robots equipped with thermal cameras can perform temperature checks at entry points, screening visitors and staff for potential symptoms of COVID-19.

It's important to note that while humanoid robots offer many potential benefits, they should not replace human healthcare professionals. Instead, they should be seen as tools to augment the capabilities of healthcare teams and improve patient care. Additionally, implementing humanoid robots in healthcare settings requires careful consideration of ethical, privacy, and safety aspects to ensure they are used responsibly and effectively.

As technology continues to advance, the capabilities of humanoid robots are likely to evolve, making them more versatile and valuable assets in managing and treating patients during public health emergencies like the COVID-19 pandemic.

8.4 CASE STUDY WITH CHALLENGES AND BENEFITS

The COVID-19 pandemic brought unprecedented challenges to healthcare systems worldwide, overwhelming medical facilities and posing significant risks to healthcare workers. In response, innovative technologies have been explored to enhance patient care and reduce the transmission of the virus. Humanoid robots have emerged as potential tools in managing and treating COVID-19 patients, offering promising solutions to address critical healthcare needs.

A hospital facing a surge in COVID-19 cases is likely to be dealing with a challenging and stressful situation. Due to the contagious nature of the virus, ensuring the safety of frontline workers while providing efficient care for patients became a top priority. The hospital decided to implement humanoid robots to assist in various aspects of patient management and treatment.

- Robot Deployment: The hospital introduced humanoid robots in several departments to augment patient care and improve infection control measures.
- Screening and Triage: Humanoid robots were stationed at the hospital's entrance to conduct initial patient screening and triage. Equipped with thermal sensors and facial recognition technology, the robots identified potential COVID-19 cases based on temperature and facial features. Patients showing symptoms or elevated temperatures were immediately directed to designated isolation areas, minimizing the risk of infection spread within the hospital premises.
- Telemedicine Consultations: To reduce direct contact between healthcare providers and COVID-19 patients, humanoid robots were utilized as telemedicine assistants. Equipped with high-definition cameras and video conferencing capabilities, the robots facilitated virtual consultations between

patients and medical staff [19]. This approach not only allowed physicians to remotely assess patients' conditions but also conserved personal protective equipment (PPE).
- Patient Monitoring: Humanoid robots were deployed in COVID-19 wards to monitor patients' vital signs and provide real-time data to healthcare providers. Integrated with advanced sensors, the robots continuously measured patients' temperature, heart rate, and oxygen levels. Alarms were triggered if any parameters deviated from the normal range, ensuring immediate attention from the medical team.
- Medication Delivery: Incorporating autonomous navigation systems, humanoid robots delivered medications and essential supplies to patients' rooms, reducing the need for direct human contact [20]. The robots followed pre-defined paths, avoiding obstacles and ensuring timely and accurate delivery of medications, thus minimizing potential cross-contamination.

The implementation of humanoid robots in the hospital will yield several significant outcomes:

- Enhanced Infection Control: The use of humanoid robots in screening, triage, and telemedicine consultations reduced the risk of virus transmission between patients and healthcare workers.
- Reduced Workload for Medical Staff: The robots' ability to conduct routine tasks, such as patient monitoring and medication delivery, alleviated the burden on medical staff, allowing them to focus on critical care activities.
- Efficient Resource Utilization: The deployment of humanoid robots optimized the utilization of PPE and medical supplies, as robots could be disinfected after each use, minimizing the consumption of protective equipment.

8.4.1 Case Study 1—Hospital

The case study of a hospital illustrates the potential role of humanoid robots in managing and treating COVID-19 patients. By assisting in screening, triage, telemedicine consultations, patient monitoring, and medication delivery, these robots proved to be valuable assets in curbing the spread of the virus, protecting healthcare workers, and enhancing patient care. The successful integration of humanoid robots in healthcare settings during the COVID-19 pandemic highlights their potential to revolutionize patient management in future infectious disease outbreaks and beyond.

8.4.1.1 The Challenges and Potential Benefits of Utilizing Humanoid Robots

The COVID-19 pandemic has led to an increased demand for medical professionals who bravely serve the community at great personal risk. However, a global shortage of medical personnel has become evident. To address this scarcity and curb the spread of the pandemic, human-like robots or humanoids are being deployed for various crucial medical tasks, including disinfection, patient care, monitoring, and even surgical assistance [21]. Additionally, in hazardous fields like mining, bomb disposal, and cleaning jobs, where human skills are required but human safety is at risk due to hazardous materials or gases, humanoids are proving to be valuable assets. Despite the significant benefits they offer, the integration of robots in public and healthcare sectors also presents several challenges that need to be carefully addressed.

8.4.1.2 The Challenges and Benefits in Diagnosis and Screening

The process of diagnosing and screening for COVID-19 is of utmost importance in preventing the spread of the pandemic. Consequently, a significant challenge in public health has emerged: increasing the number and accuracy of COVID-19 tests. To enhance testing efficiency and precision, researchers have been developing various diagnostic techniques. Notably, this pandemic stands apart from past outbreaks due to the extensive utilization of robots right from the initial stages of the outbreak. These robots have demonstrated remarkable potential in combating the pandemic by undertaking routine tasks that traditionally demand substantial human labor, including diagnosis, screening, mask-wearing checks, expanding testing capacity, and conducting screening interviews.

8.4.2 Case Study 2—Robotic Technology during COVID-19 Pandemic Challenges and Benefits

The utilization of robotic technology during the COVID-19 pandemic brings about undeniable benefits, yet it also presents several critical challenges, such as:

- Enhancing Autonomy: While robot-assisted sampling procedures can reduce the need for close contact between medical personnel and patients, achieving full autonomy remains a priority. This requires seamless integration and in-depth study of components like Human-Robot Interaction (HRI), computer vision, and artificial intelligence (AI) to enable fully automated procedures.
- Ensuring Reliability: As robots interact directly with humans in densely populated environments and are increasingly integrated into daily life, ensuring their reliability is paramount. In medical contexts, refining technologies like AI and computer vision for human detection, face recognition, and diagnosis is crucial. Any inaccuracies leading to false positives or false negatives can pose significant risks to public health.
- Enhancing a robot's overall performance requires advancements in sensor technologies and other fundamental components. Robotics is a complex field that demands careful consideration of each component's impact on the robot's capabilities. Precise temperature screening in hospitals and public

places necessitates accurate temperature measurements, posing a challenging task in extracting valuable information from sensors while achieving the required precision and sensitivity.
- Establishing a network of robots to monitor the pandemic situation throughout communities holds promising advantages. Utilizing modern infrastructures such as cloud technology, Internet of Things (IoT), artificial intelligence (AI), and 5G enables the creation of a smart network of robots. This network can provide superior guidance in managing the pandemic and making informed policy decisions, resulting in a more efficient pandemic response.

Despite the aforementioned challenges, employing humanoid robots for diagnosis and screening offers numerous benefits compared to conventional methods.

- Enhanced Safety: Robots can operate effectively in dangerous environments, reducing unnecessary contact between patients and medical staff. This not only ensures increased safety for healthcare workers but also alleviates the burden on them.
- Extended Operation: Robots can work tirelessly for extended periods, significantly reducing the physical and mental exhaustion experienced by doctors and nurses.
- Improved Efficiency: By automating various testing procedures, robots can enhance efficiency and throughput, streamlining the diagnostic process.
- Multi-Tasking Abilities: Equipped with multiple sensors such as thermal sensors, IR sensors, speakers, cameras, and microphones, many robots developed during the COVID-19 pandemic can perform multiple tasks concurrently.
- Technological Advancements: Recent technological progress has propelled the rise of robotics, enabling robots to function as independent units while also facilitating seamless information sharing. This comprehensive network approach empowers a more effective and coordinated response to combat the pandemic.

8.4.3 Case Study 3—Disinfection Challenges and Benefits

Autonomous robots excel in hazardous environments that pose risks to human safety, such as mining, nuclear plant maintenance, and underwater exploration. They also prove valuable for tasks like sanitizing areas to prevent the transmission of the SARS-CoV-2 virus [22]. Disinfecting robots can be categorized into two main types: those employing light-based methods and those using liquid agent sprays. These robots can take the form of drones, mobile units, or semi-autonomous mechanisms, often requiring a teleoperator.

- UV disinfection has limitations concerning power and reach. It may not effectively disinfect vast open spaces or concealed areas in shadows.

- Robots using spray methods may encounter difficulties reaching regions with gaps or challenging access points for spraying. Combining both technologies in a single autonomous robot could provide a more comprehensive and effective disinfection solution.

Using autonomous robots for disinfection offers several advantages:

- Continuous Operation: Autonomous robots can operate 24/7 and be designed to be self-rechargeable, eliminating the need for constant monitoring of charging requirements. They can be pre-programmed to follow specific paths for targeted disinfection, allowing healthcare workers to focus on other essential tasks.
- Enhanced Safety: Deploying autonomous robots for disinfection reduces the risk of medical staff coming into direct contact with the virus, ensuring their safety while maintaining effective disinfection protocols.

8.4.4 Case Study 4—Surgery and Telehealth Challenges and Benefits

Robots are commonly employed in minimally invasive surgeries, guided by doctors who may be present in the operating room or located remotely. Robotic surgery enhances surgical capabilities, offering a superior viewpoint, improved control, flexibility, and precision. Patients experience reduced infection risks, minimal blood loss, smaller incisions, and faster healing times. The rapid progress in telehealth since the emergence of COVID-19 has been significant in both technical and economic aspects. However, certain challenges persist, such as initial investment and development costs, as well as network-related issues outside of healthcare facilities [23].

The adoption and expansion of telehealth have accelerated in response to the widespread impact of COVID-19. Telehealth allows patients to receive medical attention while minimizing their exposure to the virus. Moreover, patients who undergo robotic surgery via teleprocedure tend to experience shorter recovery times and spend less time in the hospital after their operation. As the world continues to grapple with the pandemic and environmental challenges, the utilization of robots for telehealth and teleprocedure is expected to continue increasing.

8.4.5 Case Study 5—Social and Care Challenges and Benefits

To mitigate the spread of COVID-19, social distancing policies were implemented worldwide, leading to social isolation. This isolation can increase the risk of psychological and physical health issues, especially among the elderly, who face the challenge of staying indoors without the ability to gather with family members. In such circumstances, using robots in the form of pets or human-like assistants to aid the elderly at home can prove highly beneficial.

As the SARS virus can be transmitted through surfaces like plastic or metal, there is a risk of the virus spreading through shared robots. For example, while robotic pets may alleviate loneliness among elderly citizens, sharing them could potentially lead to virus transmission [24, 25]. Another challenge posed by robotic technology in elderly care is the potential replacement of human labor, which could result in the loss of valuable experience and expertise among human caretakers.

Social and care robots offer numerous potential benefits in assisting elderly people and enhancing healthcare service quality, both during and after the pandemic. In contrast to traditional approaches relying solely on human labor in social care and hospital settings, robotic service can significantly enhance efficiency and service quality in various ways.

- Social robots [26] can provide physical and virtual companionship, communication, and guidance, thereby reducing the risk of virus transmission while still fulfilling their primary functions.
- Robots designed in various forms or tailored to specific functions exhibit efficient task facilitation compared to humans.
- Social and care robots have proven their capability to work tirelessly in hospitals during the pandemic. The advantages of these robots have been further amplified by the efforts of numerous researchers and healthcare providers, who have been significantly impacted by the COVID-19 pandemic.

8.5 OBSERVATION

The market for surgical robots was estimated at USD 4.4 billion globally in 2022, and it is anticipated to rise at a CAGR of 18.0% from 2023 to 2030, as presented in Figure 8.1. The primary drivers of the market are the global lack of doctors and surgeons as well as the growing use of automated surgical tools. The market expansion is also being positively impacted by the rising investments made by regional and international companies in the development of new and cutting-edge surgical robots.

The market for medical robotic systems was valued at USD 21.2 billion in 2022, and from 2023 to 2030 presented in Figure 8.2, it is anticipated to increase at a CAGR of 16.9%. The elderly are more prone to injuries and other illnesses that need surgical treatment due to the pathological and psychological changes that come with aging. The primary drivers of market expansion are the increased demand for precise laparoscopic procedures and the rising incidence of trauma injuries. Additionally, rising per capita healthcare spending and advancing medical equipment technology are expected to fuel the market in emerging nations.

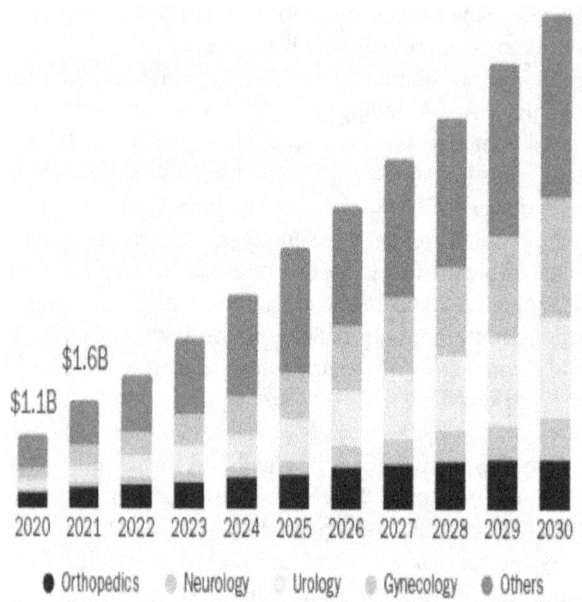

Figure 8.1 Future market prediction for surgical robots.

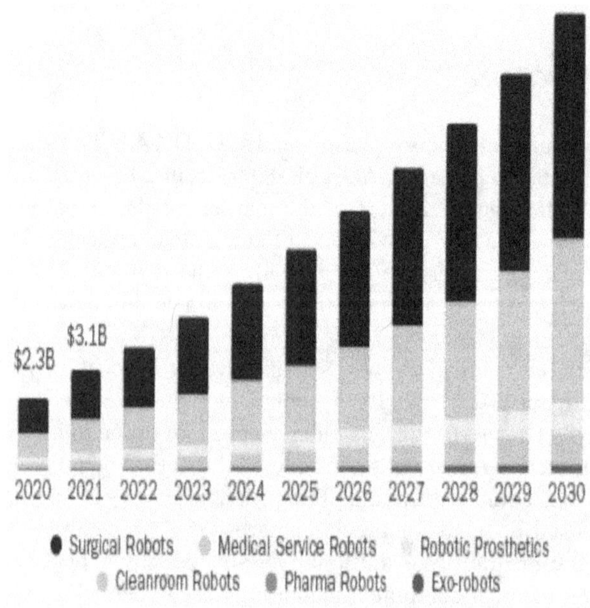

Figure 8.2 Future market prediction for medical robotic systems.

8.6 CONCLUSION

During the COVID-19 pandemic, humanoid service robots made quick headway in providing assistance to the overburdened global healthcare system. Currently battling the novel COVID-19 disease, the humanoid robot assistants can be useful in a variety of applications including distributing hand sanitizers, distributing food items and meal packets, and doing thermal imaging, among others. The purpose of this chapter is to provide a general overview of the types of assistive humanoid robots used in healthcare, with a particular focus on their role in battling the COVID-19 pandemic and responding to natural disasters.

Future scope, after the pandemic, it seems that the healthcare industry is more dependent on robots to stop human-to-human transmission. To improve healthcare and achieve financial and medical stability, many nations may therefore expand their interest in robotic breakthroughs, which would result in a sharp rise in the usage of medical robots.

REFERENCES

1. Tavakoli, M., Carriere, J., & Torabi, A. (2020). Robotics, smart wearable technologies, and autonomous intelligent systems for healthcare during the COVID-19 pandemic: An analysis of the state of the art and future vision. Advanced Intelligent Systems, 2, 2000071. https://doi.org/10.1002/aisy.202000071
2. Yang, G. Z., Nelson, B. J., Murphy, R. R., Choset, H., Christensen, H., Collins, S. H., Dario, P., Goldberg, K., Ikuta, K., Jacobstein, N., Kragic, D., Taylor, R. H., & McNutt, M. (2020). Combating COVID-19-the role of robotics in managing public health and infectious diseases. Science Robotics, 5(40), eabb5589. https://doi.org/10.1126/scirobotics.abb5589
3. How robots became essential workers in the COVID-19 response. IEEE Spectrum. Available: https://spectrum.ieee.org/how-robots-became-essential-workers-in-the-covid19-response.
4. Murphy, R. R., Babu Manjunath Gandudi, V., & Adams, J. (2020). Applications of robots for COVID-19 response. arXiv:2008.06976. Available: http://arxiv.org/abs/2008.06976.
5. Telepresence robots are helping take pressure off hospital staff. IEEE Spectrum. Available: https://spectrum.ieee.org/automaton/robotics/medical-robots/telepresence-robots-are-helping-take-pressureoff-hospital-staff.
6. We've been killing deadly germs with UV light for more than a century. IEEE Spectrum. Available: https://spectrum.ieee.org/tech-history/dawn-of-electronics/weve-been-killing-deadly-germs-with-uv-light-for-more-than-a-century.
7. Flight of the GermFalcon: How a potential coronavirus-killing airplane sterilizer was born. IEEE Spectrum. Available: https://spectrum.ieee.org/tech-talk/aerospace/aviation/germfalcon-coronavirus-airplane-ultravioletsterilizer-news.
8. Miller, G. Social distancing prevents infections, but it can have unintended consequences. Available: www.sciencemag.org/news/2020/03/we-are-social-species-how-will-social-distancing-affect-us.

9. Ghafurian, M., Ellard, C., & Dautenhahn, K. (2021). Social companion robots to reduce isolation: A perception change due to COVID-19. arXiv:2008.05382v2. Available: http://arxiv.org/abs/2008.05382.
10. Chang, T. H., Lee, C. C., Yang, C. S., Lu, Y. H., Chiang, Y. T., & Hsiao, F. Y. (2020). The use of disinfection robot to suppress the spread of COVID-19 in Taiwan. Journal of the Chinese Medical Association, 83(8), 797–798.
11. Zhang, S., Qin, L., Xu, S., Huang, L., Zhang, X., & Jiang, H. (2020). A review of the robot technology in fighting against COVID-19. IEEE/CAA Journal of Automatica Sinica, 7(6), 1591–1601.
12. Chintalapudi, N., Battineni, G., Amenta, F., Lavorgna, L., & Lippi, G. (2020). Role of Robotics in the management of COVID-19: A literature review. Journal of Medical Robotics Research, 5(3), 2030002.
13. Al-Nuaimi, Y. A., Al-Nuaimi, A. S., Al-Muharrmi, Z., & Al-Raisi, I. A. (2020). Application of humanoid robots in reducing the spread of COVID-19. International Journal of Advanced Computer Science and Applications, 11(6), 332–336.
14. Ghosal, S., Ganguly, R., & Chakraborty, S. (2021). COVID-19: The role of robotics in pandemic management—a comprehensive literature review. IEEE Robotics and Automation Magazine, 28(3), 38–48.
15. Pranav, H., Peshwani, A., & Singh, D. K. (2021). Role of social robots in the fight against COVID-19. Journal of Healthcare Engineering, 2021, 6650673.
16. Abbott, M. R., Lin, D., & Nourbakhsh, I. (2019). Human-robot interaction challenges for the COVID-19 era. arXiv:2004.01567.
17. Admoni, H., & Scassellati, B. (2017). Social eye gaze in human-robot interaction: A review. Journal of Human-Robot Interaction, 6(1), 25–63.
18. Broadbent, E., Stafford, R., & MacDonald, B. (2009). Acceptance of healthcare robots for the older population: Review and future directions. International Journal of Social Robotics, 1(4), 319–330.
19. Han, J., Guo, S., Hui, Z., Luo, D., Yang, X., & Wang, Y. (2020). The role of robots in managing the COVID-19 pandemic. Frontiers in Robotics and AI, 7, 593984.
20. Li, J., & Lu, R. (2021). An intelligent anti-epidemic robot system for COVID-19. Journal of Intelligent & Robotic Systems, 101(1), 95–104.
21. Lyons, K., & Feltovich, P. J. (2019). Telemedicine in a humanoid robot for rural remote evaluation and interactive guidance. Mayo Clinic Proceedings: Innovations, Quality & Outcomes, 3(3), 319–329.
22. Moltchanova, E., Saerbeck, M., & Sosnowski, S. (2021). Understanding the role of robotics in infectious disease outbreaks: A review. Paladyn, Journal of Behavioral Robotics, 12(1), 56–67.
23. Nalamachu, S., Croft, J. H., & Pestian, J. P. (2015). Robots in medicine. Journal of Clinical Medicine, 4(4), 665–671.
24. Sharkey, A., Sharkey, N., & Sharkey, A. (2010). The crying shame of robot nannies: An ethical appraisal. Interaction Studies, 11(2), 161–190.
25. Passi, A., & Singh, S. P. (2015). Enhancement of energy model for WiMax using bully election algorithm. Advances in Computer Science and Information Technology (ACSIT), 2(4), 317–320.
26. Tadakuma, R., Sugiura, K., & Sakamoto, D. (2016). Telexistence: A proposal for the next-generation communication tool. Journal of Advanced Computational Intelligence and Intelligent Informatics, 20(4), 629–634.

Chapter 9

Role of Robotic Healthcare Technologies during the COVID-19 Pandemic

Sampurna Panda and Rakesh Kumar

9.1 INTRODUCTION

The social and financial repercussions of a pandemic are profound. The last century has seen several devastating pandemics. The present outbreak of COVID-19 is the most recent of its type; nonetheless, lessons learned from previous pandemics and the ways in which technology helped people then can be applied to the current crisis. In this chapter, we will examine many effective responses to previous pandemics and describe their implementation [1].

It is crucial to investigate the equipment and related systems that can aid in disease detection, containment, and prevention. As a first line of defence, governments can implement a variety of cutting-edge technologies. In this section, we will look at how the Internet of Things (IoT), the Internet of Medical Things (IoMT), and other smart new technologies like drones, robots, autonomous vehicles (AVs), Bluetooth, and GPS may be used to combat this epidemic [2].

The Internet of Things (IoT) is an exciting new field in computer science that connects and communicates computing devices without requiring human involvement. IoMT has recently received a lot of interest from the medical community. It's a network of hardware and software for the healthcare industry that works together to provide better treatment. Minimising contact with potentially infectious COVID-19 patients is an added benefit of using drones, robots, and AV technologies. Using Bluetooth and GPS technologies, wearables provide another effective method of tracking one's health and stress levels in real time. Whether it's for disease prevention or for identifying and monitoring the masses, paramedical personnel, symptomatic, and asymptomatic COVID positives during the pandemic, these technologies can contribute a significant piece to the new paradigm of telemedicine [3–5].

There can be no doubt that the COVID-19 epidemic is the worst disaster of the 21st century, and perhaps the worst international crisis since World War II. Worldwide transmission of the COVID-19 pandemic began in late 2019. The pandemic poses a serious threat to the well-being of people everywhere, as well as to the efficiency and effectiveness of healthcare systems everywhere. The worldwide

DOI: 10.1201/9781003480181-9

population is under constant pressure to maintain stringent preventative measures due to the virus's quick spreading capabilities. The healthcare industry has been rocked by the virus's outcry. As a result of the pandemic, there has been a surge in demand for cutting-edge robotics and artificial intelligence-based applications, as well as for the standard healthcare equipment and pharmaceuticals needed to combat the disease. Diagnostics, risk assessments, monitoring, tele healthcare, disinfection, and other operations could all benefit greatly from the use of intelligent robot systems during this pandemic, greatly reducing the burden on frontline humans [6]. Also considerably accelerated by AI-enabled techniques is the discovery of a vaccine for this deadly virus. In addition, numerous robots and Robotics Process Automation platforms have greatly improved the delivery of the vaccine in a variety of contexts. The development of these cutting-edge technologies has also helped to alleviate the suffering of those whose mental health issues have been ignored or misdiagnosed. In this study, we look into how healthcare robotics and AI could be used to combat the ongoing COVID-19 epidemic.

Robotics technologies have sufficiently catered to public health needs, including the detection and containment of COVID-19, the lightening of the workloads of health personnel, and the streamlining of communicative operations. The development of a vaccination against COVID has been considerably aided by advances in artificial intelligence and robotics. Information on recent discoveries and investigations was processed using a number of AI-enhanced big data analysis methods. There has been successful use of unmanned aerial vehicle (UAV) and robotic process automation (RPA) platforms for managing and distributing vaccines. These are the ways in which different technologies have responded to the Covid-19 issues [7, 8].

9.2 TECHNOLOGY AND MEDICAL SCIENCE

Together, advances in medicine and technology may make for a brighter, healthier future. The medical industry has benefited greatly from technological advancements, which has helped to increase the average lifespan of humans all around the world. In addition, it has boosted people's happiness by providing a more effective means of diagnosing and treating illness. Early medical technology included the thermometer, microscope, ophthalmoscope, stethoscope, laryngoscope, and X-ray.

9.3 USE OF TECHNOLOGY DURING COVID-19

A few more names for IoT include the Internet of Things and the Industrial Internet. It's a new way of thinking about technology that involves interconnected systems of equipment and gadgets that function together effectively. Several sectors throughout the world have shown considerable interest in IoT, which they see as a crucial component of future technologies [9].

In many ways, IoT is a win-win situation. Broadband Internet's widespread availability, cheaper technology, and the massive adoption of smartphones, data-collecting wearables, and other "smart" items are at the forefront of this shift.

9.4 IoMT DEVICE CLASSIFICATION

9.4.1 Wearables

Moreover, wearables may be split into two groups:

1. Patient lifestyle gadgets, often known as fitness wearables as presented in Figure 9.1, are commonly used by those in need of continual monitoring of their physical fitness. In order to track a patient's health, built-in sensors record information about the person's movements and record it.
2. Devices for wearable clinical use the authorised and approved Internet of Things devices fall into this category. IoT based devices are used for clinical purpose as well. These kind of devices are often recommended by healthcare providers. For an example Qardio core chest trap records electrocardiogram as displayed in Figure 9.2.

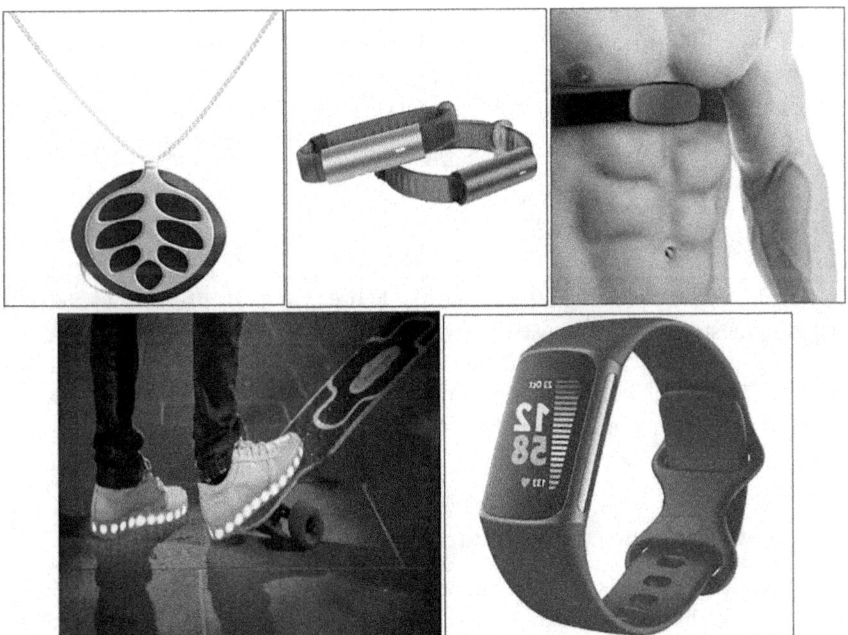

Figure 9.1 Smart lifestyle gadgets and wearables.

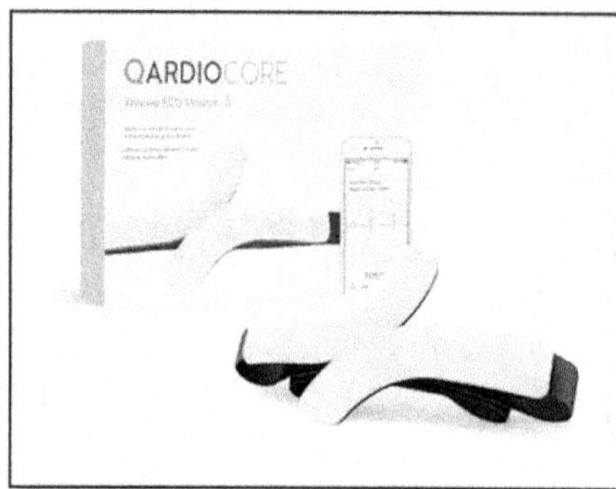

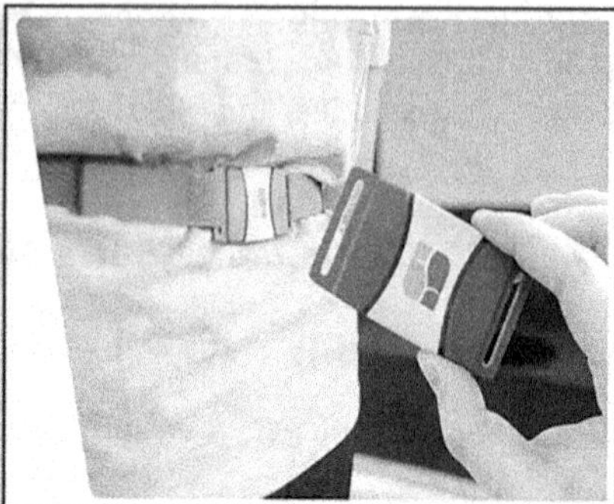

Figure 9.2 Quardiocore and smart belt.

9.4.2 Remote Patient Monitoring Devices

With the help of RPM, doctors may keep an eye on their patients from afar and handle their care in novel ways. Remote patient monitoring (RPM) is a method through which patients' medical records are compiled at a central place (such as the patient's home) and then forwarded electronically to doctors in other locations so that they may make diagnoses and recommendations.

This method is efficient since it provides care for patients without compromising on their convenience. With this data, doctors may offer patients personalised alerts and updated treatment plans based on their level of physical activity. More than four million people have been using remote health monitoring says IHS (Information Handling Services). Remote blood collection devices, continuous glucose monitoring monitors, and inexpensive surgical robots are just a few well-known examples [10]. A basic remote patient monitoring system with working has been shown in Figure 9.3.

9.4.3 Smart Pills

Smart pills, often called digital pills due to their incorporation of electronic sensors into an ingestible form, are used to monitor a patient's adherence to a prescribed drug regimen. They have medication sensors as depicted in Figure 9.4 that are triggered by stomach acids and then wirelessly communicate with external devices like tablets, cell phones, or patches. As one example of a "smart pill," Abilify MyCite has found widespread use [11].

9.4.4 Nurse Robots in Hospitals

These robots' purpose is to help doctors in the same way that nurses do in hospitals. Japan has the biggest population of elderly people (over 75 years old) compared to other OCED countries; hence, nurse robots are widely deployed there. This is becoming a bigger problem for hospitals around the country. To fulfil their social

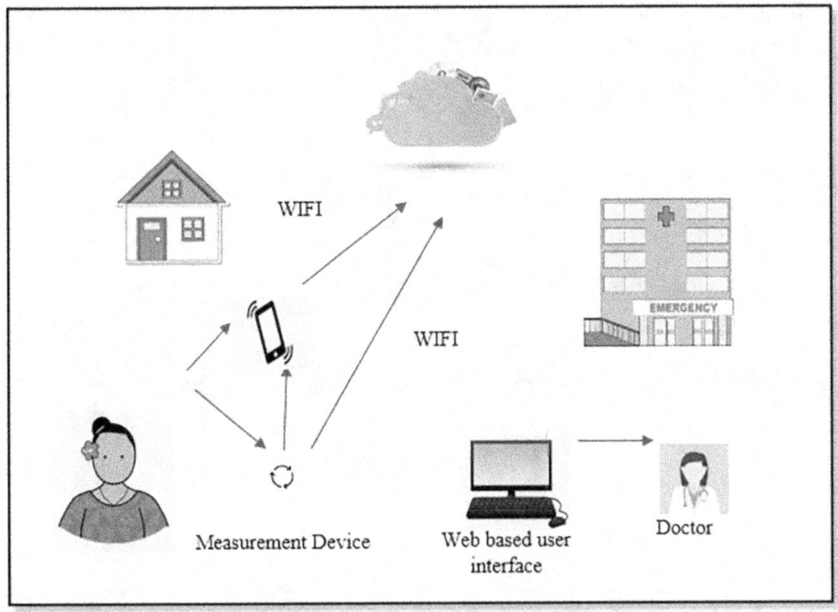

Figure 9.3 Remote patient monitoring devices.

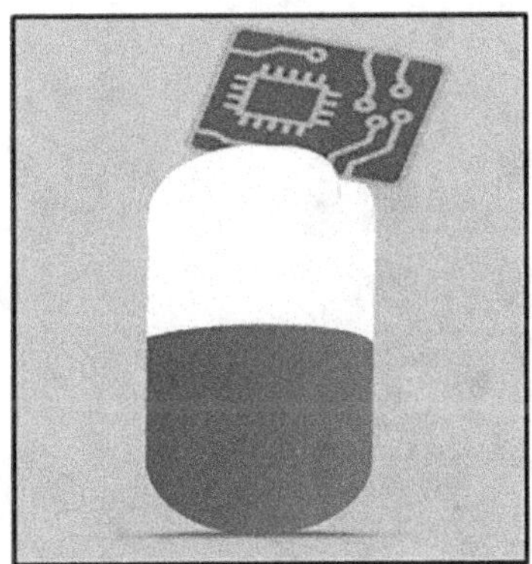

Figure 9.4 An example of a smart pill.

obligation, more Japanese people are staying at home to care for their elderly relatives because there is not enough employment in the senior care sector [12].

High levels of stress and fatigue are also experienced by the nursing and healthcare staff as a result of the large number of patients they must care for. As can be seen in Figure 9.5, the Japanese government is thus looking to technology solutions to assist in the care of the country's ageing population. Figure 9.6 shows nursing robots monitoring a patient.

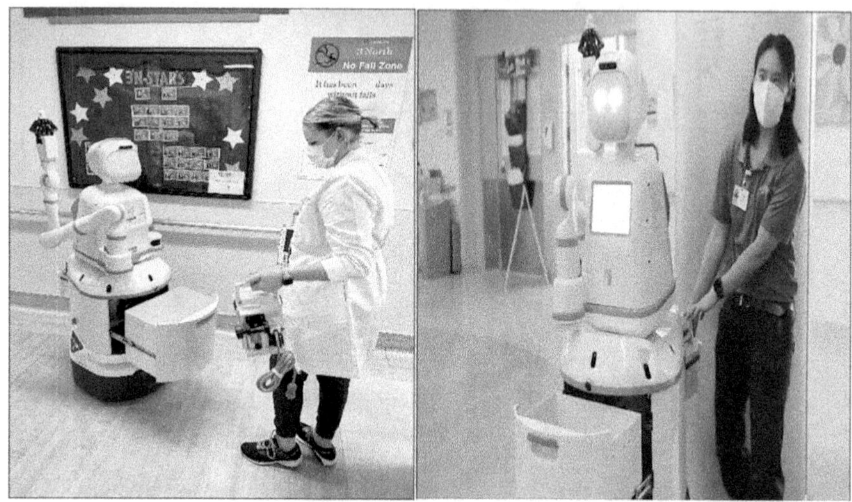

Figure 9.5 Nursing robots.

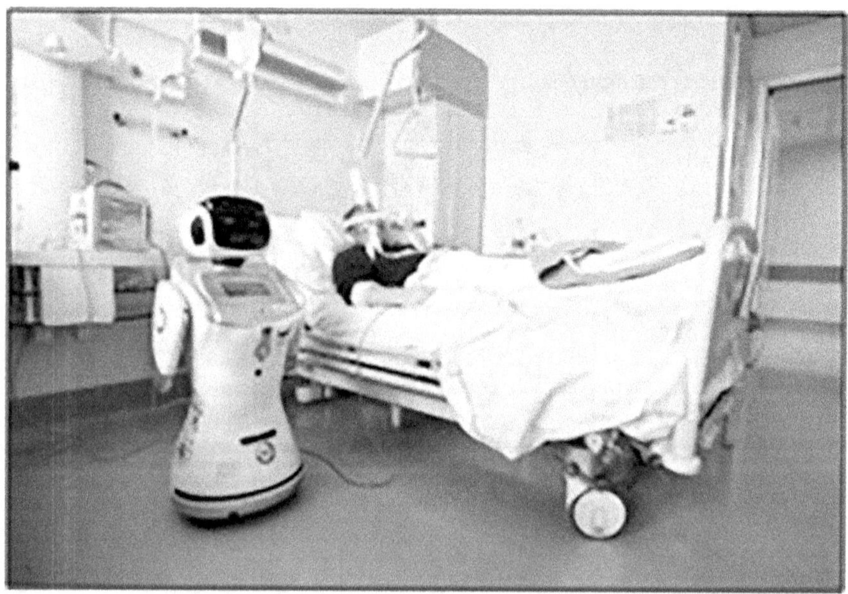

Figure 9.6 Nursing robot monitoring patient.

9.4.5 Point-of-Care Devices

Point-of-care devices are diagnostic tools used in a variety of settings, including patient homes, clinics, and hospitals. When in close proximity to the patient, they are utilised to obtain diagnostic information. Glucose monitors, cholesterol

Table 9.1 List of Robots Used in Hospitals with Work

Robot	*Make*	*Cleaning*	*Nursing*	*Lab*	*Pharmacy*	*Food Service*	*Waste removal*	*Linen*
Dinsow	CT Asia		☆	☆	☆	☆	☆	☆
Relay	Swisslog							
TUG	iRobot (USA)				☆	☆	☆	☆
RR-VITA	Aethon (USA)		☆	☆				
Roomba i7	iRobot (USA)	☆	☆		☆			
Moxi	Diligent Robots (USA)		☆		☆			
Ambubot	Thailand							
Drone Robot	TU Delft		☆		☆			

analysers, pregnancy tests, pulse oximeters, drug screening kits, etc., are all examples of diagnostic testing gadgets. The mobility, ease, and speed of these gadgets are their primary benefits [13].

9.4.6 Disease Diagnosis

Nowadays, the reverse real-time PCR assay is utilised as the gold standard for testing for COVID-19 (rRT-PCR). The usual turnaround time for this molecular-based test is between four and six hours. In addition, a well-equipped laboratory and knowledgeable professionals are necessities. The number of tests that can be performed is capped at some point, which is unacceptable under these conditions. This highlights the critical need for further quick diagnostic techniques.

Point-of-care (POC) devices using lateral flow immunoassay (LFIA) technology to detect COVID-19 in human serum are a potentially useful tool in such a scenario. This method is predicated on the observation that IgG and IgM antibodies against SARS-CoV-2 may be found in human blood following infection with COVID-19, and that the levels of these antibodies provide insight into the progression of the disease. As the number of reported cases rises throughout the world, a plethora of point-of-care LFIA devices has emerged [14].

9.4.7 Disease Monitoring

The current pandemic situation, with its worrisome rise in COVID-19 cases, necessitates a reliable monitoring and surveillance system for effective patient tracing. The Internet of Things has the potential to play a pivotal role in this pandemic, particularly in regards to contact tracing, cluster detection, and quarantine compliance.

It is crucial to identify infected individuals in crowded locations, and infrared thermometers are the primary tool for doing so at present. Nevertheless, this does not appear to be very effective for two reasons: first, the thermometer may not be able to check on everyone in the throng, and second, having a health officer examine many individuals in a queue, any one of whom may be ill, may increase the likelihood of the spread of disease. As a result, we need a new kind of technology, and the Internet of Things holds some promise in this area [15].

9.4.8 Disease Management

The rapid global spread of COVID-19 has necessitated the implementation of stringent lockdown measures throughout the world. An estimated 10 billion individuals worldwide have sequestered themselves at home. Yet, there has been a significant demand for critically important medical supplies and tools. Contrary to the efforts made for isolation and quarantine, citizens, some of whom may be potential patients, must leave their houses in order to seek medical treatment. The medical community has also advised those with mild or questionable symptoms to remain at home rather than risk being exposed to others in hospitals' inadequate isolation units. As a result, doctors are telling patients with mild or questionable symptoms to stay at home to avoid spreading the disease before they have a chance to test for it. In such a dire case, IoMT can serve as a medical podium, helping impacted individuals access the right care at home while also providing governments and healthcare organisations with a comprehensive database for illness management [16]. The stages involved in managing diseases on such a system are depicted in Figure 9.7.

Those with mild symptoms typically do not need to be hospitalised. Patients can avoid having to go to the hospital by purchasing the necessary diagnostic and healthcare equipment, such as thermometers, masks, gloves, sanitisers, and POC kits used for detecting and monitoring COVID-19 and medications, and then uploading this information regularly to the IoMT platform, where it will be broadcast to the nearest hospitals, Centre for Disease Control (CDC), and local health

Figure 9.7 Management of diseases with IoMT.

agencies. The Centres for Disease Control and other health organisations might then allocate supplies and quarantine facilities afterward.

An IoMT platform offers numerous benefits. Patients may dynamically track their disease state while also receiving the care they need without worrying about spreading their illness. In addition to being more cost-effective, such a system will also provide a more organised database for keeping track of the propagation of viruses.

9.4.9 Drone Technology

A drone is a form of unmanned aerial vehicle that flies without a human pilot on board (UAV). A ground-based controller and two-way communication system are included. It's possible to run UAV flights in a number of different ways:

a) Remote control by a human operator
b) Autonomously by onboard computers
c) Piloted by an autonomous robot

Originally, unmanned aerial vehicles (UAVs) and drones were employed for certain operations that were too risky, hazardous, or unimportant for human agents to undertake. It's not uncommon for individuals to confuse the phrases unmanned aerial vehicle (UAV) and autonomous drone. As the name suggests, many UAVs are automated; this means that they can do tasks without direct human intervention. Autonomous drones, on the other hand, are UAVs that don't need human control to function [17]. To clarify, these drones can take off, fly, accomplish the mission, and land without any human intervention (autonomously). We can conclude from this that not every unmanned aerial vehicle (UAV) is also an autonomous drone, but that autonomous drones do fall under the umbrella of UAVs.

So, any ground control system or communications management software plays a significant role in autonomous drones, and as a result, autonomous drones are also included in the category of unmanned aerial systems (Unmanned Aircraft System). Drones use a plethora of cutting-edge tools, including cloud computing, computer vision, AI, ML, DL, and heat sensors, to implement such control.

Military, commercial, scientific, agricultural, medical (in the current COVID-19 pandemic, which we will discuss in the following section), and other uses [18] include mass surveillance, aerial photography, drone racing as a hobby, infrastructure inspections, and smuggling of illegal goods and drugs as presented in Figure 9.8.

9.4.9.1 Drones as Telemedicine and Transfer Units

Access to healthcare in underserved areas can be improved with the help of drones. The absence of investment in infrastructure and reliable public transportation is a hallmark of economically depressed areas. Drones are especially useful in these areas because they may speed up the delivery of much-needed medical supplies

Figure 9.8 A modern drone.

and equipment. Drones can fly much faster than any manned vehicle, allowing them to traverse terrain that would be difficult or impossible for other vehicles to handle. As the COVID-19 patient is infectious, he or she must be kept in isolation while receiving treatment and nourishment. Beyond visual line of sight (BVLOS) [19] is an example of an autonomous drone. Protecting the human workers during a pandemic is a top priority, and these drones can fly well beyond the line of sight, maximising production while minimising costs and dangers and guaranteeing site safety and security [20]. Amazon Prime Air is just one example of how these aircraft might be utilised for consumer-facing missions like package delivery and time-sensitive medical supply drops.

9.4.9.2 Robotic Aircraft for Spying and Examining

In addition to amateur photography, surveillance is the most common use for drones equipped with cameras. In that they can quickly provide an aerial perspective of the current location, they may be particularly well suited for crowd monitoring. As a result, numerous nations are using drones to monitor crowds, especially in the midst of the current COVID-19 pandemic.

Drones equipped with temperature sensors can provide real-time information about people's core temperatures in any public space. Drone technology has been widely utilised for crowd surveillance in many countries, including China and India. The deployed drones include surveillance cameras that can keep an eye on high-risk regions of the city and alert authorities to any suspicious activity in real time [21].

9.4.9.3 *Public Service Announcement Drones*

Drones have several potential uses beyond crowd surveillance, including but not limited to the dissemination of vital information in regions with limited access to the media. Drones have been deployed by governments in California, Florida, and New Jersey to notify and warn people about social isolation. Police in Madrid, Spain, deployed a drone fitted with a loudspeaker to disseminate information about the state of emergency and its regulations to the public [22]. Drones are also useful in Europe; many countries use them to broadcast warnings to the public during pandemics or disasters, helping to stem the spread of illness or destruction.

9.4.9.4 *Disinfection by Drones*

Drones are helpful in situations where people need to avoid coming into contact with infectious pathogens. Drones can be used to spray disinfectants over polluted locations. The use of spraying drones in agricultural regions has seen a rise in popularity over the past decade. The Spanish military has begun using agricultural drones from DJI, a renowned Chinese drone manufacturer, to spray cleaning chemicals over public locations [23]. These spraying drones have a 16 L tank and can sterilise a tenth of a kilometre per hour [24].

9.5 ROBOTS IN COVID-19

Being intelligent machines, robots continued to be of use throughout the recent COVID-19 epidemic. As there is minimal likelihood of spreading contagious disease from sick patients, robots can be simply deployed as frontline warriors in medical facilities.

Furthermore, ultraviolet (UV) disinfection method (means to disinfect the regions from contagious illnesses) is easily performed with robots through pre-programmed procedures, reducing the spread of disease through contact with infected surfaces in healthcare facilities and containment wards. Rather than relying on the safety-at-risk manual disinfection methods, you should use autonomous disinfecting robots [25].

Robot technology was used in many nations to prevent the spread of the COVID-19 virus and to check in on the mental and emotional well-being of hospitalised patients and those living in seclusion. The following are a few more of the robots' many benefits during a crisis, in addition to those already listed:

a) Delivery: Medicine, medical supplies, and food service are all being delivered and served by robots in hospitals and clinics during the ongoing COVID-19 pandemic to relieve the strain on human workers. An Indian firm located in Kerala called Asimov Robotics has created a three-wheeled robot that can do all of these things and more to help patients in isolation units [26].

b) Social distancing: Camera-equipped robots can help verify, in public, whether or not the rules of social distance are being observed. More public guidance regarding preventative measures is recommended, especially in affected areas.
c) Disinfecting: As was previously mentioned, robots present less of a risk while cleaning potentially dangerous environments. Multiple disinfection robots, which use ultraviolet light to sterilise the affected area or piece of equipment, have been developed by a Danish robotics firm. UV rays are able to neutralise viruses by destroying their DNA. Robots manufactured by a company called UVD have been shipped to hospitals in China, Europe, and the United States. They claim the robots can work for 2.5 hours on a single battery and can disinfect up to 10 rooms [26].
d) Emotional support: When the pandemic hit, many nations went into lockdown for months. Psychological health suffers when people are isolated for long periods of time. To help the lonely connect with others, scientists have created robots with emotional intelligence. Doctors can monitor their patients' conditions remotely with the help of these robots.
e) Medical procedures and surgeries: Due to the infectious nature of the COVID-19 virus, many doctors and surgeons had to take extra precautions during routine procedures. Dentists, oncologists, and ear, nose, and throat (ENT) surgeons are at the forefront of the epidemic because the virus transmits so easily through the mouth and droplets. Almost every affected country delayed general-purpose processes during the pandemic, but crises still require heightened focus. Successful robotic procedures have been performed in a variety of medical settings for quite some time, well before the pandemic crisis. Although PPEs can help, the best way to stop the transmission of a virus is to keep a safe distance from infected people. Hence, nonautonomous robots can be a safer option in situations where close contact through the patient's mouth and nasal cavities is required, such as during a pandemic.

9.6 CONCLUSION

The potential of robots in medicine and associated fields is explored in this research, with a focus on its application to the containment of the COVID-19 epidemic. As was seen with the Chinese epidemic, proper control of COVID-19 may drastically cut down on both cases and fatalities. As it has become a worldwide problem, countries with more sophisticated technology might help less developed ones by providing them with donations of necessary tools and robotic infrastructure. This analysis confirms that healthcare digitalisation, which has led to the introduction of medical robots, has vastly improved the safety and quality of health management systems compared to manual methods. Medical robots, from those used for housekeeping to those designed for complex surgeries, are categorised solely according to their intended functions. Opportunities exist in the design and operation of medical robots, such as the incorporation of a cyber-physical system (CPS), the management of power using optimised algorithms and renewable sources, and the implementation of fault-tolerant control and dependable architectures to ensure the robots' continued and secure operation within hospitals and other healthcare facilities.

REFERENCES

[1] Kashani, Mostafa Haghi, Mona Madanipour, Mohammad Nikravan, Parvaneh Asghari, and Ebrahim Mahdipour. "A systematic review of IoT in healthcare: Applications, techniques, and trends." *Journal of Network and Computer Applications* 192 (2021): 103164.

[2] Javaid, Mohd, and Ibrahim Haleem Khan. "Internet of Things (IoT) enabled healthcare helps to take the challenges of COVID-19 pandemic." *Journal of Oral Biology and Craniofacial Research* 11, no. 2 (2021): 209–214.

[3] Bharadwaj, Hemantha Krishna, Aayush Agarwal, Vinay Chamola, Naga Rajiv Lakkaniga, Vikas Hassija, Mohsen Guizani, and Biplab Sikdar. "A review on the role of machine learning in enabling IoT based healthcare applications." *IEEE Access* 9 (2021): 38859–38890.

[4] Valsalan, Prajoona, Tariq Ahmed Barham Baomar, and Ali Hussain Omar Baabood. "IoT based health monitoring system." *Journal of Critical Reviews* 7, no. 4 (2020): 739–743.

[5] Tamilselvi, V., S. Sribalaji, P. Vigneshwaran, P. Vinu, and J. GeethaRamani. "IoT based health monitoring system." In *2020 6th International conference on advanced computing and communication systems (ICACCS)*, pp. 386–389. IEEE, 2020.

[6] Ratta, Pranav, Amanpreet Kaur, Sparsh Sharma, Mohammad Shabaz, and Gaurav Dhiman. "Application of blockchain and Internet of Things in healthcare and medical sector: Applications, challenges, and future perspectives." *Journal of Food Quality* 2021 (2021): 1–20.

[7] Dash, Satya Prakash. "The impact of IoT in healthcare: Global technological change & the roadmap to a networked architecture in India." *Journal of the Indian Institute of Science* 100, no. 4 (2020): 773–785.

[8] Krishnamoorthy, Sreelakshmi, Amit Dua, and Shashank Gupta. "Role of emerging technologies in future IoT-driven Healthcare 4.0 technologies: A survey, current challenges and future directions." *Journal of Ambient Intelligence and Humanized Computing* 14, no. 1 (2023): 361–407.

[9] Butt, Shariq Aziz, Jorge Luis Diaz-Martinez, Tauseef Jamal, Arshad Ali, Emiro De-La-Hoz-Franco, and Muhammad Shoaib. "IoT smart health security threats." In *2019 19th International conference on computational science and its applications (ICCSA)*, pp. 26–31. IEEE, 2019.

[10] Kumar, Rakesh, Mini Anil, Durlav Singh Parihar, Anshul Garhpale, Sampurna Panda, and Babita Panda. "A cross-sectional assessment of Gwalior residents' reports of adverse reactions to the COVID-19 immunization." *International Journal of Science &Technology* 10 (2022): 2386–2392.

[11] Kumar, Rakesh, Mini Anil, Sampurna Panda, Babita Panda, Lipika Nanda, and Chitralekha Jena. "A psychological study on accepting and rejecting COVID-19 vaccine by college students in India." In *2022 IEEE international conference on distributed computing and electrical circuits and electronics (ICDCECE)*, pp. 1–4. IEEE, 2022.

[12] Panda, Sampurna, and Rakesh Kumar Dhaka. "Application of Artificial Intelligence in medical imaging." In *Machine learning and deep learning techniques for medical science*, pp. 195–202. CRC Press, 2022.

[13] Kyrarini, Maria, Fotios Lygerakis, Akilesh Rajavenkatanarayanan, Christos Sevastopoulos, Harish Ram Nambiappan, Kodur Krishna Chaitanya, Ashwin Ramesh Babu, Joanne Mathew, and Fillia Makedon. "A survey of robots in healthcare." *Technologies* 9, no. 1 (2021): 8.

[14] Maibaum, Arne, Andreas Bischof, Jannis Hergesell, and Benjamin Lipp. "A critique of robotics in health care." *AI & Society* (2022): 1–11.

[15] Kaiser, M. Shamim, Shamim Al Mamun, Mufti Mahmud, and Marzia Hoque Tania. "Healthcare robots to combat COVID-19." *COVID-19: Prediction, Decision-Making, and Its Impacts* (2021): 83–97.

[16] Khan, Zeashan Hameed, Afifa Siddique, and Chang Won Lee. "Robotics utilization for healthcare digitization in global COVID-19 management." *International Journal of Environmental Research and Public Health* 17, no. 11 (2020): 3819.

[17] Van Wynsberghe, Aimee, and Shuhong Li. "A paradigm shift for robot ethics: From HRI to human–robot–system interaction (HRSI)." *Medicolegal and Bioethics* (2019): 11–21.

[18] Mukati, Naveen, Neha Namdev, R. Dilip, N. Hemalatha, Viney Dhiman, and Bharti Sahu. "Healthcare assistance to COVID-19 patient using Internet of Things (IoT) enabled technologies." *Materials Today: Proceedings* 80 (2023): 3777–3781.

[19] Firouzi, Farshad, Bahar Farahani, Mahmoud Daneshmand, Kathy Grise, Jaeseung Song, Roberto Saracco, Lucy Lu Wang, et al. "Harnessing the power of smart and connected health to tackle COVID-19: IoT, AI, robotics, and blockchain for a better world." *IEEE Internet of Things Journal* 8, no. 16 (2021): 12826–12846.

[20] Sarfraz, Zouina, Azza Sarfraz, Hamza Mohammad Iftikar, and Ramsha Akhund. "Is COVID-19 pushing us to the fifth industrial revolution (society 5.0)?" *Pakistan Journal of Medical Sciences* 37, no. 2 (2021): 591.

[21] Nasajpour, Mohammad, Seyedamin Pouriyeh, Reza M. Parizi, Mohsen Dorodchi, Maria Valero, and Hamid R. Arabnia. "Internet of Things for current COVID-19 and future pandemics: An exploratory study." *Journal of Healthcare Informatics Research* 4 (2020): 325–364.

[22] Jain, Rachna, Meenu Gupta, Kashish Garg, and Akash Gupta. "Robotics and drone-based solution for the impact of COVID-19 worldwide using AI and IoT." *Emerging Technologies for Battling COVID-19: Applications and Innovations* (2021): 139–156.

[23] Hussain, Khalid, Xingsong Wang, Zakarya Omar, Muhanad Elnour, and Yang Ming. "Robotics and artificial intelligence applications in manage and control of COVID-19 pandemic." In *2021 international conference on computer, control and robotics (ICCCR)*, pp. 66–69. IEEE, 2021.

[24] Castiglione, Aniello, Muhammad Umer, Saima Sadiq, Mohammad S. Obaidat, and Pandi Vijayakumar. "The role of Internet of Things to control the outbreak

of COVID-19 pandemic." *IEEE Internet of Things Journal* 8, no. 21 (2021): 16072–16082.

[25] Leila, Ennaceur, Soufiene Ben Othman, and Hedi Sakli. "An internet of robotic things system for combating coronavirus disease pandemic (COVID-19)." In *2020 20th international conference on sciences and techniques of automatic control and computer engineering (STA)*, pp. 333–337. IEEE, 2020.

[26] Mohammed, M. N., I. S. Arif, S. Al-Zubaidi, S. H. K. Bahrain, A. K. Sairah, and Y. Eddy. "Design and development of spray disinfection system to combat coronavirus (COVID-19) using IoT based robotics technology." *Revista Argentina de Clínica Psicológica* 29, no. 5 (2020): 228.

Chapter 10

Innovating Healthcare Delivery

Harnessing Drones for Medical Services

Shalom Akhai

10.1 INTRODUCTION TO THE RISE OF DRONE-ASSISTED MEDICAL SERVICES

Healthcare delivery has undergone a significant transformation over the past few decades, with technological advancements playing a pivotal role in this shift. Among the most exciting innovations in healthcare delivery is the integration of drone technology, which is generating widespread excitement and presenting unprecedented opportunities for improved patient outcomes and efficient healthcare delivery [1–5].

Drones, or unmanned aerial vehicles, are being increasingly used in healthcare for a wide range of applications, from delivering medical supplies to conducting remote consultations and performing medical procedures. The benefits of drone-assisted health delivery are numerous and varied, including increased access to care, improved patient outcomes, and reduced costs [6–10].

One of the most significant advantages of drone-assisted health delivery is its ability to reach remote and underserved populations. Drones can overcome geographical barriers and deliver medical supplies to areas that are difficult to access, such as remote villages or disaster-stricken areas. They can also bypass infrastructure challenges, such as poor roads or limited transportation options, by delivering medical supplies and equipment directly to patients [10–13].

Moreover, drone-assisted health delivery can significantly improve patient outcomes by reducing the time it takes for patients to receive critical care. Drones can deliver life-saving medications and equipment quickly and efficiently, allowing for early intervention and preventing the worsening of medical conditions. Additionally, drones can be equipped with sensors, cameras, and other medical devices, making them ideal for conducting remote consultations and monitoring patients' health remotely [14–16].

However, despite the tremendous potential of drone-assisted health delivery, there are also significant challenges and limitations that must be considered. These include technical limitations, such as limited flight time and payload capacity, as well as regulatory and ethical considerations, such as privacy and safety concerns. To fully realize the potential of drone-assisted health delivery, it is crucial to

DOI: 10.1201/9781003480181-10

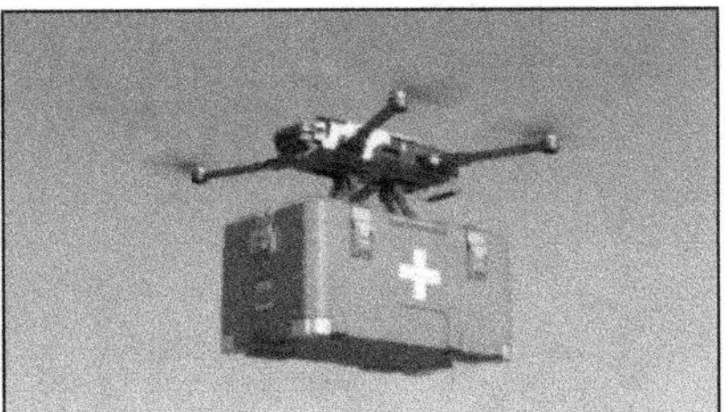

Figure 10.1 Drone assisted service.

address these challenges and limitations while continuing to innovate and explore new applications for this technology. Moreover, it is important to ensure that the use of drones in healthcare is regulated and ethical, with a focus on protecting patient privacy and safety [17–20].

Overall, the emergence of drone-assisted health delivery as represented in Figure 10.1 is a significant step forward in the delivery of healthcare services. With its potential to increase access to care, improve patient outcomes, and reduce costs, this technology has the potential to transform the way healthcare is delivered in the years to come. As such, it is an exciting time to be involved in the rapidly evolving field of drone-assisted health delivery.

10.2 DRONE TECHNOLOGY AND DESIGN

Recent years have witnessed remarkable advancements in drone technology, enabling their utilization in various sectors, including healthcare. These unmanned aerial vehicles (UAVs), commonly referred to as drones, are essentially flying robots that can be remotely operated by a ground-based controller. They can operate autonomously using GPS technology or be manually controlled by an operator.

Figure 10.2 shows some drone designs used widely. The following section discusses the key features of drone technology and design [21–25]:

- There are several components to a typical drone design, including the airframe, propulsion system, control system, and payload. The airframe is the structure of the drone that gives it its shape and provides a platform for attaching other components. It can be made from a variety of materials, including carbon fiber, aluminum, or plastic.
- The propulsion system is what allows the drone to move through the air. Most drones use electric motors and propellers to generate lift and move forward, backward, and sideways. The control system includes the hardware and software that allows the operator to control the drone. This can include a remote controller with joysticks and buttons, or it can be done using a computer or mobile device.

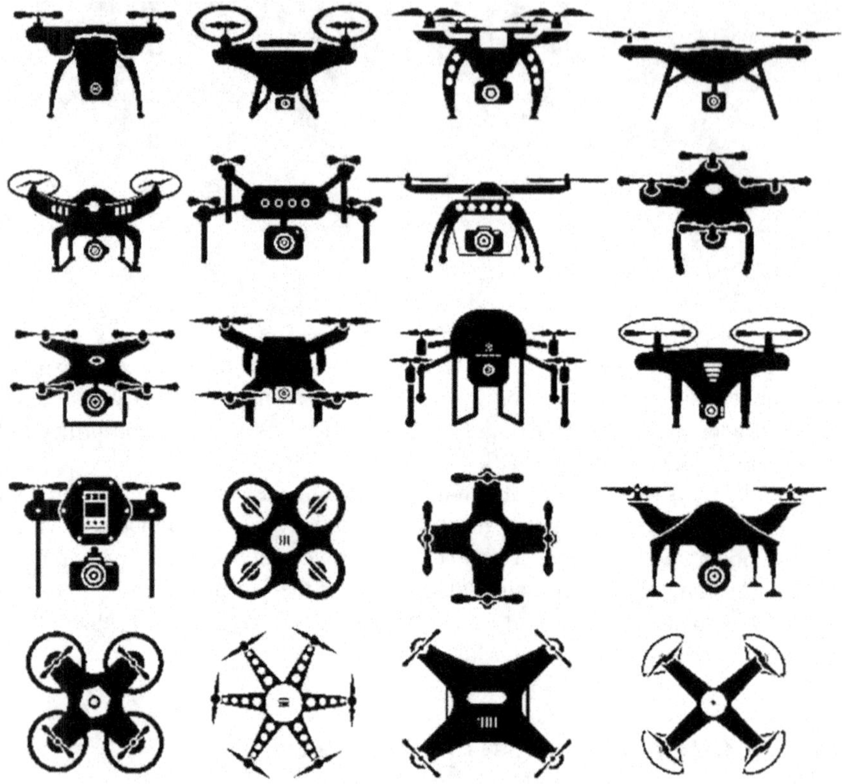

Figure 10.2 Drone designs.

- Finally, the payload is the equipment or cargo that the drone is designed to carry. In healthcare applications, this can include medical supplies, equipment, or even patients in some cases. Drones can be designed to carry payloads of various sizes and weights, depending on the intended application.
- One of the key advantages of drone technology is its versatility. Drones can be designed for a wide range of applications, from surveillance and monitoring to search and rescue operations. In healthcare, drones can be used for delivering medical supplies and equipment to remote areas, conducting medical procedures, and providing remote consultations and telemedicine services.
- To be effective in healthcare applications, drones must be designed with several key features in mind. First and foremost, they must be reliable and safe to operate. They must also be able to carry payloads of various sizes and weights, depending on the intended application. This requires careful attention to the drone's design and construction, including the airframe, propulsion system, and control system.

- In addition to these features, healthcare drones must also be able to operate in a variety of environments and weather conditions. This can include flying in high winds, rain, and even snow in some cases. To be effective in these environments, drones must be designed with durable materials that can withstand harsh conditions and still function reliably.
- Another important consideration in drone design is the range and endurance of the drone. Healthcare drones may need to travel long distances to reach remote areas or deliver medical supplies to areas that are difficult to access by traditional means. This requires drones with long-range capabilities and extended endurance to ensure that they can complete their missions successfully.
- Finally, drone design must take into account the regulations and guidelines governing drone operations in healthcare settings. This includes considerations such as privacy and security, as well as the need to ensure that drones do not interfere with other medical equipment or procedures.

Overall, drone technology has the potential to revolutionize healthcare delivery, providing improved access to care and more efficient and cost-effective services. However, to fully realize this potential, drones must be designed and constructed with careful attention to the unique needs and requirements of healthcare applications. As technology continues to advance, it is likely that we will see even more innovative uses for drones in healthcare and other industries in the years to come.

10.3 BENEFITS OF DRONE-ASSISTED HEALTH DELIVERY

The integration of drones in the healthcare industry has the potential to revolutionize the way healthcare is delivered, improving patient outcomes and increasing access to care. This section explores the benefits of drone-assisted health delivery in greater detail [26–31]:

- **Improved Response Time and Efficiency**—One of the main benefits of using drones in healthcare is the ability to deliver medical supplies and equipment quickly and efficiently. In emergency situations, every second counts, and drones can significantly reduce response times. They can quickly transport critical medical supplies such as blood, vaccines, and medications to remote or hard-to-reach areas, where traditional transportation methods may not be feasible. Drones can also deliver medical supplies to disaster areas where roads and other infrastructure may be damaged or inaccessible.
- **Increased Access to Healthcare**—Drones can be especially useful in delivering healthcare services to underserved areas where access to care may be limited. Remote communities, such as those in rural or isolated regions, often lack healthcare infrastructure and resources. In these areas, drones can be used to deliver medical supplies, conduct remote consultations, and even

perform medical procedures. By providing access to healthcare services in these areas, drones can help improve health outcomes and reduce healthcare disparities.

- **Reduced Costs**—Drones can also help reduce the costs associated with healthcare delivery. They are less expensive to operate than traditional transportation methods, such as helicopters or planes, and require less infrastructure, such as runways or landing pads. Drones can also reduce the costs associated with transporting medical supplies and equipment by providing a more efficient and cost-effective mode of transportation. Additionally, by providing access to healthcare services in remote or underserved areas, drones can help reduce the costs associated with transporting patients to larger healthcare facilities for treatment.
- **Improved Patient Outcomes**—Drones can help improve patient outcomes by providing quicker access to medical care and supplies. In emergency situations, such as natural disasters or medical emergencies, the timely delivery of medical supplies and equipment can mean the difference between life and death. Drones can also be used to monitor patients remotely, providing real-time data on vital signs and other health indicators. This information can be used to make more informed decisions about patient care and improve treatment outcomes.
- **Enhanced Safety and Security**—Using drones to transport medical supplies and equipment can enhance safety and security. Drones can navigate through challenging terrain, such as mountains or forests, without putting human pilots at risk. They can also be used to transport medical supplies and equipment to areas that may be unsafe for human transportation, such as war zones or disaster areas. Drones can also be equipped with advanced sensors and cameras, which can be used to monitor and detect potential threats to patient safety and security.
- **Reduced Environmental Impact**—Drones have a smaller environmental footprint than traditional transportation methods, such as cars or planes. They require less fuel to operate and produce fewer emissions. Drones can also reduce the amount of waste generated by healthcare delivery by providing a more efficient mode of transportation. This can help reduce the environmental impact of healthcare delivery, making it more sustainable in the long run.
- **Improved Healthcare System Resilience**—Drones can help improve the resilience of healthcare systems by providing backup transportation and communication capabilities in emergency situations. In the event of a natural disaster or other emergency, traditional transportation and communication infrastructure may be damaged or destroyed. Drones can be used to transport medical supplies and equipment to affected areas and provide communication capabilities, helping to ensure that healthcare services continue to be delivered.

In conclusion, the benefits of drone-assisted health delivery are significant and far-reaching. Drones have the potential to improve patient outcomes, increase access to care, reduce costs, enhance safety and security, and reduce the environmental impact of healthcare delivery. By leveraging the latest technological advancements, drones can transform the way healthcare is delivered, improving lives.

10.4 CHALLENGES AND LIMITATIONS OF DRONE TECHNOLOGY IN HEALTHCARE

While drone technology offers immense promise in healthcare delivery, it is not without its challenges and limitations. This section will explore some of the key obstacles that must be addressed to ensure the successful integration of drones in healthcare. This section explores the hurdles, obstacles, and restrictions of using drone technology in healthcare [32–39]:

- **Regulatory Hurdles**—One of the biggest challenges to the widespread use of drones in healthcare is navigating the complex regulatory environment. The Federal Aviation Administration (FAA) oversees the use of drones in the United States, and any commercial operation of drones must be approved by the agency. This includes healthcare-related applications of drone technology, such as medical deliveries and emergency response.

  The FAA has established rules for the operation of drones, including restrictions on the altitude, distance, and location of flights. Drones must also be operated within the line of sight of the operator and cannot fly over people or beyond certain distances. Meeting these regulations can be particularly difficult in healthcare settings, where drones may need to fly over densely populated areas or outside of the line of sight of the operator.

  Additionally, healthcare providers must comply with HIPAA regulations that protect patient privacy and the security of patient information. This means that healthcare-related drone operations must ensure that patient information is not exposed or breached during the delivery process.
- **Technical Limitations**—Drone technology is still in its infancy, and there are several technical limitations that must be overcome to make it a viable option for healthcare delivery. One of the biggest challenges is the limited range of drones. Currently, most drones can only travel a few miles before needing to recharge or refuel. This makes it difficult to use drones for long-distance medical deliveries or emergency response. Another technical limitation is the weight capacity of drones. Most commercial drones have a weight capacity of a few pounds, which limits their ability to deliver larger medical supplies, such as oxygen tanks or defibrillators. Additionally, drones are vulnerable to weather conditions, such as high winds or

rain, which can interfere with their flight and make it difficult to complete deliveries.

- **Cost**—While drones have the potential to reduce the cost of healthcare delivery, the initial investment required to implement drone technology can be significant. Drones themselves can be expensive, and healthcare providers must also invest in training for operators, maintenance, and infrastructure to support drone operations. Additionally, healthcare providers must consider the cost of potential liability issues that may arise from drone operations. Accidents involving drones can result in damage to property or injury to individuals, which can lead to costly lawsuits.
- **Public Perception**—Drones have a negative public perception due to their association with military operations and surveillance. This perception can create a significant barrier to the adoption of drone technology in healthcare, as patients may be hesitant to receive medical supplies or undergo medical procedures delivered by a drone. Additionally, some patients may be uncomfortable with the idea of their medical information being transmitted via drone or may have concerns about the safety and reliability of drone deliveries.
- **Security Concerns**—Drone technology raises security concerns that must be addressed to ensure the safety of both patients and healthcare providers. Drones are vulnerable to hacking and interference, which can result in unauthorized access to patient information or interference with medical deliveries. Additionally, drones may be vulnerable to theft, which can lead to the loss of expensive medical supplies and equipment.
- **Environmental Concerns**—Finally, the use of drones in healthcare delivery raises environmental concerns, particularly in terms of their carbon footprint. Drones are powered by batteries or fuel, and their operation can contribute to air pollution and carbon emissions.

While drone technology offers tremendous potential for improving healthcare delivery, it is not without its challenges and limitations. Healthcare providers must navigate complex regulatory environments, address technical limitations, and manage costs to successfully implement drone technology in their operations. Additionally, healthcare providers must address public perception and security concerns to ensure patient trust in the safety and reliability of drone deliveries.

10.5 APPLICATIONS OF DRONE TECHNOLOGY IN HEALTHCARE

The use of drone technology in the healthcare industry has opened up new avenues for medical delivery, remote consultations, and emergency response. Here are some of the applications of drone technology in healthcare [40–49]:

10.5.1 Medical Supply Delivery

One of the primary applications of drone technology in healthcare is the delivery of medical supplies to remote and hard-to-reach areas. Drones equipped with GPS and other navigational technologies can fly over mountains, forests, and other challenging terrains to deliver essential medicines, vaccines, and blood products to patients in need. This can be especially useful in disaster-stricken areas, where traditional modes of transportation may be disrupted. In 2019, the government of Ghana partnered with Silicon Valley startup Zipline to launch the world's largest medical drone delivery network. The initiative aims to deliver essential medical supplies to over 12 million people living in remote and hard-to-reach areas of the country. The drones can carry up to 1.75 kg of cargo and deliver supplies to health facilities within a 75 km radius in less than 30 minutes.

10.5.2 Remote Consultations and Telemedicine

Drones can be equipped with cameras and other medical devices to conduct remote consultations and telemedicine sessions. This can be particularly useful for patients who are unable to visit a healthcare facility due to distance, mobility issues, or other factors. Drones can fly to a patient's location and provide a live video feed to a healthcare professional, who can remotely assess the patient's condition and provide medical advice. In 2017, a team of researchers from the University of Maryland and the University of Michigan developed a drone equipped with an automated external defibrillator (AED) that can be dispatched to a cardiac arrest patient's location. The drone can fly at speeds of up to 60 mph and reach a patient within an average of 5.21 minutes, which is faster than the average response time of an ambulance. The AED can then be used by a bystander to deliver an electric shock to the patient's heart, potentially saving their life.

10.5.3 Environmental Monitoring

Drones equipped with sensors and cameras can be used to monitor environmental conditions that may impact public health. For instance, drones can be used to detect air pollution levels, water quality, and other environmental hazards that can affect human health. The data collected by drones can be used by public health officials to develop targeted interventions and policies that can improve public health outcomes. In 2020, researchers from the University of South Australia used drones to monitor the water quality of the Murray River, one of Australia's largest rivers. The drones were equipped with sensors that measured water temperature, pH levels, and other factors that can impact water quality. The data collected by the drones was used to develop a water quality index that can help identify potential sources of pollution and guide environmental management strategies.

10.5.4 Search and Rescue Operations

Drones can be used in search and rescue operations to locate missing persons or victims of natural disasters. Drones equipped with thermal cameras and other sensors can fly over disaster-stricken areas to detect the presence of survivors or victims. This can help rescue teams locate and provide medical assistance to those in need. In 2019, the Los Angeles Fire Department deployed drones to survey a wildfire that had broken out in the city. The drones were equipped with thermal cameras that could detect the presence of hot spots, which helped firefighters locate and contain the fire more efficiently. The drones also provided live video feeds that allowed firefighters to monitor the fire's progress in real time.

10.5.5 Medical Waste Management

Drones can be used to transport medical waste from healthcare facilities to treatment facilities. Medical waste, such as syringes, needles, and other biohazardous materials, needs to be disposed of safely to prevent the spread of infectious diseases and protect public health. Traditional methods of medical waste disposal, such as incineration, can be costly, time-consuming, and harmful to the environment. Drone-assisted medical waste management can provide a faster, more efficient, and safer solution to this problem. Drones can be equipped with specially designed containers to transport medical waste securely and hygienically. These containers are leak-proof, puncture-resistant, and capable of withstanding extreme temperatures. They can also be fitted with GPS tracking devices to monitor their location and ensure secure transport. Using drones for medical waste management can offer several benefits. First, it can reduce the risk of infection for healthcare workers who handle and transport medical waste. By eliminating the need for physical contact with hazardous materials, drones can minimize the risk of exposure to infectious diseases such as HIV, Hepatitis B and C, and COVID-19. Second, drone-assisted medical waste management can save time and money. Drones can transport medical waste faster than traditional methods, reducing the time required for waste disposal and minimizing the associated costs. Moreover, drones can transport medical waste directly from healthcare facilities to treatment facilities, eliminating the need for intermediate steps such as transportation by road or air. Third, drone-assisted medical waste management can be more environmentally friendly than traditional methods. Incineration, for example, can emit harmful pollutants into the air, contributing to air pollution and climate change. By contrast, drones can transport medical waste using renewable energy sources, such as solar power or electric batteries, reducing their carbon footprint and environmental impact. Several organizations are already using drones for medical waste management. In Rwanda, a startup called Zipline is using drones to transport medical supplies, including blood and vaccines, as well as medical waste from remote clinics to treatment facilities. The company has been successful in reducing the time required to transport medical supplies and waste and improving access to healthcare services in remote areas. In the United States, the Federal Aviation Administration (FAA) has granted permission to several companies to test the use of drones for medical waste management. In 2019, a company called Sharps Compliance, Inc. tested a drone-assisted medical waste transport system in Texas, demonstrating the feasibility and safety of using drones for this purpose. In addition to medical waste management, drones can also be used for the delivery of medication to patients in remote or inaccessible areas. This can be particularly beneficial for patients with chronic conditions who require regular medication but have limited access to healthcare services. For example, in 2018, a startup called Vayu, in collaboration with the government of Madagascar, conducted a pilot project to deliver medical supplies and medication to remote villages in the country using drones. The project was successful in improving access to healthcare services and reducing the time required to transport medical supplies. Another potential application of drone technology in healthcare is the delivery of emergency medical supplies and equipment to disaster-stricken areas. Natural disasters, such as earthquakes, hurricanes, and floods, can disrupt healthcare services and cause shortages of essential medical supplies and equipment. Drones can be used to transport emergency medical supplies, such as medicines, bandages, and first aid kits, to disaster-stricken areas quickly and efficiently. Drones can also be used to transport medical equipment, such as ventilators and defibrillators, to disaster sites, enabling healthcare workers to provide immediate care to those in need. In 2017, in the aftermath of Hurricane Maria, a startup called Flirtey partnered with the American Red Cross to deliver emergency medical supplies, including insulin, to residents in Puerto Rico using drones. The project was successful in providing critical medical supplies to those in need and improving access to healthcare services in the aftermath of the hurricane.

In conclusion, drones have the potential to revolutionize healthcare delivery, providing faster, more efficient, and safer solutions for medical waste

10.5.6 Delivery of Medical Supplies and Equipment

Another promising application of drone technology in healthcare is the delivery of medical supplies and equipment. This is particularly relevant in areas with poor infrastructure or limited access to healthcare facilities, where traditional methods of transportation may be inefficient or unavailable. Drones can be used to transport medical supplies such as vaccines, blood products, medications, and medical equipment to remote or hard-to-reach areas quickly and efficiently. In Rwanda, Zipline, a California-based drone delivery company, has partnered with the Rwandan government to deliver medical supplies across the country. Zipline uses autonomous fixed-wing drones to transport blood products, vaccines, and other medical supplies from its distribution centers to health facilities. The company claims that its drones have made more than 20,000 deliveries to date and have helped to reduce the time it takes to deliver medical supplies to remote areas from several days to just a few hours. In the United States, UPS has also been exploring the use of drones for medical supply deliveries. In 2019, the company received Federal Aviation Administration (FAA) approval to operate a drone airline and began delivering medical supplies to a hospital in North Carolina. The program has since expanded to include deliveries to other healthcare facilities and is expected to grow in the coming years.

10.5.7 Drone-Assisted Medical Procedures

In addition to delivering medical supplies and conducting remote consultations, drones can also be used to perform medical procedures. This is particularly relevant in areas with limited access to healthcare facilities or specialists, where drones can be used to transport medical equipment and specialists to the patient. One example of this is the use of drones to transport defibrillators to people experiencing cardiac arrest. In Sweden, the Swedish Transportation Agency has been testing the use of drones to transport defibrillators to cardiac arrest patients. The drones are equipped with a defibrillator and a camera that allows a specialist to provide instructions to the person on how to use it. The aim of the project is to reduce the response time and increase the chances of survival for people experiencing cardiac arrest. Another example is the use of drones to transport medical samples for testing. In Madagascar, the government has partnered with Vayu, a drone delivery company, to transport medical samples for testing. The drones are used to transport blood samples from remote health clinics to a central laboratory, where they can be tested for diseases such as HIV, tuberculosis, and malaria. The use of drones has reduced the time it takes to transport the samples from several days to just a few hours, enabling faster diagnosis and treatment.

10.6 ETHICAL AND REGULATORY CONSIDERATIONS IN DRONE-ASSISTED HEALTH DELIVERY

The integration of drone technology in healthcare delivery presents unique ethical and regulatory considerations. While drones have the potential to revolutionize healthcare delivery, there are concerns surrounding their use, including patient privacy, safety, and security. This section discusses some of the ethical and regulatory considerations associated with drone-assisted health delivery [50–58]:

- **Patient Privacy and Security**—The use of drones in healthcare delivery can raise concerns about patient privacy and security. Drones can capture and transmit images and data, raising concerns about the protection of sensitive patient information. Patient privacy is protected by the Health Insurance Portability and Accountability Act (HIPAA) in the United States, which mandates that healthcare providers take steps to protect patient information. HIPAA requirements apply to all forms of electronic communication, including drone technology.

To protect patient privacy, drones should be equipped with secure transmission technologies, such as encryption, to prevent unauthorized access to sensitive patient data. Drone operators must also ensure that they comply with all applicable privacy regulations and that all data collected by the drone is stored and transmitted securely.

- **Safety**—Safety is another significant concern when it comes to the use of drones in healthcare. Drones can be hazardous to people and property if they malfunction or crash. For this reason, drone operators must follow strict safety guidelines and adhere to all applicable regulations.

  In the United States, the Federal Aviation Administration (FAA) regulates drone operations, and drone operators must obtain a license to operate commercially. The FAA also mandates that drones must fly below 400 feet and that drone operators must maintain visual contact with the drone at all times. These regulations help ensure the safety of people and property in the vicinity of the drone.
- **Regulatory Compliance**—Drone operators must comply with all applicable regulations when using drones in healthcare delivery. In addition to complying with FAA regulations, drone operators must also comply with regulations governing healthcare delivery, such as HIPAA and the Clinical Laboratory Improvement Amendments (CLIA).

  HIPAA regulates the privacy and security of patient information, while CLIA regulates laboratory testing. Healthcare providers using drones for laboratory testing must comply with all CLIA regulations, including quality control, personnel qualifications, and record-keeping requirements.
- **Ethical Considerations**—There are also ethical considerations associated with the use of drones in healthcare delivery. Drones can be used for a variety of purposes in healthcare, including delivering medical supplies, conducting remote consultations, and performing medical procedures. While these applications can improve patient outcomes and access to care, they also raise ethical questions about the role of technology in healthcare.

  One ethical concern is the potential for drones to reduce the human element in healthcare delivery. While drones can provide access to care in remote or underserved areas, they may also reduce the personal connection between patients and healthcare providers. Patients may feel uncomfortable receiving care from a drone, particularly if they are accustomed to receiving care from a human healthcare provider.

  Another ethical concern is the potential for drones to exacerbate existing healthcare disparities. Drones can provide access to care in remote or underserved areas, but they may also reinforce existing disparities by providing care only to those who can afford it. Drones can also exacerbate existing disparities by limiting access to specialized care, particularly in rural areas.

The integration of drone technology in healthcare delivery presents unprecedented opportunities for improved patient outcomes and efficient healthcare delivery.

However, it also presents unique ethical and regulatory considerations. To ensure the safe and responsible use of drone technology in healthcare, drone operators must comply with all applicable regulations, including those governing privacy, safety, and security. They must also consider the ethical implications of using drones in healthcare delivery and strive to balance the benefits of this technology with the need to provide compassionate, human-centered care.

10.7 RESULTS

The utilization of drone technology within the healthcare sector has yielded a multitude of significant findings and insights.

Drone Design: Healthcare drones are equipped with essential components, including an airframe, propulsion system, control system, and payload. In the realm of healthcare applications, it is imperative to prioritize the fundamental aspects of reliability, payload capacity, and durability.

Versatility: Unmanned aerial vehicles (UAVs), commonly known as drones, have the capacity to transport essential medical resources, facilitate remote medical consultations, and augment search and rescue operations within the healthcare sector. These instruments possess a high degree of flexibility, rendering them suitable for utilization across a wide range of healthcare settings.

Benefits: Drones offer numerous benefits within the healthcare sector. The advantages encompass enhanced timeliness and efficacy of response, heightened healthcare accessibility in marginalized regions, diminished financial burdens, improved patient outcomes, fortified safety measures, reduced environmental impact, and bolstered resilience of the healthcare system.

Challenges: The utilization of drone technology in the healthcare sector presents certain limitations. New technologies commonly encounter challenges. The factors encompassing legislative barriers, technological limitations pertaining to range and weight capacity, financial implications, public sentiment, security considerations, and environmental consequences are among the key considerations. In order to achieve successful integration, it is imperative to effectively address the aforementioned challenges.

Applications: Drones have been extensively employed within the healthcare sector to fulfill various functions, including the transportation of medical supplies to geographically isolated regions, enabling remote consultations and telemedicine services, monitoring environmental factors, executing search and rescue operations, and overseeing the management of medical waste. The aforementioned examples demonstrate the potential of Unmanned Aerial Vehicles (UAVs) to enhance the provision of healthcare services.

Ethical and Regulatory Considerations: The utilization of drones within the healthcare sector gives rise to a range of ethical and regulatory concerns. These concerns encompass matters such as safeguarding patient privacy, ensuring safety, and maintaining a patient-centered approach to treatment. In order to safeguard patient privacy, maintain data integrity, and ensure the safety of unmanned aerial vehicles (UAVs), it is imperative to adhere to the regulations set forth by the Health Insurance Portability and Accountability Act (HIPAA) and the Federal Aviation Administration (FAA).

10.8 CONCLUSIONS: THE FUTURE OF DRONE-ASSISTED HEALTH DELIVERY AND ITS POTENTIAL FOR IMPROVING PATIENT OUTCOMES

In conclusion,

- Drone-assisted health delivery is a rapidly growing field that holds great potential for improving patient outcomes, increasing access to care, and reducing costs in the healthcare industry. As this technology continues to develop, it will be important to balance the benefits of drone-assisted health delivery with the need to protect patient privacy and safety, as well as to address regulatory and ethical considerations.
- Drone technology has the potential to revolutionize the delivery of healthcare services, especially in remote and underserved areas. It can be used for remote consultations, delivering medical supplies and equipment, transporting patient samples and medical waste, and even performing medical procedures. The use of drones can lead to faster response times, more efficient use of resources, and improved access to healthcare services for patients who live in remote areas or areas with limited healthcare infrastructure.
- There are also challenges and limitations associated with the use of drone technology in healthcare, such as limited battery life, restricted air space regulations, and concerns about privacy and safety. These challenges need to be addressed through continued innovation and responsible implementation of drone technology.
- Furthermore, ethical and regulatory considerations must be taken into account when implementing drone-assisted health delivery. Healthcare providers must ensure that patient privacy is protected and that they are compliant with local regulations governing the use of drones in healthcare. Additionally, healthcare providers must ensure that drone operators are properly trained and certified to operate drones safely and effectively.
- Despite the challenges and limitations, drone-assisted health delivery is poised to transform the way healthcare services are delivered. As this technology continues to develop, it has the potential to improve patient outcomes, increase access to care, and reduce costs in the healthcare industry. The future of healthcare delivery is exciting and drone technology will undoubtedly play a significant role in shaping it.

In conclusion, drone-assisted health delivery has the potential to revolutionize healthcare delivery and improve patient outcomes, but it must be implemented responsibly, taking into account ethical and regulatory considerations, and addressing the challenges and limitations associated with this technology. As this field continues to evolve, it is important for healthcare professionals, policymakers, and other stakeholders to stay informed and engage in ongoing discussions about the responsible use of drone technology in healthcare.

10.9 FUTURE SCOPE

The utilization of drones in healthcare delivery has the potential to revolutionize the industry. The rapidly expanding sector possesses the capacity to enhance accessibility, patient outcomes, and pricing. In order to maximize the advantages of technological advancements, it is imperative to strike a delicate equilibrium between the progress made and the preservation of patient privacy and safety.

The utilization of drones has the potential to revolutionize the provision of healthcare services, particularly in underserved rural areas. Drones have been found to enhance response times, optimize resource allocation, and enhance accessibility to medical treatment through various applications such as remote consultations, delivery of medical supplies, transportation of samples, and even execution of medical procedures. Nevertheless, in order to successfully navigate the landscape of technological advancements, it is imperative for innovation and judicious utilization to surmount obstacles such as the constraints imposed by limited battery life, airspace limitations, and concerns regarding privacy.

The utilization of drones in health delivery is subject to the influence of both ethical considerations and legal regulations. The preservation of patient confidentiality, adherence to regional legal frameworks, and the acquisition of adequate training and certification for drone pilots are of utmost importance.

Despite the numerous challenges that exist, it is plausible that the utilization of drones has the potential to revolutionize the delivery of healthcare services. As this technology continues to develop, it has the potential to improve patient outcomes, increase accessibility to healthcare services, and reduce costs. The utilization of drones is expected to have a transformative impact on the delivery of healthcare services. It is imperative for healthcare professionals, decision-makers, and stakeholders to maintain active involvement, stay well-informed, and adopt a proactive approach in promoting the safety of drones in the healthcare sector.

REFERENCES

1. Jazieh, A. R., & Kozlakidis, Z. (2020). Healthcare transformation in the post-coronavirus pandemic era. *Frontiers in Medicine*, *7*, 429.

2. Nyaaba, A. A., & Ayamga, M. (2021). Intricacies of medical drones in healthcare delivery: Implications for Africa. *Technology in Society*, *66*, 101624.
3. Euchi, J. (2021). Do drones have a realistic place in a pandemic fight for delivering medical supplies in healthcare systems problems? *Chinese Journal of Aeronautics*, *34*(2), 182–190.
4. Ackerman, E., & Koziol, M. (2019). The blood is here: Zipline's medical delivery drones are changing the game in Rwanda. *IEEE Spectrum*, *56*(5), 24–31.
5. Scott, J. E., & Scott, C. H. (2019). Models for drone delivery of medications and other healthcare items. In *Unmanned aerial vehicles: Breakthroughs in research and practice* (pp. 376–392). IGI Global.
6. Zailani, M. A. H., Sabudin, R. Z. A. R., Rahman, R. A., Saiboon, I. M., Ismail, A., & Mahdy, Z. A. (2020). Drone for medical products transportation in maternal healthcare: A systematic review and framework for future research. *Medicine*, *99*(36).
7. Gera, U. K., Saini, D. K., Singh, P., & Siddharth, D. (2021). IoT-based UAV platform revolutionized in smart healthcare. In *Unmanned aerial vehicles for Internet of Things (IoT) concepts, techniques, and applications* (pp. 277–293). Wiley Online Library.
8. Awad, A., Trenfield, S. J., Pollard, T. D., Ong, J. J., Elbadawi, M., McCoubrey, L. E.,... & Basit, A. W. (2021). Connected healthcare: Improving patient care using digital health technologies. *Advanced Drug Delivery Reviews*, *178*, 113958.
9. Zailani, M. A. H., Sabudin, R. Z. A. R., Rahman, R. A., Saiboon, I. M., Ismail, A., & Mahdy, Z. A. (2021). Drone technology in maternal healthcare in Malaysia: A narrative review. *The Malaysian Journal of Pathology*, *43*(2), 251–259.
10. Bahrainwala, L., Knoblauch, A. M., Andriamiadanarivo, A., Diab, M. M., McKinney, J., Small, P. M.,... & Grandjean Lapierre, S. (2020). Drones and digital adherence monitoring for community-based tuberculosis control in remote Madagascar: A cost-effectiveness analysis. *PLos One*, *15*(7), e0235572.
11. Flemons, K., Baylis, B., Khan, A. Z., Kirkpatrick, A. W., Whitehead, K., Moeini, S.,... & Hawkins, W. (2022). The use of drones for the delivery of diagnostic test kits and medical supplies to remote First Nations communities during COVID-19. *American Journal of Infection Control*, *50*(8), 849–856.
12. Balasingam, M. (2017). Drones in medicine—the rise of the machines. *International Journal of Clinical Practice*, *71*(9), e12989.
13. Bhattacharya, S., Hossain, M. M., Hoedebecke, K., Bacorro, M., Gökdemir, Ö., & Singh, A. (2020). Leveraging unmanned aerial vehicle technology to improve public health practice: Prospects and barriers. *Indian Journal of Community Medicine: Official Publication of Indian Association of Preventive & Social Medicine*, *45*(4), 396.
14. Sharma, K., Singh, H., Sharma, D. K., Kumar, A., Nayyar, A., & Krishnamurthi, R. (2021). Dynamic models and control techniques for drone delivery of medications and other healthcare items in COVID-19 hotspots. In *Emerging technologies for battling COVID-19: Applications and innovations* (pp. 1–34). Springer.
15. Rosser Jr, J. C., Vignesh, V., Terwilliger, B. A., & Parker, B. C. (2018). Surgical and medical applications of drones: A comprehensive review. *JSLS: Journal of the Society of Laparoendoscopic Surgeons*, *22*(3).
16. Masud, U., Saeed, T., Akram, F., Malaikah, H., & Akbar, A. (2022). Unmanned aerial vehicle for laser based biomedical sensor development and examination of device trajectory. *Sensors*, *22*(9), 3413.

17. Ayamga, M., Tekinerdogan, B., & Kassahun, A. (2021). Exploring the challenges posed by regulations for the use of drones in agriculture in the African context. *Land*, *10*(2), 164.
18. Pathak, H., Kumar, G. A. K., Mohapatra, S. D., Gaikwad, B. B., & Rane, J. (2020). Use of drones in agriculture: Potentials, problems and policy needs. *ICAR-National Institute of Abiotic Stress Management*, 4–5.
19. Tatsidou, E., Tsiamis, C., Karamagioli, E., Boudouris, G., Pikoulis, A., Kakalou, E., & Pikoulis, E. (2019). Reflecting upon the humanitarian use of unmanned aerial vehicles (drones). *Swiss Medical Weekly*, *149*(1314), w20065.
20. Vergouw, B., Nagel, H., Bondt, G., & Custers, B. (2016). Drone technology: Types, payloads, applications, frequency spectrum issues and future developments. In *The future of drone use: Opportunities and threats from ethical and legal perspectives* (pp. 21–45). Springer.
21. Garg, P. K. (2021). *Unmanned aerial vehicles: An introduction*. Mercury Learning and Information.
22. Valavanis, K. P., & Vachtsevanos, G. J. (Eds.). (2015). *Handbook of unmanned aerial vehicles* (Vol. 1). Springer.
23. Valavanis, K. P. (Ed.). (2008). *Advances in unmanned aerial vehicles: State of the art and the road to autonomy*. Springer.
24. Fahlstrom, P. G., Gleason, T. J., & Sadraey, M. H. (2022). *Introduction to UAV systems*. John Wiley & Sons.
25. Tsach, S., Tatievsky, A., & London, L. (2010). Unmanned Aerial Vehicles (UAVs). In *Encyclopedia of Aerospace Engineering*. Wiley Online Library.
26. Sanz-Martos, S., Lopez-Franco, M. D., Álvarez-García, C., Granero-Moya, N., López-Hens, J. M., Cámara-Anguita, S.,... & Comino-Sanz, I. M. (2022). Drone applications for emergency and urgent care: A systematic review. *Prehospital and Disaster Medicine*, 1–7.
27. Sabino, H., Almeida, R. V., de Moraes, L. B., da Silva, W. P., Guerra, R., Malcher, C.,... & Passos, F. G. (2022). A systematic literature review on the main factors for public acceptance of drones. *Technology in Society*, 102097.
28. Zailani, M. A. H., Sabudin, R. Z. A. R., Rahman, R. A., Saiboon, I. M., Ismail, A., & Mahdy, Z. A. (2020). Drone for medical products transportation in maternal healthcare: A systematic review and framework for future research. *Medicine*, *99*(36).
29. Ullah, H., Nair, N. G., Moore, A., Nugent, C., Muschamp, P., & Cuevas, M. (2019). 5G communication: An overview of vehicle-to-everything, drones, and healthcare use-cases. *IEEE Access*, *7*, 37251–37268.
30. Moshref-Javadi, M., & Winkenbach, M. (2021). Applications and research avenues for drone-based models in logistics: A classification and review. *Expert Systems with Applications*, *177*, 114854.
31. Moshref-Javadi, M., & Lee, S. (2017). Using drones to minimize latency in distribution systems. In *IIE Annual Conference. Proceedings* (pp. 235–240). Institute of Industrial and Systems Engineers (IISE).
32. Ullah, H., Nair, N. G., Moore, A., Nugent, C., Muschamp, P., & Cuevas, M. (2019). 5G communication: An overview of vehicle-to-everything, drones, and healthcare use-cases. *IEEE Access*, *7*, 37251–37268.
33. Gupta, R., Kumari, A., & Tanwar, S. (2021). Fusion of blockchain and artificial intelligence for secure drone networking underlying 5G communications. *Transactions on Emerging Telecommunications Technologies*, *32*(1), e4176.

34. Jeyabalan, V., Nouvet, E., Meier, P., & Donelle, L. (2020). Context-specific challenges, opportunities, and ethics of drones for healthcare delivery in the eyes of program managers and field staff: A multi-site qualitative study. *Drones*, *4*(3), 44.
35. Ullah, S., Kim, K. I., Kim, K. H., Imran, M., Khan, P., Tovar, E., & Ali, F. (2019). UAV-enabled healthcare architecture: Issues and challenges. *Future Generation Computer Systems*, *97*, 425–432.
36. Sony, S., Laventure, S., & Sadhu, A. (2019). A literature review of next-generation smart sensing technology in structural health monitoring. *Structural Control and Health Monitoring*, *26*(3), e2321.
37. Alzahrani, B., Oubbati, O. S., Barnawi, A., Atiquzzaman, M., & Alghazzawi, D. (2020). UAV assistance paradigm: State-of-the-art in applications and challenges. *Journal of Network and Computer Applications*, *166*, 102706.
38. Yaacoub, J. P., Noura, H., Salman, O., & Chehab, A. (2020). Security analysis of drones systems: Attacks, limitations, and recommendations. *Internet of Things*, *11*, 100218.
39. Wazid, M., Das, A. K., & Lee, J. H. (2018). Authentication protocols for the internet of drones: Taxonomy, analysis and future directions. *Journal of Ambient Intelligence and Humanized Computing*, 1–10.
40. Hiebert, B., Nouvet, E., Jeyabalan, V., & Donelle, L. (2020). The application of drones in healthcare and health-related services in North America: A scoping review. *Drones*, *4*(3), 30.
41. Javaid, M., Haleem, A., Khan, I. H., Singh, R. P., Suman, R., & Mohan, S. (2022). Significant features and applications of drones for healthcare: An overview. *Journal of Industrial Integration and Management*, 2250024.
42. Jeyabalan, V., Nouvet, E., Meier, P., & Donelle, L. (2020). Context-specific challenges, opportunities, and ethics of drones for healthcare delivery in the eyes of program managers and field staff: A multi-site qualitative study. *Drones*, *4*(3), 44.
43. Carrillo-Larco, R. M., Moscoso-Porras, M., Taype-Rondan, A., Ruiz-Alejos, A., & Bernabe-Ortiz, A. (2018). The use of unmanned aerial vehicles for health purposes: A systematic review of experimental studies. *Global Health, Epidemiology and Genomics*, *3*, e13.
44. Nyaaba, A. A., & Ayamga, M. (2021). Intricacies of medical drones in healthcare delivery: Implications for Africa. *Technology in Society*, *66*, 101624.
45. Zailani, M. A., Azma, R. Z., Aniza, I., Rahana, A. R., Ismail, M. S., Shahnaz, I. S.,... & Mahdy, Z. A. (2021). Drone versus ambulance for blood products transportation: An economic evaluation study. *BMC Health Services Research*, *21*, 1–10.
46. Truog, S., Maxim, L., Matemba, C., Blauvelt, C., Ngwira, H., Makaya, A.,... Defawe, O. (2020). Insights before flights: How community perceptions can make or break medical drone deliveries. *Drones*, *4*(3), 51.
47. Comtet, H. E., & Johannessen, K. A. (2022). A socio-analytical approach to the integration of drones into health care systems. *Information*, *13*, 62.
48. Ciasullo, M. V., Orciuoli, F., Douglas, A., & Palumbo, R. (2022). Putting Health 4.0 at the service of Society 5.0: Exploratory insights from a pilot study. *Socio-Economic Planning Sciences*, *80*, 101163.
49. Geronel, R. S., Begnini, G. R., Botez, R. M., & Bueno, D. D. (2022). An overview on the use of unmanned aerial vehicles for medical product transportation: Flight dynamics and vibration issues. *Journal of the Brazilian Society of Mechanical Sciences and Engineering*, *44*(8), 349.

50. Stadler, F., & Tatham, P. (2022). Drone-assisted medicinal maggot distribution in compromised healthcare settings. In *A complete guide to maggot therapy: Clinical practice, therapeutic principles, production, distribution, and ethics* (pp. 383–402). Open Book Publishers.
51. Akhai, S. (2023). From Black Boxes to transparent machines: The quest for explainable AI. Social Science Research Network. Available at: http://dx.doi.org/10.2139/ssrn.4390887.
52. Bone, E., & Bolkcom, C. (2003, April). *Unmanned aerial vehicles: Background and issues for congress*. Library of Congress, Congressional Research Service.
53. Geer, H., & Bolkcom, C. (2005, November). *Unmanned aerial vehicles: Background and issues for congress*. Library of Congress, Congressional Research Service.
54. Merkert, R., & Bushell, J. (2020). Managing the drone revolution: A systematic literature review into the current use of airborne drones and future strategic directions for their effective control. *Journal of Air Transport Management*, *89*, 101929.
55. Pepper, T. (2012). Drones–ethical considerations and medical implications. *Journal of the Royal Naval Medical Service*, *98*(1).
56. Tarr, A. A., Perera, A. G., Chahl, J., Chell, C., Ogunwa, T., & Paynter, K. (2021). Drones—healthcare, humanitarian efforts and recreational use. In *Drone LAW AND POLICY* (pp. 35–54). Routledge.
57. Bhatt, K., Pourmand, A., & Sikka, N. (2018). Targeted applications of unmanned aerial vehicles (drones) in telemedicine. *Telemedicine and e-Health*, *24*(11), 833–838.
58. Johnson, A. M., Cunningham, C. J., Arnold, E., Rosamond, W. D., & Zègre-Hemsey, J. K. (2021). Impact of using drones in emergency medicine: What does the future hold? *Open Access Emergency Medicine*, 487–498.

Chapter 11

Development, Deployment, and Management of IoT Systems

A Software Hypothesis

Shaik Himam Saheb

11.1 INTRODUCTION

The modern world cannot function without software. Computerized systems govern national infrastructures and utilities, and the majority of electrical devices include a computer and control system. Industrial manufacturing and the financial system are also computerized. The entertainment sector, including the music industry, computer games, movies, and television, is heavily reliant on software. As a result, software engineering is critical to the operation of the society. As new software engineering approaches emerge, the demands change, allowing us to create larger and more advanced systems. There is a need to build and deliver systems at a faster pace than before, while also incorporating capabilities that were once considered impossible. Existing software engineering approaches cannot meet these increased expectations, and new software techniques must be developed. The dependence of individuals and society on intricate software systems is growing. Organizations should be able to produce reliable and trustworthy software in a cost-effective and timely manner. The software business is continuously looking for efficient and adaptable solutions to create high-quality software quickly and at a low cost. Agile techniques strive to consistently deliver excellent products and services to their consumers, ensuring a quick return on investment and the flexibility to adapt to changes in client needs over time and permanent environment changes.

Due to the high level of competitiveness in the software industry, software development companies are currently seeking to increase their productivity. These organizations often embrace and establish agile and simplified procedures that enable them to swiftly adapt to business changes. Despite advances in the design of solutions and models, these alone will not yield the optimum outcomes in highly productive environments that require consumers to be served constantly [1].

DevOps has recently acquired prominence in the software development process. DevOps as its name suggests is Developers and Operations teams who work together to develop software. Trust is an important term used by companies when working with their team and making the software build a successful model. Be

DOI: 10.1201/9781003480181-11

it Amazon, Google, or Netflix, everybody looks for trust. Hence to implement DevOps, "Trust" is important. Ever since its inception, several software development trends have been growing, always striving for better practices that ensure the delivery of high-quality products while fulfilling industry standards and customer expectations. Within this evolution, traditional frameworks are continually evolving as a result of the homogeneity of their distinctions, comparison, and integration of best practices. The primary goal of DevOps is to integrate software development and operations into a unified, integrated, automated process, to break down boundaries between Dev and Ops. It is intended from a DevOps perspective to leverage the expertise and knowledge of reliable processes, and technology to foster collaboration and creativity throughout the SDLC process. This is accomplished fast, frequently, and consistently, without diminishing the quality and value supplied to clients, which are distinguishing aspects in agile methodologies, while also boosting some of its benefits [2–5].

IoT (Internet of Things) applications have several objectives that drive their development and implementation across various industries and domains. These objectives are centered around leveraging the connectivity and data-sharing capabilities of IoT devices to achieve specific goals and outcomes. One of the primary objectives of IoT applications is to collect data from sensors, devices, and other sources in real time. This data is then analyzed to extract meaningful insights, patterns, and trends that can inform decision-making, optimize processes, and drive improvements.

The IoT applications aim to automate processes and tasks by enabling devices to communicate and interact with each other without human intervention. This automation leads to increased efficiency, reduced errors, and streamlined operations. IoT devices are often used to monitor the condition of equipment, machinery, and infrastructure. The objective here is to predict when maintenance is needed before a failure occurs, thus minimizing downtime and reducing maintenance costs. IoT applications strive to enhance user experiences by providing personalized, convenient, and seamless interactions with technology. Examples include smart homes, wearable devices, and personalized recommendations based on user data. IoT allows for remote monitoring and control of devices and systems, which is particularly useful in scenarios where physical access is limited or challenging.

This objective is crucial in fields like healthcare, agriculture, and industrial operations. IoT applications aim to optimize the use of resources such as energy, water, and raw materials. By collecting and analyzing data, businesses and organizations can make informed decisions to reduce waste and conserve resources. IoT devices contribute to safety and security objectives by providing surveillance, monitoring, and threat detection capabilities. These devices can enhance physical security, detect anomalies, and provide alerts in case of emergencies. IoT applications can support environmentally sustainable practices by facilitating better resource management and reducing the carbon footprint. Smart grids, waste management systems, and pollution monitoring are examples of how IoT contributes to this objective. IoT-generated data enables organizations to gain insights into customer behavior, operational performance, and market trends. This data-driven approach helps in making informed strategic decisions and staying competitive.

IoT applications aim to provide real-time data that enables timely decision-making. This is crucial in scenarios where immediate action is required, such as in supply chain management or emergency response situations. IoT can lead to the development of innovative products, services, and business models. Organizations can create value by offering IoT-enabled solutions that address new market demands and customer needs. IoT applications often integrate with existing digital systems, such as enterprise resource planning (ERP) and customer relationship management (CRM) systems. This integration improves data flow, coordination, and communication within an organization. The objectives of IoT applications revolve around harnessing the capabilities of connected devices to gather data, improve efficiency, enhance user experiences, enable informed decision-making, and achieve specific outcomes that drive benefits for individuals, businesses, and society as a whole.

11.2 OVERVIEW OF THE HISTORY OF IoT

The history of the IoT is traced back to several key milestones and developments over the years. While the concept of connecting devices and enabling communication between them has been around for decades, the term "Internet of Things" was officially coined in the late 1990s.

Early Concepts (1970s–1980s): The idea of interconnected devices and machines was first explored in the 1970s and 1980s. Researchers and engineers experimented with connecting appliances and industrial equipment to share data and automate processes.

The Birth of the Term "Internet of Things" (1999): The term "Internet of Things" is coined by scientist Kevin Ashton in 1999 while working at Procter & Gamble. He used the term to describe a system where objects could be uniquely identified and tracked through the internet using RFID.

RFID and Wireless Communication (2000s): In the early 2000s, RFID technology gained traction, allowing devices to communicate wirelessly over short distances. This marked a significant step towards the development of IoT applications.

IPv6 Adoption (2000s): The adoption of IPv6, which allows for a vastly expanded number of unique IP addresses, was a critical enabler for IoT growth. With IPv4 addresses running out, IPv6 provided the necessary address space to accommodate the massive number of IoT devices.

Smart Home Appliances and Wearables (2010s): The 2010s saw the rise of consumer-oriented IoT devices such as smart home appliances. These devices offered greater connectivity and interaction with users through mobile apps and cloud services.

Industrial IoT and Industry 4.0 (2010s): The concept of Industrial IoT (IIoT) emerged, focusing on IoT technologies in industrial processes and manufacturing. Industry 4.0, a related concept, highlighted the automation and data exchange in manufacturing technologies.

IoT Standardization Efforts (2010s): Various organizations and alliances worked on standardizing IoT protocols and communication frameworks to ensure interoperability and security. Examples include MQTT, CoAP, and AllSeen Alliance (later merged into the Open Connectivity Foundation).

5G Connectivity (2010s–2020s): The rollout of 5G networks offers higher bandwidth, lower latency, and improved connectivity, which is expected to further accelerate the adoption of IoT applications.

Edge Computing (2010s–2020s): The concept of edge computing gained prominence, enabling data processing and decision-making closer to IoT devices, reducing latency and bandwidth requirements.

Current Trends (2020s): The IoT continues to evolve and expand into various industries, including healthcare, agriculture, transportation, and smart cities. It is becoming an integral part of digital transformation efforts.

The history of IoT is an ongoing narrative, and it is expected to shape the future of technology and connectivity as more devices become connected and intelligent, driving innovation and transforming various aspects of our lives and industries.

11.3 KEY COMPONENTS OF IoT

- Devices and Sensors: IoT devices are attached with various sensors that can gather data related to temperature, humidity, motion, light, and more. These sensors play a crucial role in capturing real-world information.
- Connectivity: IoT devices utilize various technologies like Wi-Fi, Bluetooth, cellular networks, Zigbee, LoRaWAN, and more to connect to the internet and communicate with each other and central systems.
- Data Processing and Data Analytics: Data collected from IoT devices is processed and analyzed either at the edge (on the devices themselves) or in the cloud to derive meaningful insights and trigger appropriate actions.
- Cloud Services: Cloud computing is often used to store and process the vast amounts of data generated by IoT devices. Cloud services enable scalability and provide a centralized platform for data analysis and management.
- Actuators: In addition to sensors, IoT devices may have actuators that allow them to take actions based on the data they receive.
- Machine learning: Machine learning enriches IoT applications by providing the capability to extract valuable insights, make informed decisions, and improve efficiency based on the data generated by IoT devices. The components of IoT are shown in Figure.11.1.

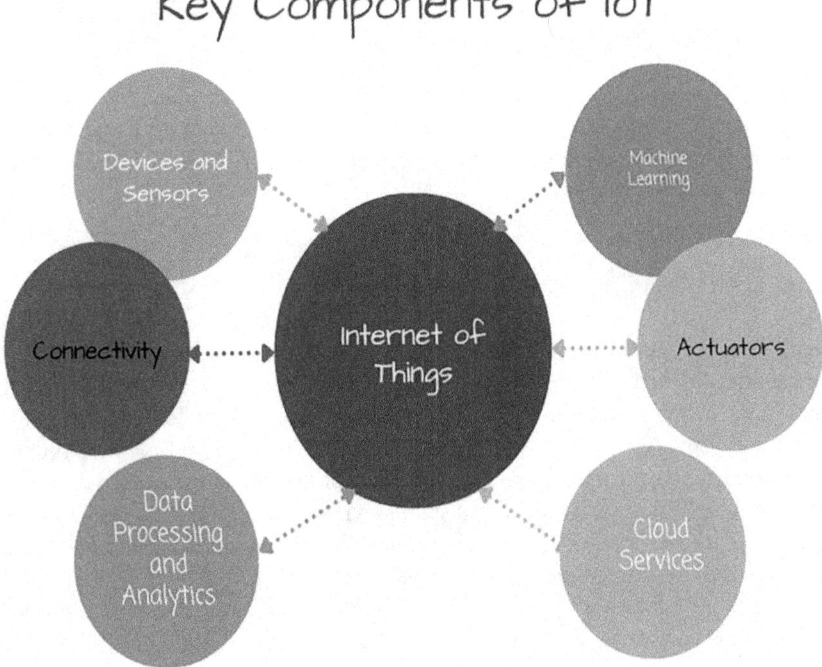

Figure 11.1 Key components of IoT.

11.3.1 Benefits of IoT

- Real-Time Insights: IoT provides real-time data that helps businesses and individuals make data-driven decisions promptly.
- Improved User Experience: IoT devices offer personalized and convenient experiences for users, enhancing comfort and convenience.
- Enhanced Monitoring and Control: IoT helps in remote monitoring and control of devices, assets, and environments, leading to improved safety and security.
- Optimized Resource Management: IoT applications in industries such as agriculture and energy help optimize resource utilization and reduce waste.

11.3.2 Challenges of IoT

- Security and Privacy: The interconnected nature of IoT devices raises concerns about data security, privacy, and potential vulnerabilities.
- Data Overload: The enormous amount of data generated by IoT devices can overwhelm systems and require efficient data management and analysis.

Power Consumption: Many IoT devices operate on batteries, and managing power consumption is essential to prolong battery life and reduce maintenance efforts.

IoT has immense potential to transform industries, improve quality of life, and drive innovation across various domains. However, addressing the challenges and ensuring responsible deployment will be crucial to fully realize the benefits of this rapidly evolving technology.

11.4 DEPLOYMENT OF IoT SYSTEMS

Deploying IoT applications involves making your application available for use by users and devices in a production environment. Here's a step-by-step guide to deploying IoT applications. Choose whether to deploy your IoT application on cloud platforms (e.g., AWS, Azure, Google Cloud) or on-premises infrastructure. Determine the required compute resources, storage, databases, and networking components. Set up cloud resources such as virtual machines, databases, storage buckets, and IoT-specific services. Configure security groups, access controls, and firewall rules to ensure the security of your deployment.

Prepare the hardware infrastructure required for your IoT application, including servers, gateways, and networking equipment. Install and configure necessary software components on the on-premises hardware. Configure IoT devices to connect to the appropriate network and communication protocols. Set up device IDs, authentication credentials, and network settings to enable secure communication. Set up data routing rules to direct incoming data streams from IoT devices to the appropriate processing and storage components. Ensure that data is transmitted securely and efficiently between devices and the cloud/on-premises infrastructure.

Data Storage and Processing deals with the database configuration and data storage solutions to store the incoming data from IoT devices. Set up data processing pipelines to transform, filter, and analyze the collected data. Deploy the backend APIs and services that handle device communication, data processing, and user interactions. Ensure that the APIs are accessible and secure for both devices and users. Deploy web and mobile applications that users will use to interact with IoT devices and access data. Make sure that the interfaces are responsive, accessible, and compatible with various devices and browsers. Conduct thorough testing of your deployed application to ensure that it functions correctly in the production environment. Test the application's performance, scalability, and reliability under real-world conditions.

Implement monitoring tools to track the health, performance, and availability of your IoT application and infrastructure. Set up alerts to notify you of any issues or anomalies that require attention. Plan for scalability by having strategies in place to handle increased device connections, data volume, and user traffic. Consider implementing auto-scaling mechanisms to dynamically allocate resources based on demand. Documentation and user support helps to provide comprehensive documentation for users, administrators, and developers on how to use, maintain, and troubleshoot the deployed IoT application. Offer customer support channels to assist users with any issues they encounter.

Regularly monitor the performance and security of your deployed IoT application. Deploy updates, patches, and new features as needed based on user feedback and changing requirements. Deploying IoT applications requires careful planning, coordination, and expertise in various domains including hardware, software, networking, and security. Continuous monitoring and improvement are essential to ensuring the reliable and effective operation of your IoT deployment.

11.5 MANAGEMENT OF IoT SYSTEMS

Managing IoT applications systems involves overseeing the ongoing operation, maintenance, monitoring, and optimization of your deployed IoT solutions. Effective management ensures that your IoT applications continue to function reliably, securely, and efficiently while meeting user needs. Here's a comprehensive guide to managing IoT applications systems:

11.5.1 Monitoring and Analytics

Implement monitoring tools to track the performance, health, and availability of IoT devices, cloud services, and applications. Set up real-time alerts to notify you of any anomalies or issues that require immediate attention. Utilize analytics to gain insights into device behavior, user interactions, and system usage patterns.

11.5.2 Data Management

Ensure data integrity, accuracy, and compliance with data privacy regulations.

Implement data retention policies and manage data storage to prevent unnecessary accumulation of data.

11.5.3 Device Management

Monitor the health and status of IoT devices, including connectivity, firmware version, and battery levels. Implement remote device management features such as remote configuration, diagnostics, and firmware updates. User Experience Optimization is to collect user feedback and usage data to identify areas for improvement in user interfaces and interactions. Continuously iterate and update UI/UX based on user needs and preferences.

11.5.4 Scalability Performance and Customer Support

Monitor system performance and scalability to ensure that the application can handle increased loads and device connections. Implement auto-scaling mechanisms to dynamically allocate resources as needed. Regularly update software components, APIs, and services to incorporate bug fixes, security patches, and new features. Plan maintenance windows for minimal disruption to users. Provide customer support channels for users to report issues, ask questions, and seek assistance. Establish efficient workflows for issue tracking, resolution, and communication. Ensure that your IoT application complies with relevant industry regulations and data privacy laws. Regularly review and update your application to remain compliant with changing requirements.

11.5.5 Training Collaborations and Documentation

Provide training materials and documentation for users and administrators on how to use and manage the IoT application effectively. Keep documentation up to date as the application evolves. Foster collaboration between development, operations, and support teams to ensure smooth communication and alignment. Regularly share updates and insights about the IoT application's performance and improvements.

11.5.6 Continuous Improvement and Cost Management

Use data-driven insights to identify opportunities for optimization and enhancement in various aspects of the application. Continuously gather user feedback to inform future updates and enhancements. Monitor resource utilization to optimize costs related to cloud services, data storage, and other infrastructure components. Adjust resource allocation based on actual usage patterns to avoid unnecessary expenses. Managing IoT applications systems is an ongoing effort that requires a multidisciplinary approach involving technical expertise, communication skills, and a deep understanding of user needs. Regular monitoring, proactive maintenance, and continuous improvement are key to ensuring the long-term success and value of your IoT deployment.

11.6 MICROSERVICES

Microservices in DevOps and IoT have revolutionized the way software is developed, deployed, and managed. This approach aligns perfectly with DevOps principles, enabling organizations to achieve faster release cycles, improved scalability, and increased flexibility. One of the primary benefits of using microservices in DevOps is the ability to foster agility and continuous delivery. By decomposing an application into smaller, loosely coupled services, development teams can work on different services simultaneously. This parallel development allows for faster iteration and deployment of new features or bug fixes. Each microservice can have its own dedicated development, testing, and deployment pipelines. The detailed microservice architecture is depicted in Figure 11.2, which enables CI/CD practices [6, 7]. DevOps teams can automate the build, test, and deployment processes for each microservice, ensuring that changes can be deployed quickly and reliably. Microservices also promote scalability and resilience in DevOps environments. Microservices, on the other hand, allow for independent scaling of individual services. This means that resources can be allocated specifically to the services experiencing high demand, resulting in improved resource utilization and cost-effectiveness. Moreover, if one microservice fails, the impact is limited to that specific service, as other services continue to operate unaffected. This fault isolation enhances the overall resilience of the system. Another advantage of microservices in DevOps is the flexibility they offer in technology choices.

Each microservice can be developed using different technologies, frameworks, or programming languages that are most suitable for the specific business logic it handles. This flexibility allows development teams to adopt the best tools and technologies for each microservice, optimizing performance and productivity. It also enables organizations to leverage existing systems or integrate with third-party services seamlessly, as microservices can be designed to communicate through well-defined APIs. Despite the benefits, microservices in DevOps also pose challenges. Managing and coordinating multiple servs can become complex, especially when dealing with inter-service communication and data consistency. Organizations need to invest in robust service discovery mechanisms, load balancing, and monitoring solutions to ensure effective management of microservices at scale. Additionally, testing and deploying changes across multiple services

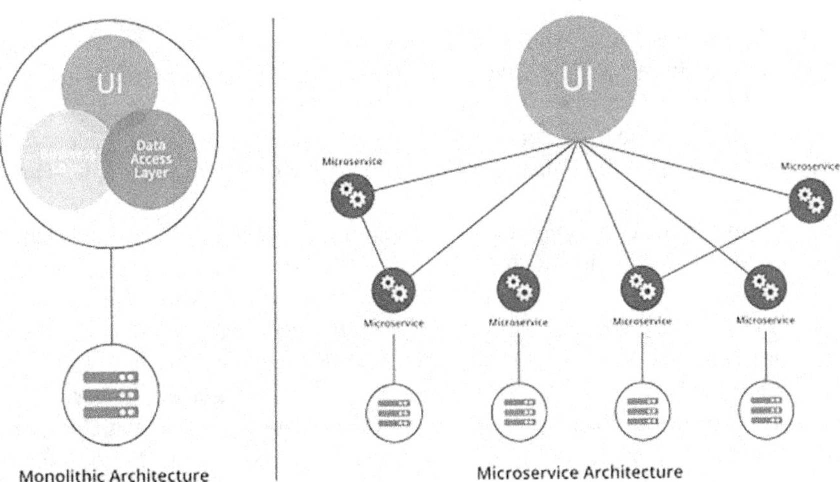

Figure 11.2 Microservice architecture.

require a well-defined strategy to ensure compatibility and avoid regressions. In conclusion, microservices in DevOps have transformed software development and delivery practices. By embracing microservices, organizations can achieve faster release cycles, improved scalability, and increased flexibility. The combination of microservices and DevOps enables teams to work independently, deploy changes rapidly, and build resilient systems. While challenges exist, with proper planning, tooling, and practices, microservices in DevOps can unlock significant benefits for organizations.

11.7 THE RELATIONSHIP BETWEEN IoT AND DevOps

IoT projects often involve complex systems with multiple components, including hardware, embedded software, communication protocols, cloud services, and user interfaces. To effectively manage these diverse elements, a DevOps approach can be beneficial in the following ways [8, 9].

11.7.1 CI/CD

IoT applications typically require frequent updates and improvements. CI/CD practices can automate the building, testing, and deployment of IoT software, making it easier to deliver updates and new features more quickly and reliably. It's a set of practices and principles used in software development to automate and streamline the process of building, testing, and deploying software applications. CI/CD aims to improve the speed, quality, and reliability of software development by automating repetitive tasks, reducing manual intervention, and enabling more frequent releases.

- **Continuous Integration (CI)**: CI involves the process of integrating code changes from multiple developers into a shared repository multiple times a day. Each integration triggers an automated build and testing process to catch integration issues and bugs early. The primary goal is to ensure that code changes are compatible with the existing codebase and do not introduce conflicts or regressions.
- **Continuous Deployment (CD)**: Continuous Deployment is the practice of automatically deploying code changes to production environments after passing through the CI pipeline. This process aims to ensure that validated code changes are quickly and reliably delivered to users without manual intervention. Automated tests, quality checks, and approval processes are typically included in the CD pipeline to maintain high-quality releases.
- **Continuous Delivery (CD)**: Continuous Delivery is closely related to Continuous Deployment but involves a manual decision point before releasing code changes to production. In Continuous Delivery, code changes are automatically built, tested, and prepared for deployment, but they are not

automatically deployed to production. Once the code changes have passed through the pipeline, a human decision-maker can choose when to initiate the deployment to production.

CI/CD practices enable rapid feedback on code changes, helping developers catch and fix bugs early in the development process. Automated testing and deployment reduce the risk of introducing defects and inconsistencies in the codebase. CI/CD encourages smaller, more frequent releases, which can lead to quicker delivery of new features and improvements to users. Automation ensures that the same process is followed for each code change, reducing human error and maintaining consistency. CI/CD encourages collaboration among developers, testers, and operations teams by providing a standardized process for integration and deployment. When issues arise, pinpointing the problematic code changes becomes easier due to the smaller scope of recent changes. CI/CD supports scaling development and deployment processes, making it easier to accommodate larger and more complex projects. Automated testing ensures that code changes meet quality standards and reduces the likelihood of regression bugs. CI/CD tools and platforms, such as Jenkins, Travis CI, CircleCI, GitLab CI/CD, and others, help teams automate and manage these processes, making it easier to integrate, test, and deliver software efficiently and reliably.

11.7.2 Automated Testing and Monitoring

In the IoT realm, there can be numerous devices with different configurations and behaviors. Automated testing is crucial to validate the functionality and compatibility of IoT applications across various devices, ensuring a consistent experience for users. IoT devices generate large amounts of data, and monitoring these devices in real time is essential. DevOps practices can help set up monitoring systems and alerts to identify and respond to issues promptly. IoT systems often involve a combination of cloud services, edge computing, and on-premises infrastructure. IoT allows the infrastructure to be managed and provisioned programmatically, making it easier to maintain consistency and scalability.

11.7.3 Version Control and Security

IoT projects involve multiple stakeholders, including hardware and software teams. Using version control systems helps manage changes to software and hardware designs effectively. Security is a significant concern in the IoT landscape, given the potential vulnerabilities of connected devices. DevOps practices can help embed security into the development process, ensuring that security considerations are addressed from the outset.

IoT and DevOps share a symbiotic relationship, where DevOps practices can enhance the development, deployment, and management of IoT applications, leading to more efficient and reliable IoT systems.

11.8 PROS AND CONS OF DEVOPS PRACTICES IN IOT

- **Faster Time-to-Market**: DevOps fosters a culture of continuous integration and deployment, enabling faster development cycles. This agility is crucial in the rapidly evolving IoT landscape, where timely product releases are essential to stay competitive.
- **Improved Collaboration**: DevOps encourages closer collaboration between development and operations teams. In the context of IoT, this collaboration is vital, as it helps align product requirements with operational considerations and ensures a smoother development process.
- **Automated Testing and Deployment (ATD)**: Automation is a core principle of DevOps, and it plays a significant role in IoT. ATD can help detect and resolve issues more efficiently, reducing the risk of errors and downtime in IoT applications.
- **Enhanced Scalability**: IoT deployments often involve a high number of devices and data streams. DevOps practices can help scale IoT applications efficiently, ensuring they can handle increasing workloads and device connectivity.
- **Continuous Monitoring and Improvement**: DevOps promotes continuous monitoring of applications in production. This ongoing feedback loop allows for timely identification of performance issues or anomalies, leading to quicker resolution and continuous improvement of IoT solutions.
- **Security Integration**: Security is a critical concern in IoT, and DevOps practices can help integrate security measures into the development process from the outset. This proactive approach can minimize vulnerabilities and reduce the risk of security breaches.
- **Skillset Requirements**: DevOps practices often require a diverse skillset that combines development, operations, and automation expertise. Organizations may need to invest in upskilling their teams or hiring individuals with the necessary skills, which can be time-consuming and costly.
- **Complexity and Tooling**: The adoption of DevOps practices involves the implementation of various tools and technologies for automation, continuous integration, deployment, and monitoring. Managing and integrating these tools can introduce complexity and require additional effort in terms of setup, maintenance, and training.

11.8.1 Cons of Adopting DevOps Practices in IoT

- **Complex Ecosystem**: IoT solutions often involve a complex ecosystem of hardware, firmware, software, and cloud services. Coordinating the integration of these diverse components within a DevOps workflow can be challenging.
- **Resource Intensive**: Implementing DevOps in IoT requires investments in automation tools, infrastructure, and skilled personnel. Small IoT projects

or organizations with limited resources may find it challenging to adopt full-scale DevOps practices.

- **Hardware Constraints**: Many IoT devices have limited computing power and memory. Implementing complex DevOps tools and processes on resource-constrained devices can be difficult and may impact performance.
- **Security Risks**: While DevOps can enhance security, it also introduces new risks if not properly managed. Continuous integration and deployment could lead to the unintended introduction of vulnerabilities if security testing and validation are not adequately integrated.
- **Standardization Challenges**: The lack of standardized protocols and platforms in the IoT space can make it challenging to implement consistent DevOps practices across different IoT projects and ecosystems.
- **Regulatory Compliance**: IoT applications may have to comply with specific regulations and standards, such as data privacy and safety requirements. Aligning DevOps practices with compliance needs can be complex and time-consuming. Adopting DevOps practices in IoT can lead to faster development, improved collaboration, and scalability. However, it also presents challenges related to complexity, resource requirements, security, and standardization. Organizations considering DevOps in IoT should carefully assess their specific needs, resources, and potential risks before implementation.
- **Security Considerations**: With the increased emphasis on automation and rapid deployments, security can be a concern. It is crucial to ensure that security practices are embedded throughout the DevOps pipeline and that vulnerabilities are identified and addressed early in the process.

It's significant to note that the pros and cons of adopting DevOps practices can vary depending on the organization, its specific context, and the maturity of its software delivery processes. Successful adoption needs careful planning, strong leadership, and a commitment to continuous improvement [10].

11.9 WHY USE DevOps INSTEAD OF THE TRADITIONAL METHODS

DevOps is often preferred over conventional software development and operations methods due to several compelling reasons. By bringing these teams together, DevOps promotes shared responsibilities and fosters a culture of transparency and accountability. This collaborative approach leads to faster decision-making, reduced friction between teams, and improved overall efficiency.

Moreover, DevOps places a strong emphasis on software quality, an aspect often overlooked.

Secondly, DevOps leverages automation to streamline and accelerate the software development lifecycle. Traditional methods often involve manual and time-consuming processes, such as manual testing, deployment, and configuration management.

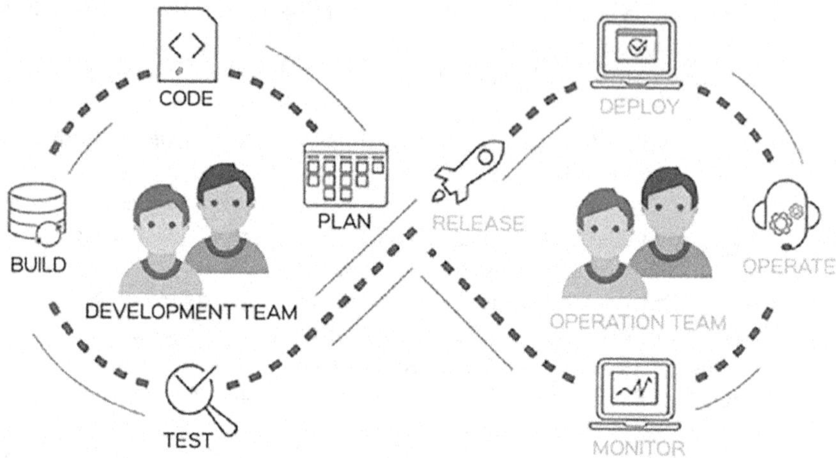

Figure 11.3 DevOps methods vs traditional methods.

DevOps practices automate these processes, enabling rapid and repeatable deployments, automated testing, and infrastructure provisioning. Automation reduces human error, improves consistency, and allows teams to focus on higher-value tasks. Ultimately, DevOps empowers organizations to deliver software faster, with improved quality and stability Figure 11.3 explains how DevOps differs from traditional IT and why. DevOps is preferred over traditional methods due to its emphasis on collaboration, communication, and automation. By breaking down silos and promoting cross-functional teams, DevOps fosters a culture of collaboration and accelerates decision-making. The automation of repetitive tasks and the implementation of CI/CD pipelines enable faster, more reliable software delivery. With these advantages, DevOps helps organizations stay competitive in today's fast-paced and rapidly changing software development landscape.

11.10 AI AND IoT

Artificial intelligence (AI) and the Internet of Things (IoT) are two transformative technologies that are often used together to enhance the capabilities of IoT applications and systems. When AI and IoT are combined, they create powerful, intelligent, and data-driven ecosystems that can generate valuable insights, automate processes, and improve overall efficiency [11].

11.10.1 Data Processing and Predictive Maintenance

IoT devices generate massive amounts of data from sensors and connected devices. AI algorithms can analyze and process this data in real time to extract meaningful patterns and insights. AI can identify anomalies, predict trends, and detect patterns that may not be apparent through traditional data analysis. AI can enable predictive maintenance in IoT systems. By analyzing historical data and sensor readings, AI algorithms can anticipate potential equipment failures or maintenance needs. This proactive approach helps avoid costly downtime and optimize maintenance schedules.

11.10.2 Machine Learning

AI-powered machine learning models can be used to learn the behavior of IoT devices over time. These models can detect deviations from normal behavior, indicating possible malfunctions or security breaches. AI enables intelligent automation in IoT applications. It can process data and make decisions in real time, allowing IoT systems to respond to changing conditions without human intervention. AI-driven NLP allows IoT devices to understand and respond to voice commands, enabling seamless interaction with smart home devices, virtual assistants, and other IoT applications. AI can analyze user data from IoT devices to offer personalized experiences, such as customized recommendations or tailored user interfaces.

11.10.3 Edge AI

Edge computing, which involves processing data closer to the foundation (on the IoT devices themselves or on edge servers), combined with AI enables faster response times and reduces the need for constant data transmission to the cloud. AI can optimize energy consumption in IoT applications by intelligently managing devices based on usage patterns and demand. AI can bolster IoT security by identifying abnormal behavior and potential security threats in real time, helping prevent cyberattacks.

11.10.4 Data Fusion and Traffic Optimization

AI can fuse data from multiple IoT devices and sources to gain a comprehensive view of a situation or environment. AI can optimize traffic management in smart cities by analyzing data from connected vehicles and infrastructure to improve traffic flow and reduce congestion.

The integration of AI and IoT enables a symbiotic relationship, where AI augments the capabilities of IoT devices and IoT provides valuable data to train and improve AI models. As both technologies continue to advance, their synergy will lead to even more innovative and transformative applications in various industries, from healthcare and manufacturing to transportation and agriculture. However, ethical considerations, data privacy, and ensuring secure AI-driven IoT applications are essential aspects to address as these technologies become more pervasive in our daily lives.

11.11 MANAGEMENT OF IoT APPLICATIONS

Managing IoT applications involves a range of activities aimed at ensuring the successful development, deployment, operation, and maintenance of these applications. Due to the complex nature of IoT systems, managing them requires careful planning, coordination, and ongoing monitoring. Here are some key aspects of managing IoT applications including the objectives and requirements of your IoT application. Identify the devices, sensors, and components that will be part of the IoT ecosystem. Design the architecture, data flow, communication protocols, and security measures. Set up a system to provision and onboard new IoT devices. Monitor the health and status of devices to detect anomalies or failures. Implement remote device management for software updates, configuration changes, and troubleshooting. Develop a strategy for collecting, storing, and processing the data generated by IoT devices. Implement data analytics to extract valuable insights and trends from the collected data. Ensure data security, privacy, and compliance with relevant regulations.

11.11.1 Security and Privacy

Implement strong security measures to protect IoT devices and the data they transmit. Use encryption, authentication, and authorization mechanisms to prevent unauthorized access. Regularly update and patch devices to address security vulnerabilities. Choose the appropriate communication protocols and technologies for connecting devices. Manage network infrastructure to ensure reliable and efficient communication. Plan for scalability as the number of connected devices grows. Develop and test applications that make use of the IoT data and capabilities. Implement automated testing and continuous integration practices to maintain code quality. Test how applications behave in different scenarios and conditions. Ensure that IoT devices and applications can seamlessly integrate with other systems and platforms. Implement standards and protocols that enable interoperability between different devices and vendors.

11.11.2 Deployment and Operations

Plan the deployment strategy, considering factors such as device placement and coverage. Monitor the performance, uptime, and health of the IoT application and devices. Set up alerting mechanisms to notify teams about issues or anomalies. Design the IoT system to handle increased loads as the number of devices and users grows. Optimize performance to ensure responsive user experiences and timely data processing. Regularly update and patch both the IoT devices and the software applications. Perform periodic maintenance to prevent degradation of performance or security. Prioritize user experience by designing intuitive interfaces and responsive applications. Provide user support for troubleshooting, inquiries, and assistance with using the IoT application. Managing IoT applications requires collaboration among cross-functional teams, including software developers, hardware engineers, data scientists, security experts, and operations personnel. It's an ongoing process that involves continuous improvement and adaptation to technological advancements and changing user needs.

11.12 THE INTERNET OF THINGS ADVANTAGES

IoT offers a wide range of advantages and benefits that have the potential to transform industries, improve efficiency, and enhance the quality of life. IoT enables automation of various tasks and processes, reducing the need for manual intervention. This leads to increased efficiency, cost savings, and optimized resource utilization. IoT devices continuously collect and transmit data in real time. This data can be analyzed to gain valuable insights, enabling data-driven decision-making and better understanding of processes and environments. IoT allows for remote monitoring and control of devices and systems. This capability is particularly valuable in industries like healthcare, where remote patient monitoring can improve patient outcomes and reduce hospitalization. IoT facilitates predictive maintenance by monitoring the condition of equipment and machinery in real time. This helps in identifying potential issues early and scheduling maintenance before a breakdown occurs, reducing downtime and costs.

IoT applications can provide personalized and seamless user experiences. Smart home devices, wearables, and personalized recommendations are some examples of how IoT enhances user experiences. IoT technologies can optimize energy consumption by monitoring and controlling energy usage in buildings, factories, and homes. This leads to reduced energy wastage and lower utility costs. IoT enables precision agriculture by collecting data on soil conditions, weather, and crop health. This data-driven approach optimizes irrigation, fertilization, and pest control, leading to increased crop yields and sustainable farming practices. IoT plays a significant role in creating smart cities. Connected sensors and devices in urban infrastructure enhance traffic management, waste management, public safety, and energy distribution. IoT improves supply chain management by providing real-time tracking and visibility of goods and shipments. This helps in reducing delays, minimizing losses, and improving overall efficiency. IoT applications in healthcare lead to better patient care and health outcomes. Wearable devices and remote monitoring solutions enable continuous health monitoring and early detection of health issues.

IoT can be used to monitor and manage environmental factors, such as air quality, water usage, and waste management. This data-driven approach helps in sustainable environmental practices. IoT enhances safety and security in various settings. Smart surveillance systems, connected alarms, and monitoring devices help prevent accidents and improve overall security. IoT data provides valuable insights into customer behavior, market trends, and product performance, enabling businesses to make data-driven strategic decisions. The advantages of IoT are vast and diverse, with the potential to revolutionize industries, enhance quality of life, and address global challenges. As IoT technologies continue to evolve and become more prevalent, their impact is expected to grow significantly, touching almost every aspect of modern living.

11.12.1 Conclusions

IoT application development involves designing and implementing software solutions that connect, collect data from, and interact with IoT devices. Integration of hardware, firmware, and software components requires collaboration between multidisciplinary teams, including software developers, hardware engineers, and UI/UX designers. Deploying IoT applications involves setting up the necessary infrastructure, cloud services, and connectivity for IoT devices to operate effectively. Consideration should be given to scalability, security, and performance when deploying IoT applications to ensure they can handle increasing device connections and data volume. Effective management of IoT applications is essential for ensuring their continued operation, security, and optimization. Ongoing monitoring of device health, data flow, and system performance helps identify and address issues promptly. IoT applications generate large volumes of data that can provide valuable insights when analyzed using data analytics and machine learning. Implementing robust authentication, encryption, and access controls is crucial to prevent breaches and maintain user trust. IoT applications may need to adhere to industry-specific regulations and data privacy laws, necessitating careful consideration of compliance requirements. Effective collaboration between development, operations, support, and business teams is essential for successful development, deployment, and management of IoT applications.

REFERENCES

1. Evans, D. S., Hagiu, A., & Schmalensee, R. (2016). Invisible Engines: How Software Platforms Drive Innovation and Transform Industries. Available at: https://ssrn.com/abstract=2747032
2. Hamzehloui, M. S., Sahibuddin, S., & Ashabi, A. (2019). A study on the most prominent areas of research in microservices. International Journal of Machine Learning and Computing, 9(2), 242–247.
3. Chen, S., Xu, H., Liu, D., Hu, B., & Wang, H. (2014). A vision of IoT: Applications, challenges, and opportunities with China perspective. IEEE Internet of Things Journal, 1(4), 349–359.
4. Al-Saqqa, S., Sawalha, S., & AbdelNabi, H. (2020). Agile software development: Methodologies and trends. International Journal of Interactive Mobile Technologies, 14(11).
5. Kaim, R., Härting, R. C., & Reichstein, C. (2019). Benefits of agile project management in an environment of increasing complexity—a transaction cost analysis. In Intelligent Decision Technologies 2019: Proceedings of the 11th KES International

Conference on Intelligent Decision Technologies (KES-IDT 2019), Vol. 2, pp. 195–204. Springer, Singapore.

6. Alnafessah, A., Gias, A. U., Wang, R., Zhu, L., Casale, G., & Filieri, A. (2021). Quality-aware DevOps research: Where do we stand? IEEE Access, 9, 44476–44489.
7. Baptista, G., & Abbruzzese, F. (2020). Software Architecture with C# 9 and. NET 5: Architecting software solutions using microservices, DevOps, and design patterns for Azure. Packt Publishing.
8. Szmeja, P., Fornés-Leal, A., Lacalle, I., Palau, C. E., Ganzha, M., Pawłowski, W.,... & Schabbink, J. (2023). ASSIST-IoT: A modular implementation of a reference architecture for the next generation Internet of Things. Electronics, 12(4), 854.
9. Russell, B., & Van Duren, D. (2016). Practical Internet of Things security. Packt Publishing.
10. Igor Muzetti Pereira, Tiago Carneiro, and Eduardo Figueiredo. 2021. A systematic review on the use of DevOps in internet of things software systems. In Proceedings of the 36th Annual ACM Symposium on Applied Computing (SAC '21). Association for Computing Machinery, New York, NY, USA, 1569–1571. https://doi.org/10.1145/3412841.3442126.
11. Sarker, I. H. (2022). Smart City Data Science: Towards data-driven smart cities with open research issues. Internet of Things, 19, 100528.

Chapter 12

Software-Defined Network's Security Concerns

A Meticulous Synopsis

Ashvini Pradeep Shende, Vijaya Kumbhar, Parag Tamhankar, Yudhishthir Raut, and Anirudh Mangore

12.1 INTRODUCTION

Software-defined networking is a type of ***network design***. ***Control plane*** is one of the important components of SDN, and grants ***centralized network management*** and ***programmability***. The control plane is abstracted from the underlying infrastructure and installed in a software-based controller, while the data plane remains in the devices of the network such as switches and routers [1–6].

Figure 12.1 depicts the SDN in pictorial form.

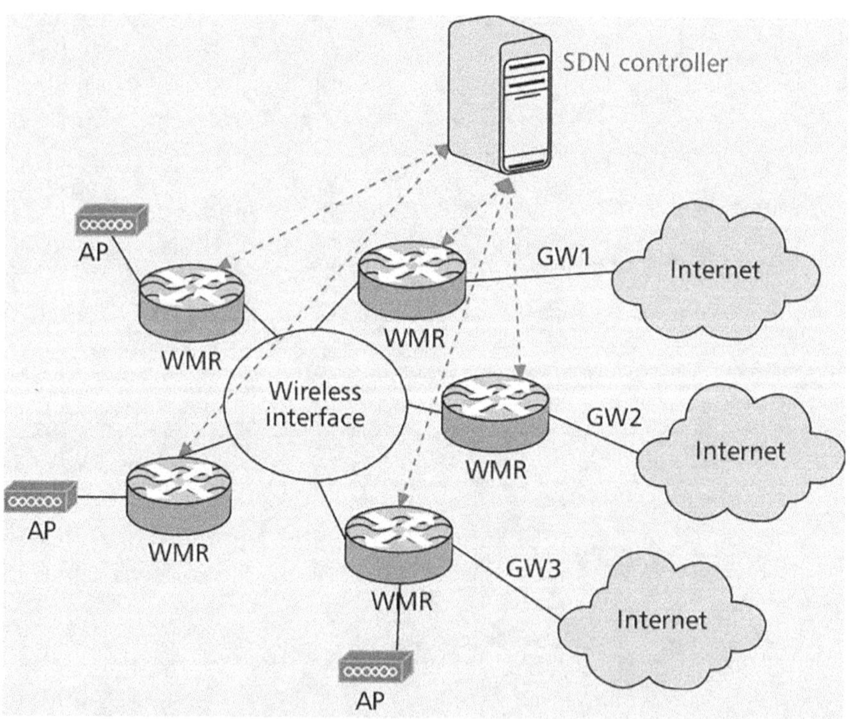

Figure 12.1 Software-defined network [7].

DOI: 10.1201/9781003480181-12

SDN provides several benefits, including ***improved network agility***, ***scalability***, and ***automation***, as well as ***better network visibility*** and ***control***. The technology has found applications in various domains, including ***data centres***, ***cloud computing***, and ***telecommunications*** [4, 5].

12.2 EVOLUTION OF SDN

Over the years, there has been significant development in software-defined networking (SDN) to overcome the drawbacks of conventional network architectures [8–12]. The progression of SDN can be categorized into three main stages: the ***initial*** stage, the ***advanced*** stage, and the ***forthcoming*** stage.

In the ***initial stage of SDN***, the primary emphasis was on exploring the core principles of SDN, which involved separating the control and data planes, enabling programmability, and centralizing control. Researchers dedicated their efforts to designing architectures, protocols, and tools that could effectively realize these concepts. Notable examples of early SDN research projects include OpenFlow, NOX, and ONOS.

During the ***advanced stage of SDN***, the primary objective was to put SDN into practical use within real network environments. Extensive research efforts were dedicated to creating solutions that addressed scalability, reliability, security, and interoperability challenges. Additionally, this phase witnessed the rise of network function virtualization (NFV) as a complementary technology to SDN. NFV played a significant role in enhancing the capabilities and functionalities of SDN during this period.

Examples of mature SDN research include Open Daylight, Floodlight, and Open vSwitch.

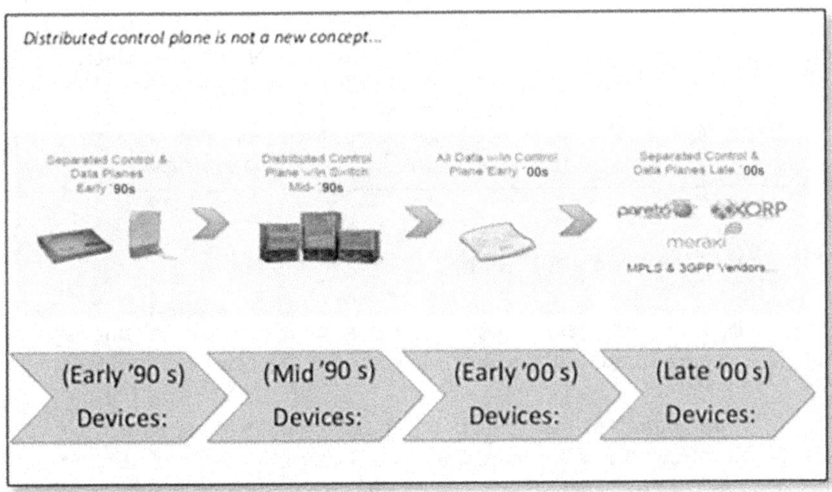

Figure 12.2 Evolution of SDN [13].

The ***forthcoming stage of SDN*** is characterized by the integration of SDN and NFV, and the adoption of SDN in emerging technologies such as edge computing and 5G networks. Research efforts in this phase are aimed at developing solutions for network slicing, service chaining, mobility management, and security. Examples of future SDN research include SONATA, 5G-TRANSFORMER, and 5G-EmPOWER.

12.3 TIMELINE OF SDN

SDN has its origins in the early 2000s, but it wasn't until the development of the OpenFlow protocol in 2008 that SDN began to take shape as a distinct technology. Since then, SDN has evolved rapidly, and its timeline can be roughly divided into several phases [9, 13–17]:

Table 12.1 Timeline and evolution of SDN [Compiled by Author]

SDN Evolution Phase	*Time Period*	*Technological Evolution in SDN*
Early SDN	2008–2012	• Development of the OpenFlow protocol • Introduction of the concept of SDN • Focus on developing proof-of-concept implementations
Maturing SDN	2012–2016	• SDN technology matured and more widely adopted • Focus on developing more sophisticated SDN controllers • Improved security and scalability of SDN networks • SDN usage in real-world applications started
SDN and NFV Integration	2016–2020	• Integration of SDN with network function virtualization (NFV) • Development of new architectures and standards to support their use together
Future SDN	2020 and beyond	• Further integration of SDN with other emerging technologies such as artificial intelligence, blockchain, and 5G networks • SDN support for new use cases like edge computing, Internet of Things (IoT), and smart cities

12.4 COMPONENTS OF SDN

The process of SDN is carried out at three levels: data plane, control, and application plane. Each of these fulfils a vital role within an SDN system [14, 15, 19–21].

- The ***data plane's*** primary function is to facilitate the forwarding of network traffic. It comprises switches and routers that are configured to carry out forwarding tasks according to instructions received from the control plane.

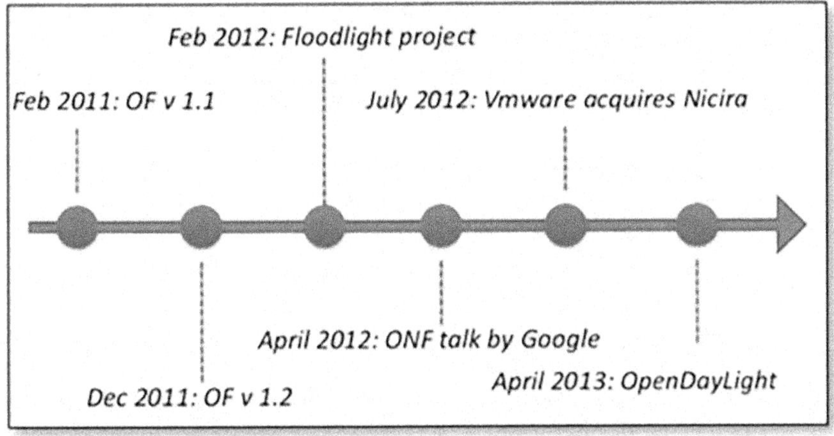

Figure 12.3 Timeline of SDN [18].

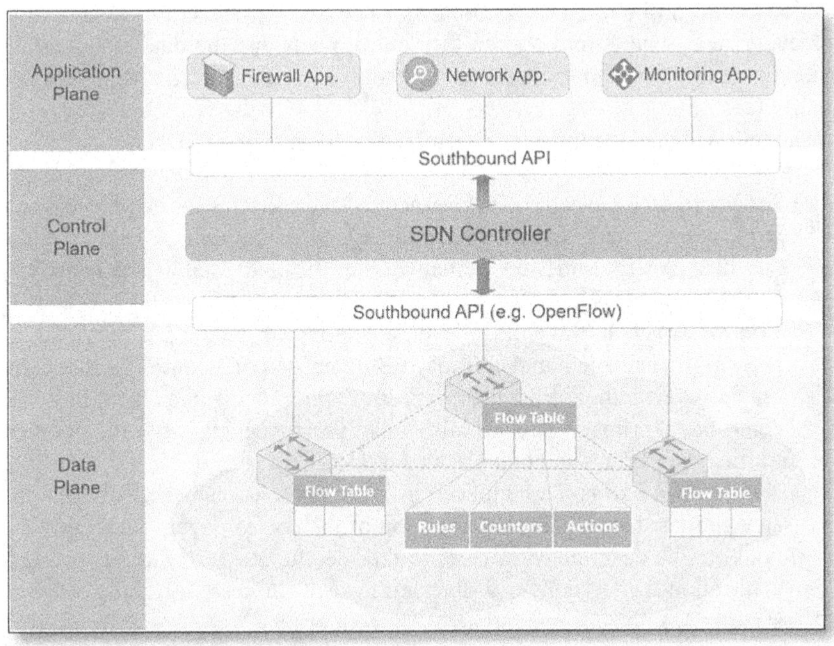

Figure 12.4 Components of SDN [19].

Typically, the data plane is executed using standard hardware, such as Ethernet switches.

- The ***control plane's*** primary role is to oversee and regulate the operations of the data plane. It encompasses a centralized controller that takes charge of programming the switches and routers within the data plane. Communication between the controller and data plane occurs through the utilization of the OpenFlow protocol, enabling the controller to define the specific forwarding behaviour for each packet. Tasks handled by the control plane include routing, traffic engineering, and quality of service (QoS) management.
- The ***application plane*** offers an elevated interface to the SDN, comprising applications that are constructed upon the control plane. These applications can deliver diverse network services like load balancing, security, and network monitoring. The application plane can be tailored and adapted to suit the specific requirements of individual network applications.

12.5 ARCHITECTURE OF SDN

The architectural perspective of SDN pertains to the arrangement and interconnection of the different SDN components. The SDN architecture is devised to achieve a clear separation between the control plane and the data plane, establishing a centralized and programmable interface for network management and control [15, 21].

The following are the main architectural components of an SDN:

a) ***Infrastructure Layer***: The infrastructure layer pertains to the physical network devices, such as switches, routers, and access points. These devices facilitate the transmission of network traffic and establish connectivity between end hosts.
b) ***Control Layer***: In the control layer, a centralized interface is provided for network management and control. It is composed of a controller that communicates with the infrastructure layer using protocols like OpenFlow. The controller's primary responsibility involves managing network policies, traffic engineering, and other related tasks.
c) ***Application Layer***: The application layer involves network services and applications that are developed on top of the control layer. These applications can be customized to cater to the specific needs of various network applications. They deliver a diverse range of services, including network security, load balancing, and network monitoring.

Understanding the architectural perspective of SDN is crucial as it enables the creation of networks that possess greater flexibility, scalability, and manageability compared to traditional networks. Through the separation of the control and data

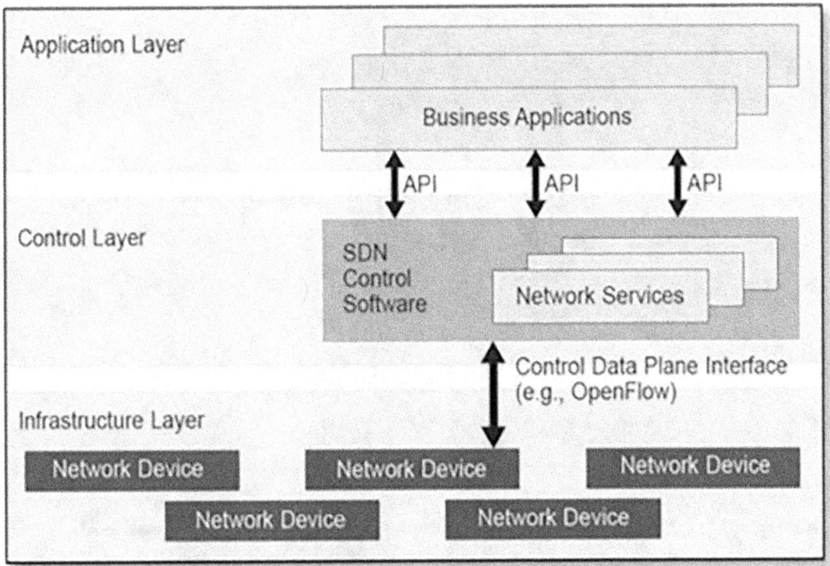

Figure 12.5 Architectural view of SDN [22].

planes, SDN facilitates centralized network management and control, simplifying the deployment of new applications and services.

12.6 WORKING OF SDN

The operation of SDN is centred around the segregation of the control plane and data plane within a network. This division facilitates centralized control and management of the network, thereby enhancing its flexibility and scalability [15, 21, 23].

The following outlines the fundamental steps involved in the functioning of SDN:

a) ***Traffic Forwarding***: Within the SDN architecture, the data plane consists of network devices such as switches and routers that are responsible for forwarding network traffic between end hosts.
b) ***Control Plane Communication***: The control plane comprises a centralized controller that communicates with the network devices using protocols like OpenFlow. The controller assumes the responsibility of managing network policies and forwarding rules.
c) ***Network Policies***: The controller governs network policies that dictate how traffic should be directed between different segments of the network. These

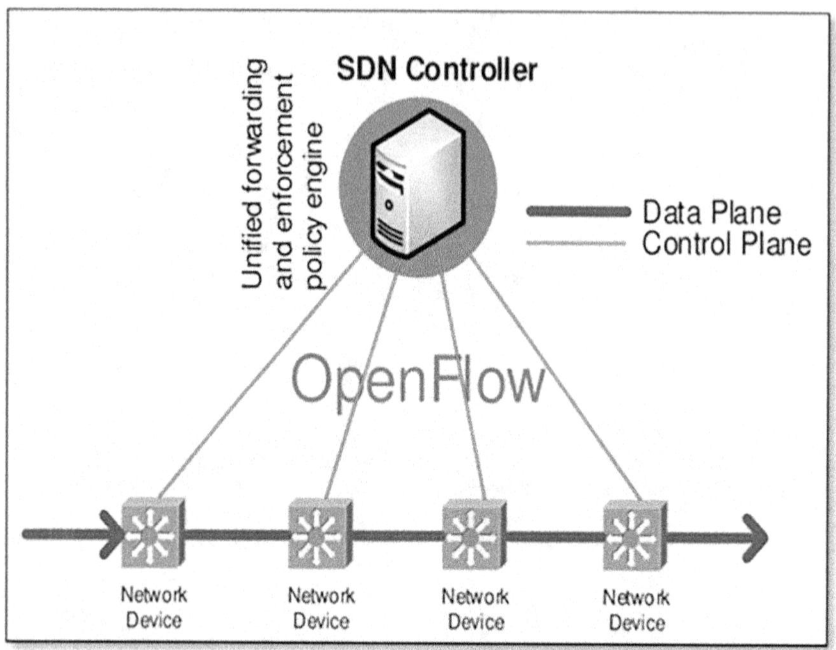

Figure 12.6 Working of SDN [24].

policies are defined using a high-level programming language specific to the SDN controller.

d) ***Flow Tables***: The controller communicates with the network devices to install forwarding rules in their flow tables. These rules determine how traffic should be directed based on the network policies.

e) ***Traffic Steering***: When network traffic reaches a network device, the device consults its flow table to determine the appropriate forwarding path for the traffic. In the absence of a matching rule in the flow table, the device forwards the traffic to the controller for further instructions.

f) ***Network Management***: The SDN architecture facilitates centralized network management and control, simplifying the deployment of new network applications and services. The controller can be leveraged for tasks such as network traffic monitoring, network policy management, and detection and mitigation of network security threats.

Hence, the functioning of SDN is designed to establish a more adaptable and scalable network architecture capable of accommodating evolving network conditions and supporting new network applications and services.

12.7 SOFTWARE-DEFINED NETWORK SECURITY

SDN has separate control plane and data plane. It enables a more flexible network for programmable network management. However, this separation introduces new security considerations. Here are some security concerns related to SDN:

- ***Controller Security***: The centralized controller is a critical component of the SDN architecture, and any compromise of the controller can potentially lead to a security breach. Controllers can be susceptible to attacks like denial of service, SQL injection, and buffer overflow [25].
- ***Data Plane Security***: As control plane and data plane are split, it can give rise to security issues within the data plane. Attackers may exploit vulnerabilities in network devices such as switches and routers to bypass security policies set by the controller [26].
- ***Virtualization Security***: SDN allows for the virtualization of network functions, introducing new security considerations. Virtual switches and network functions can be vulnerable to attacks like hypervisor attacks and side-channel attacks [27].
- ***Inter-controller Communication Security***: In a multi-controller SDN environment, the communication between controllers can be susceptible to attacks such as man-in-the-middle attacks and eavesdropping [28].
- ***Policy Management Security***: SDN enables fine-grained policy management, but this also introduces security concerns such as policy conflicts, misconfiguration, and policy manipulation [29].

It is essential to address these security concerns to ensure the robustness and integrity of SDN deployments. Implementing appropriate security measures, conducting regular vulnerability assessments, and employing strong access controls are some of the strategies to mitigate these risks and safeguard SDN environments.

12.7.1 Research Perspectives of SDN Security

Adoption and deployment of SDN have been hindered by security concerns.

Following are some research perspectives on SDN's security:

- ***Threat Detection and Prevention***: One of the key challenges in SDN security is detecting and preventing attacks. Researchers are developing techniques such as machine learning, deep learning, and artificial intelligence to detect and prevent attacks in real time [30].
- ***Security Policies and Management***: Researchers are working on developing more efficient and effective security policies and management frameworks for SDN. This includes techniques such as automated policy generation, policy verification, and policy enforcement [31].

- ***Secure Communication Protocols***: The communication protocols used in SDN need to be secure and reliable to prevent attacks such as eavesdropping and man-in-the-middle attacks. Researchers are working on developing more secure communication protocols for SDN, such as the use of Transport Layer Security (TLS) and Secure Socket Layer (SSL) [32].
- ***Authentication and Access Control***: SDN networks require authentication and access control mechanisms to prevent unauthorized access and malicious activities. Researchers are developing techniques such as Public Key Infrastructure (PKI), Lightweight Directory Access Protocol (LDAP), and Role-Based Access Control (RBAC) for secure authentication and access control in SDN [33].
- ***Resilience and Survivability***: SDN networks need to be resilient and able to survive attacks and failures. Researchers are working on developing techniques such as fault tolerance, redundancy, and backup and recovery mechanisms to ensure the resilience and survivability of SDN networks [34].

Overall, the research perspectives on SDN's security are focused on developing effective techniques for threat detection and prevention, security policies and management, secure communication protocols, authentication and access control, and resilience and survivability. These techniques will be crucial for ensuring the security and reliability of SDN networks.

12.7.2 Research Gap Analysis in SDN Security

SDN has emerged as a promising technology for network management, but its security is still a major concern. Here is some research gap analysis in SDN's security:

- ***SDN Security Threat Modelling***: The current threat modelling techniques for SDN do not consider all possible attack scenarios and do not provide effective countermeasures. Researchers need to develop a comprehensive threat model for SDN networks that can capture all possible attack scenarios and provide effective countermeasures [35].
- ***SDN Security Testing***: The testing of SDN security mechanisms is a challenging task, and there is a lack of standard testing frameworks for SDN security. Researchers need to develop standard testing frameworks for SDN security that can ensure the effectiveness of SDN security mechanisms [36].
- ***SDN Security Management***: The current SDN security management approaches are limited in their ability to detect and respond to attacks in real time. Researchers need to develop more effective and efficient SDN security management frameworks that can provide real-time detection and response to attacks [37].
- ***SDN Security Metrics***: The current metrics for evaluating SDN security are limited and do not provide a comprehensive view of SDN security.

Researchers need to develop new metrics that can provide a comprehensive view of SDN security, including both network-level and application-level security metrics [38].

- ***SDN Security Integration***: The integration of SDN security with other security mechanisms, such as intrusion detection and prevention systems, is still a challenging task. Researchers need to develop more effective and efficient approaches for integrating SDN security with other security mechanisms [39].

In summary, the research gaps in SDN's security include the need for a comprehensive threat model, standard testing frameworks, real-time security management frameworks, new security metrics, and effective integration with other security mechanisms. Addressing these research gaps is essential for improving the security of SDN networks.

12.7.3 Attack Surface in SDN

The attack surface of SDN refers to the potential vulnerabilities or entry points in the network that can be exploited by attackers to compromise the system. The introduction of SDN architecture, which involves centralized control and programmable network configurations, brings forth new security challenges and attack vectors [40, 41].

Common attack surfaces in SDN include the following [23, 24, 40, 41]:

- ***SDN Controller***: The controller serves as the central control point in an SDN architecture and is often a prime target for attackers. Gaining access to the controller can allow an attacker to manipulate network configurations, steal sensitive data, or launch further attacks.
- ***Network Infrastructure***: The switches, routers, and other devices comprising the data plane in an SDN architecture can also be targeted by attackers. Exploiting vulnerabilities in these devices can lead to data theft, network traffic disruption, or unauthorized access to other parts of the network.
- ***OpenFlow Protocol***: It is a common interface used for connection between the controller and data plane devices in SDN architectures. Vulnerabilities in the OpenFlow protocol can be exploited by attackers to intercept or manipulate network traffic, launch denial-of-service attacks, or gain unauthorized network access.
- ***Applications***: SDN applications are software programs that run on top of the controller and are responsible for managing network policies and configurations. Attackers can exploit vulnerabilities in these applications to manipulate network configurations or steal sensitive data.

To mitigate the attack surface of SDN, it is essential to implement robust security measures such as firewalls, intrusion detection systems, and access controls. Regular security audits and vulnerability assessments should be conducted to identify and address potential security issues effectively.

12.7.4 SDN Attacks Examples

Here are some examples of SDN attacks [23, 42]:

- ***Flow Table Overflow Attack***: In this attack, the attacker sends many flow table modification messages to the controller, causing the table to overflow and disrupting network traffic.
- ***Fabrication Attack***: In this attack, the attacker fabricates flow table entries to redirect network traffic to a malicious destination.
- ***Denial-of-Service (DoS) Attack***: In this attack, the attacker floods the controller with many requests, causing it to crash and disrupt network traffic.
- ***Malware Injection Attack***: In this attack, the attacker injects malware into the SDN infrastructure to steal sensitive data or disrupt network traffic.
- ***Man-in-the-Middle (MitM) Attack***: In this attack, the attacker intercepts and modifies network traffic between the controller and data plane devices to steal sensitive data or manipulate network configurations.
- ***Eavesdropping Attack***: In this attack, the attacker monitors network traffic between the controller and data plane devices to steal sensitive data or gain unauthorized access to the network.

To protect against these attacks, various security mechanisms such as access controls, encryption, and network monitoring can be implemented. Additionally, regular security audits and vulnerability assessments should be conducted to identify and remediate potential security issues.

12.7.5 Issues and Prevailing Solutions in SDN Security

SDN has many potential benefits for network management, but it also poses some significant security challenges. Here are some of the key issues and solutions in SDN's security:

- ***Issue: Lack of Visibility and Control***
  Solution: Researchers have proposed various techniques to enhance the visibility and control of SDN networks, including traffic monitoring and analysis, access control mechanisms, and intrusion detection and prevention systems [43].
- ***Issue: Vulnerabilities in the SDN Controller***
  Solution: Researchers have proposed various solutions to mitigate the vulnerabilities in the SDN controller, including secure coding practices, runtime verification, and runtime intrusion detection and prevention [44].
- ***Issue: Security Threats to the SDN Data Plane***
  Solution: Researchers have proposed various solutions to mitigate the security threats to the SDN data plane, including flow-level encryption, secure routing, and secure forwarding mechanisms [20].

- ***Issue: Lack of Security Mechanisms for SDN Applications***
  Solution: Researchers have proposed various solutions to enhance the security mechanisms for SDN applications, including access control mechanisms, policy-based security, and security-aware resource allocation [45].
- ***Issue: Lack of Standardization and Interoperability***
  Solution: Researchers have proposed various solutions to address the lack of standardization and interoperability in SDN security, including the development of standard security protocols and interfaces, and the use of open-source software and platforms [46].

In summary, the key issues in SDN's security include lack of visibility and control, vulnerabilities in the SDN controller, security threats to the SDN data plane, lack of security mechanisms for SDN applications, and lack of standardization and interoperability. Researchers have proposed various solutions to address these issues, including traffic monitoring and analysis, secure coding practices, flow-level encryption, access control mechanisms, and the development of standard security protocols and interfaces.

12.7.6 Current Research Challenges in SDN Security

SDN is a promising approach to network management, but it also poses significant security challenges. Here are some of the current research challenges of SDN's security:

- *Challenge 1*: **Security Threats in Multi-Tenant Environments**
  SDN allows for multi-tenancy, where multiple customers share a common physical infrastructure. However, this introduces security challenges such as isolation and access control. Researchers are working on developing secure multi-tenancy mechanisms, including virtualization and access control [47].
- *Challenge 2*: **Distributed Denial of Service (DDoS) Attacks**
  SDN's centralized control plane can make it vulnerable to DDoS attacks, where many malicious packets flood the network. Researchers are working on developing DDoS defence mechanisms for SDN, including traffic filtering and diversion [48].
- *Challenge 3*: **Insider Attacks SDN's Centralized Control Plane**
  SDN's centralized control panel can also make it vulnerable to insider attacks, where an authorized user intentionally or unintentionally causes harm to the network. Researchers are working on developing access control and monitoring mechanisms to detect and prevent insider attacks [49].
- *Challenge 4*: **Privacy Concerns SDN's Centralized Control Plane and Programmability**
  SDN can raise privacy concerns, as network operators can potentially access sensitive user data. Researchers are working on developing

privacy-enhancing mechanisms, including data anonymization and secure data exchange protocols [50].

- *Challenge 5*: **Verification and Validation of Security Policies**
  SDN's programmability allows for the creation of custom security policies, but verifying and validating these policies can be challenging. Researchers are working on developing automated verification and validation tools for SDN security policies [51].

In summary, the current research challenges of SDN's security include security threats in multi-tenant environments, DDoS attacks, insider attacks, privacy concerns, and verification and validation of security policies. Researchers are working on developing various mechanisms to address these challenges, including virtualization and access control, traffic filtering and diversion, access control and monitoring mechanisms, data anonymization and secure data exchange protocols, and automated verification and validation tools.

12.7.7 Future Research Challenges of SDN's Security

As SDN continues to evolve, it is expected to face new and more complex security challenges. Here are some of the future research challenges of SDN's security:

- *Challenge 1*: **Machine Learning-Based Attacks:**
  As machine learning (ML) becomes more prevalent in SDN, it can also be exploited by attackers to launch attacks, such as adversarial attacks or poisoning attacks. Researchers are exploring the development of ML-based defence mechanisms to detect and prevent such attacks [52].
- *Challenge 2*: **Quantum Computing-Based Attacks:**
  As quantum computing continues to advance, it can also pose a threat to SDN's security, as it can break many of the encryption mechanisms used in SDN. Researchers are exploring the development of post-quantum cryptography and quantum safe SDN architectures [53].
- *Challenge 3*: **Secure Inter-Domain Communication SDN:**
  It allows for inter-domain communication, where different domains share information and resources. However, this introduces security challenges such as data confidentiality and integrity. Researchers are exploring the development of secure inter-domain communication mechanisms, including encryption and authentication [54].
- *Challenge 4*: **Secure SDN Orchestration:**
  It is the process of coordinating the different elements of an SDN infrastructure to achieve a specific goal. However, this can also introduce security challenges, such as unauthorized access and malicious orchestration. Researchers are exploring the development of secure SDN orchestration mechanisms, including access control and trust management [55].
- *Challenge 5*: **Compliance with Regulations and Standards:**

As SDN becomes more widely adopted, compliance with regulations and standards, such as GDPR and NIST, becomes increasingly important. Researchers are exploring the development of SDN security mechanisms that comply with such regulations and standards [56].

In summary, the future research challenges of SDN's security include machine learning-based attacks, quantum computing-based attacks, secure inter-domain communication, secure SDN orchestration, and compliance with regulations and standards. Researchers are exploring the development of various mechanisms to address these challenges, including ML-based defence mechanisms, post-quantum cryptography and quantum-safe SDN architectures, encryption and authentication for inter-domain communication, access control and trust management for SDN orchestration, and SDN security mechanisms that comply with regulations and standards.

12.7.8 Future Scope of SDN

The future scope of SDN's security is vast and encompasses a wide range of areas. Here are some of the future scopes of SDN's security:

- ***Improved Threat Intelligence and Security Analytics***:
  As SDN environments become more complex, traditional security approaches may no longer be sufficient. Researchers are exploring the development of advanced threat intelligence and security analytics solutions to provide real-time, proactive threat detection and response [57].
- ***Blockchain-Based Security Mechanisms***:
  Blockchain technology offers a decentralized and tamper-proof way of storing and sharing information. Researchers are exploring the use of blockchain technology to enhance the security of SDN, such as for secure control plane communication and distributed security policy enforcement [58].
- ***Security Automation and Orchestration***:
  As SDN environments become more complex, security automation and orchestration can help to reduce the workload on security teams and improve response times. Researchers are exploring the development of automated security policies and workflows that can adapt to changing network conditions [59].
- ***Application-Specific Security Policies***:
  Different applications may have different security requirements, and traditional network-wide security policies may not be sufficient. Researchers are exploring the development of application-specific security policies that can be enforced at the application level [60].
- ***Integration With Emerging Technologies***:
  SDN will need to integrate with emerging technologies, such as 5G and IoT, which will bring new security challenges. Researchers are exploring

the development of SDN security mechanisms that can integrate with these emerging technologies and address their unique security challenges [61].

In summary, the future scope of SDN's security includes improved threat intelligence and security analytics, blockchain-based security mechanisms, security automation and orchestration, application-specific security policies, and integration with emerging technologies. Researchers are exploring the development of various mechanisms to address these future scopes, including advanced threat intelligence and security analytics solutions, blockchain technology, automated security policies and workflows, application-specific security policies, and SDN security mechanisms that can integrate with emerging technologies.

12.8 CONCLUSION

- SDN is a paradigm shift in ***network architecture*** that separates the ***control plane*** from the ***data plane.***
- While SDN offers many benefits, it also introduces new security concerns that need to be addressed.
- Some of the primary security concerns in SDN include unauthorized access to the control plane, data plane attacks, and lack of confidentiality, integrity, and availability in network operations. In addition, SDN's dynamic nature and reliance on software make it susceptible to various types of attacks, such as malware and denial of service (DoS) attacks.
- To address these security concerns, researchers are exploring various solutions, such as encryption and authentication mechanisms, firewalls, intrusion detection and prevention systems (IDPS), and secure network virtualization. Additionally, security automation and orchestration can help to reduce the workload on security teams and improve response times.
- Looking to the future, researchers are exploring the development of advanced threat intelligence and security analytics solutions, blockchain-based security mechanisms, application-specific security policies, and integration with emerging technologies such as 5G and IoT.
- In summary, while SDN offers many benefits, it also introduces new security concerns that need to be addressed. Researchers are exploring various solutions and approaches to mitigate these security concerns and ensure the security and resilience of SDN-based networks.

REFERENCES

[1] Open Networking Foundation. (n.d.). SDN Definition. Open Networking Foundation. Retrieved from www.opennetworking.org/sdn-resources/sdn-definition/

[2] Nadeem, T., Howarth, M. P., Pavlou, G., & Li, K. (2016). Software-defined networking: A comprehensive survey. IEEE Communications Surveys & Tutorials, 18(1), 551–586. doi: 10.1109/COMST.2015.2477041

[3] SDxCentral. (n.d.). What is Software-Defined Networking (SDN)? SDxCentral. Retrieved from www.sdxcentral.com/sdn/definitions/what-is-software-defined-networking-sdn/

[4] Cisco. (n.d.). Software-Defined Networking (SDN). Cisco. Retrieved from www.cisco.com/c/en/us/solutions/data-center-virtualization/software-defined-networking-sdn.html

[5] VMware. (n.d.). Software-Defined Networking (SDN). VMware. Retrieved from www.vmware.com/topics/glossary/content/software-defined-networking-sdn

[6] Nisar, K., Jimson, E. R., Hijazi, M. H. A., Welch, I., Hassan, R., Aman, A. H. M.,... & Khan, S. (2020). A survey on the architecture, application, and security of software defined networking: Challenges and open issues. *Internet of Things*, *12*, 100289.

[7] Pinterest. (n.d.). Software-Defined Networking (SDN). Pinterest. Retrieved from www.pinterest.com/pin/671880838149306287/

[8] Shenoy, N., & Prabhu, J. (2016). Software-defined networking: A comprehensive survey. Journal of Network and Computer Applications, 68, 157–183. doi: 10.1016/j.jnca.2015.11.010

[9] Nikaein, N., Pentikousis, K., & Tyan, H.-Y. (2016). Software-defined networking: A survey. IEEE Communications Surveys and Tutorials, 18(1), 27–51. doi: 10.1109/COMST.2015.2477041

[10] Ahmed, M., AlSayed, A., & Nasser, N. (2018). Software-Defined Networking (SDN) and Network Function Virtualization (NFV) integration: A survey. Journal of Network and Computer Applications, 108, 33–57. doi: 10.1016/j.jnca.2017.08.017

[11] Ahmadinejad, S. H., Fotouhi, A., & Arabnia, H. R. (2018). SDN, NFV, and network virtualization: An overview, benefits, and challenges. International Journal of Network Management, 28(5), e2036. doi: 10.1002/nem.2036

[12] Herrería-Alonso, S., Chirivella-Pérez, E., & García-Dorado, J. L. (2018). Software-defined networking and network function virtualization: A tutorial. IEEE Communications Magazine, 56(8), 80–87. doi: 10.1109/MCOM.2018.1700134

[13] www.slideshare.net/isaurabh/understanding-sdn

[14] Casado, M., Koponen, T., Shenker, S., & Tootoonchian, A. (2014). Fabric: A retrospective on evolving SDN. Communications of the ACM, 57(10), 54–59. doi: 10.1145/2643132

[15] Feamster, N., Rexford, J., & Zegura, E. (2013). The road to SDN: An intellectual history of programmable networks. ACM SIGCOMM Computer Communication Review, 43(4), 87–98. doi: 10.1145/2534169.2485925

[16] Jarraya, Y., Samet, M., & Ghamri-Doudane, Y. (2018). SDN and NFV integration: A systematic literature review. Journal of Network and Computer Applications, 119, 50–64. doi: 10.1016/j.jnca.2018.04.017

[17] Rotsos, C., & Moore, A. W. (2016). A brief history of SDN: The evolution of the network as an application. Proceedings of the IEEE, 104(8), 1384–1406.

[18] www.slideshare.net/sebastiengoasguen/sdn-network-agility-in-the-cloud

[19] Latah, M., & Toker, L. (2019). Artificial intelligence enabled software-defined networking: A comprehensive overview. IET Networks, 8(2), 79–99.
[20] Kreutz, D., Ramos, F. M. V., Verissimo, P. E., Rothenberg, C. E., Azodolmolky, S., & Uhlig, S. (2015). Software-defined networking: A comprehensive survey. Proceedings of the IEEE, 103(1), 14–76.
[21] McKeown, N., Anderson, T., Balakrishnan, H., Parulkar, G., Peterson, L., Rexford, J.,... & Shenker, S. (2008). OpenFlow: Enabling innovation in campus networks. ACM SIGCOMM Computer Communication Review, 38(2), 69–74.
[22] www.electronicsforu.com/technology-trends/tech-focus/software-defined-networking-revolution
[23] Kumar, S., Chowdhury, N. M., Jukan, A., & Boutaba, R. (2016). A survey of software-defined networking (SDN) security. Journal of Network and Computer Applications, 72, 112–134.
[24] www.cjcheema.com/2018/05/18/what-is-openflow-how-it-is-beneficial-in-sdn/
[25] Sheng, S., Li, Q., Zhang, T., & Wang, C. (2019). Security threats and solutions of software-defined networking: A survey. Journal of Network and Computer Applications, 137, 82–98.
[26] Gürbüz, O. F., & Kocabaş, Ü. (2018). Security challenges and solutions in software-defined networking: A survey. Computer Networks, 139, 126–146.
[27] Dhiman, G., & Kumar, N. (2019). Software-defined networking security challenges and solutions. Wireless Personal Communications, 104(3), 1201–1222.
[28] Zhang, Y., Liu, Y., Liu, Y., & Chen, J. (2019). Security challenges and solutions in software-defined networking: A survey. Journal of Ambient Intelligence and Humanized Computing, 10(2), 659–672.
[29] Barik, R. K., & Mohanty, S. P. (2020). Security challenges and solutions in software-defined networking: A review. Journal of Ambient Intelligence and Humanized Computing, 11(5), 1665–1680.
[30] Wang, Q., Li, B., He, Y., & Zhang, C. (2021). Survey on the security of software-defined networking. Journal of Network and Computer Applications, 175, 102981.
[31] Zaki, M. H., Alsheikh, M. A., & Abdel-Maguid, M. (2020). A survey on software-defined networking security challenges and solutions. Journal of Network and Computer Applications, 149, 102464.
[32] Alotaibi, F., & Alharthi, M. (2020). Security issues in software-defined networking: A survey. Journal of Network and Computer Applications, 149, 102469.
[33] Liu, Y., Zhang, L., & Wang, W. (2019). Software-defined networking security: Challenges and opportunities. Journal of Network and Computer Applications, 131, 1–16.
[34] Raza, S., Imran, M., & Tahir, M. A. (2020). A comprehensive review on software-defined networking security. Journal of Network and Computer Applications, 149, 102467.
[35] Yuan, X., Li, L., Li, H., & Liu, Y. (2019). A survey on security threat modeling for software-defined networks. Journal of Network and Computer Applications, 124, 93–104.
[36] Kim, D., & Lee, J. (2020). A survey on testing methodologies for software-defined networks. Journal of Network and Computer Applications, 154, 102530.
[37] Lee, D. H., Kim, D. Y., Kim, J., & Han, K. S. (2019). Towards an SDN security management framework for real-time security management. Future Generation Computer Systems, 100, 749–761.

[38] Zhang, C., Li, B., Wang, Q., & He, Y. (2021). Metrics for software-defined networking security: A survey. IEEE Communications Surveys & Tutorials, 23(1), 464–492.

[39] Khan, S. U., Guizani, M., Khan, M. K., & Ur Rehman, M. (2019). Software-defined networking security: Prospects, potentialities, and challenges. IEEE Communications Magazine, 57(7), 90–96.

[40] Nguyen, T. M., Li, B., Chen, J., Li, X., & Li, W. (2017). Software-defined networking security: Overview of security threats and challenges. 2017 IEEE Conference on Communications and Network Security (CNS).

[41] Bello, D. D., Han, X., & Liu, Y. (2016). Security concerns and solutions for software-defined networking: A survey. Computer Networks, 109, 233–246.

[42] Zhu, L., Cheng, X., Hu, X., & Fu, X. (2017). Software-defined networking security: Status and research directions. IEEE Communications Surveys & Tutorials, 19(4), 1842–1872.

[43] Alqahtani, S., & Zincir-Heywood, A. N. (2019). SDN-based security: A survey. IEEE Communications Surveys & Tutorials, 21(2), 1652–1682.

[44] Sezer, S., Scott-Hayward, S., Chouhan, P. K., Fraser, B., & Mcbride, R. (2013). SDN security: A survey. Future Internet, 5(4), 475–487.

[45] Karim, A., & Kim, D. (2019). Survey on security challenges of software-defined networking and its applications. IEEE Access, 7, 9619–9640.

[46] Mavromoustakis, C. X., Mastorakis, G., & Batalla, J. M. (2016). Software defined networks security: Requirements, classification and open issues. Ad Hoc Networks, 47, 3–13.

[47] Kim, J. K., & Kim, D. (2019). Software-defined networking security in multi-tenant environment: A survey. IEEE Communications Surveys & Tutorials, 21(2), 1452–1476.

[48] Liao, Y., & Wang, W. (2020). A survey of DDoS defense mechanisms in software-defined networking. IEEE Access, 8, 52806–52822.

[49] Tunc, C., & Yilmaz, O. (2020). Insider threat in software defined networking: A survey. IEEE Communications Surveys & Tutorials, 22(3), 1946–1972.

[50] Rostami, M., & Boreli, R. (2016). Privacy in software-defined networks: A survey. Computer Networks, 109, 128–147.

[51] Canini, M., Venzano, D., Perešíni, P., Kuzniar, M., & Giordano, S. (2016). A survey of network verification and testing techniques: From debugging to provable security. Proceedings of the IEEE, 105(11), 2186–2209.

[52] Wu, J., Yang, M., Li, X., & Zhang, W. (2021). Machine learning-based security threats in software-defined networking: A survey. IEEE Transactions on Network and Service Management, 18(3), 1199–1215.

[53] Zhang, Y., Dong, L., Chen, X., Chen, J., & Lin, K. J. (2020). SDN security challenges and countermeasures under quantum computing environment. IEEE Transactions on Network and Service Management, 17(4), 1848–1863.

[54] Hong, Y., Wang, H., Zhang, L., & Cheng, H. (2019). A survey on security of software-defined inter-domain communication. IEEE Communications Surveys & Tutorials, 21(4), 3483–3503.

[55] Liu, Y., Gong, X., & Zhao, J. (2021). Security challenges and solutions in software-defined networking orchestration: A survey. IEEE Communications Surveys & Tutorials, 23(2), 954–981.

[56] Sabri, A. A., Hashim, H. M., & Alobaidy, H. M. (2021). SDN security framework: Compliance with regulations and standards. IEEE Access, 9, 33160–33175.
[57] Shafiee, S., & St-Hilaire, M. (2021). SDN security: Current trends, challenges, and future research directions. Journal of Network and Computer Applications, 176, 102912.
[58] Liu, Q., Li, J., Cheng, X., Huang, Y., & Luo, X. (2020). A blockchain-based security architecture for software-defined networking. IEEE Network, 34(2), 276–282.
[59] Zhang, Y., Hong, W., & Chen, J. (2020). Software-defined networking security automation and orchestration: A survey. IEEE Communications Surveys & Tutorials, 22(3), 1733–1759.
[60] Zhang, X., Zhu, M., Liu, Q., & Cheng, X. (2018). An application-specific security policy enforcement approach for software-defined networks. IEEE Transactions on Network and Service Management, 15(4), 1631–1644.
[61] Wang, X., Sun, S., Hu, J., & Liu, Y. (2021). A survey on security issues and solutions of SDN in 5G and beyond networks. IEEE Transactions on Network and Service Management, 18(1), 302–319.

Chapter 13

Analyzing the Internet of Things and Cloud Computing

Multimedia Integration on the Internet of Things (IoT)

Rubaid Ashfaq

13.1 INTRODUCTION

With the significant technological advancements of recent years, modern technologies in computer systems, communications, robotics, and virtual reality have emerged. Exploring the integration of these diverse areas of work becomes especially important. Among these advancements, cloud computing and the Internet of Things (IoT) stand out as crucial technologies that can lead to the development of new systems improving people's quality of life (Hwang et al., 2013). The IoT enables users to connect billions of smart machines and exchange information, allowing monitoring and control of various services, such as home automation, healthcare, agriculture, security monitoring, power grids, and critical services. Controlling the IoT infrastructure represents the contemporary frontier, with the digitization of physical systems providing value-added services for mobile devices and blurring the lines between artificial and real environments (Aazam et al., 2014).

This technological advancement has led to the creation of interconnected objects in a connected world. It enables surgeons to perform operations remotely, empowers users to monitor their homes, and aids energy providers in efficient management. The number of smart devices connected to the IoT is expected to exceed 30 million by the end of 2023, highlighting the widespread adoption of IoT technology. However, managing the diverse IoT devices and data generation efficiently is a challenge, which is where the integration of cloud computing and IoT becomes indispensable. Cloud computing offers scalable and flexible resources, allowing for the efficient handling and analysis of the vast amounts of data generated by IoT devices. This integration, often referred to as the Cloud IoT paradigm, enables more sophisticated data analytics, improved device management, and enhanced decision-making processes. By leveraging cloud computing's power, the IoT ecosystem can achieve greater efficiency in processing, storing, and retrieving data, facilitating real-time insights and actions. This synergy not only enhances operational efficiency but also fosters innovation in developing new applications and services tailored to specific needs and contexts. As such, the convergence of cloud

DOI: 10.1201/9781003480181-13

computing and IoT holds the potential to transform industries, revolutionize how we interact with technology, and significantly improve the quality of life by making smarter environments a reality (Alzakholi et al., 2020). Connecting the cloud to IoT enables the generation of media content, but this also demands effective management of cloud resources due to the multimedia's processing, storage, and space requirements.

As IoT services with mission-critical tasks require high processing and response capacity, the integration of cloud computing proves beneficial. Both IoT and cloud computing have been extensively studied and applied in various fields, providing intelligent perception, M2M connection (man to man, machine, and machine to machine), and efficient exchange of resources. These technologies open up new possibilities for on-demand usage and enhance resource efficiency (Doukas & Maglogiannis, 2012).

Figure 13.1 illustrates an idea of how cloud computing and the IoT would collaborate. When these technologies function independently, they yield impressive results and contribute to new research domains. However, when these two technologies collaborate, the anticipated results become extraordinary, paving the way for unimaginable solutions that would greatly benefit the fields of science and research.

Cloud computing is poised to gain increasing significance in application execution and as a world-class computing model. Concurrently, the ever-expanding IoT will take center stage from various angles, encompassing ubiquitous computing and cutting-edge technologies, including the latest-generation printers,

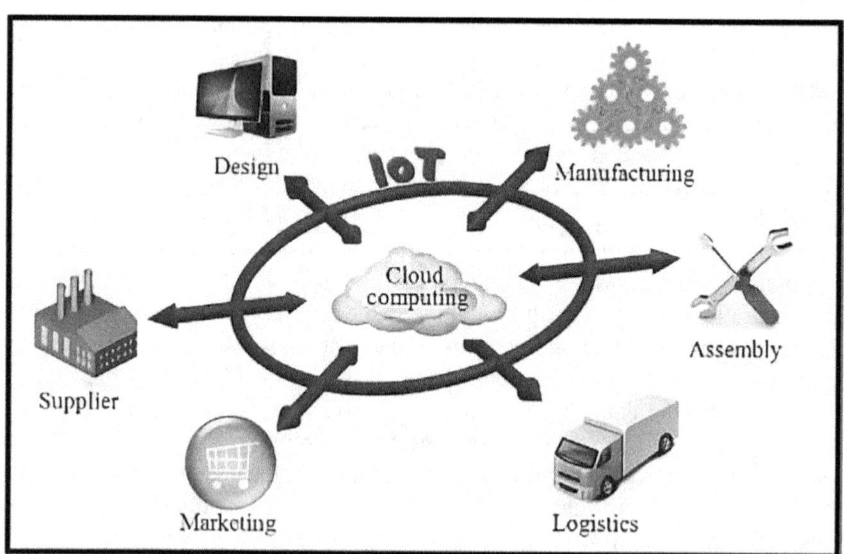

Figure 13.1 IoT and cloud computing.

among others, all within the realm of the IoT universe. An investigation into these emerging technologies has led to the creation of an article titled "IoT and Cloud Computing in Automation of Assembly Modelling Systems," authored by Zhang et al., 2010. This article proposes the enhancement of conventional assembly modelling systems into advanced systems capable of autonomously handling complexity and adapting to changes.

Similarly, the article "Bringing IoT and Cloud Computing to Pervasive Healthcare" by Charalampos Doukas and Ilias Maglogiannis introduces a cloud computing-based platform for managing mobile and portable health sensors. This platform demonstrates the application of the IoT paradigm to healthcare. Integrated health applications utilizing networks of body sensors generate a substantial amount of data, necessitating efficient management and storage for processing and future utilization (Doukas & Maglogiannis, 2012).

Despite the significance of these two modern technologies, there exists a limited body of research on their integration. Therefore, a comprehensive analysis is essential to gain insights into each of these innovative technologies, which can be achieved by exploring their collaborative potential. This work is structured as follows: Section 13.2 provides a detailed overview of the IoT, Section 13.3 addresses all aspects of cloud computing and proposes an integration of these modern technologies, followed by the conclusions and future directions, and finally, the references utilized in this investigation.

13.2 METHODOLOGY

Cloud computing services are implemented in numerous fields related to the IoT, encompassing genome data processing, teaching and learning, SME services, e-learning methods, augmented reality, manufacturing, emergency recovery, smart cities, remote forensic investigation, hospitality, business, government email, human resources management, Internet of Things for automobiles, among others (Perera et al., 2013)

The challenges posed by cloud computing and IoT, both as separate entities and within a unified application environment, have grown rapidly, giving rise to new study scenarios for researchers. The primary difficulty in studying the integration of IoT and cloud computing lies in the realm of uncertain discovery. The global proliferation of cloud computing and IoT has occurred at a rapid pace in recent years (Zhang et al., 2010). When combined, the exhibited characteristics become particularly noteworthy, holding a special and vital place among them (Aazam et al., 2014). Researchers have organized a series of applications related to the coordination between cloud and IoT to develop and accumulate data, leveraging the capabilities of cloud computing and storage. The layers of knowledge depicted in Figure 13.2 are elucidated in the following. Notably, the application, network, and detection layers are interlinked.

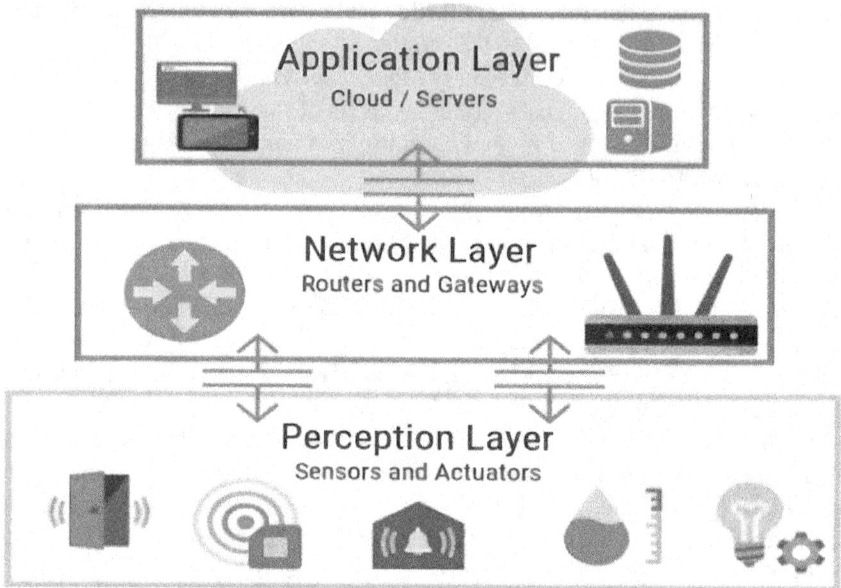

Figure 13.2 Cloud-IoT architecture.

13.3 INTERNET OF THINGS (IoT)

The IoT is a rapidly emerging paradigm that is becoming the cornerstone of modern wireless communications. The fundamental concept underlying this paradigm is the presence of various elements around us, such as radio frequency identification (RFID), sensors, actuators, cell phones, and more, which, through collaboration, can achieve common goals (Atzori et al., 2010).

Undoubtedly, the key strength of the IoT lies in its profound impact on various aspects of daily life and human behavior. For individual users, the introduction of IoT will manifest in various ways, both at work and at home. Home automation, assisted living, eHealth, and enhanced learning represent just a few application scenarios where this new paradigm will play a pivotal role soon. Similarly, from a business perspective, prominent consequences will be evident in domains like automation and industrial manufacturing, process management and coordination, intelligent transportation of people and goods (Gubbi et al., 2013).

The multitude of definitions of IoT found within the research community attests to the significant interest in the subject and the lively discussions surrounding it. While exploring the literature, an interested reader may encounter challenges in comprehending the true essence of IoT—its core concepts, as well as the social, economic, and technical implications that will arise upon its full deployment (Perera et al., 2013). Figure 13.3 vividly illustrates this understanding.

Figure 13.3 Internet of Things definition.

In short, IoT allows people and things to connect to the internet, anywhere and anytime. The IoT is a promising technology that has the potential to revolutionize and link the world by allowing heterogeneous smart devices to communicate with one another in a seamless manner. To bring the new vision of IoT to life, the demand for machine type communication (MTC) has resulted in numerous communication technologies with multiple service requirements. Long Term Evolution (LTE) for mobile devices is incompatible with IoT devices, which have low power consumption and low data rates. Several IoT standards are brought together (Ashfaq, 2021).

13.4 CLOUD COMPUTING

Cloud computing can be defined as both a technological concept and a business model that offers services for storing, accessing, and utilizing computing resources, all rooted in the network. This paradigm revolutionizes the manner in which computer resources and applications are employed and delivered, essentially transforming them into services. The primary resources provided as services encompass computing power, storage, and network infrastructure. Cloud computing entails furnishing these resources as services accessible via the internet to the public or within organizations for private use (Armbrust et al., 2010). There exist three types of cloud services, each distinct in how they offer resources. The initial approach involves presenting hardware infrastructure as a service, termed Infrastructure as a Service (IaaS).

The subsequent approach entails supplying a platform (comprising the operating system, essential software, frameworks, and tools) atop the hardware infrastructure,

Figure 13.4 Cloud computing services.

denoted as Platform as a Service (PaaS). The third approach encompasses delivering applications as services, designated as Software as a Service (SaaS) (Emeras et al., 2016). Figure 13.4 visually depicts the core concept of these services.

13.4.1 Infrastructure as a Service (IaaS)

IaaS involves the delivery of hardware as a service, representing an evolution in accommodation that necessitates no long-term commitment and permits users to provision resources on demand. The IaaS provider's role is primarily limited to data center operation. Users are responsible for implementing and managing software services themselves, similar to managing their own data centers. Examples of IaaS offerings include Amazon Web Services' Elastic Compute Cloud (EC2) and Secure Storage Service (S3) (Rajan, 2013).

13.4.2 Platform as a Service (PaaS)

PaaS furnishes a computing platform that utilizes cloud infrastructure. It encompasses all necessary applications required by the client, eliminating the need for the customer to acquire and install the software and hardware components. Developers using PaaS gain access to all the systems and environments needed throughout the software development lifecycle—spanning development, testing, deployment, and web application hosting. Notable examples include Google App Engine (GAE) and Microsoft's Azure (Jadeja & Modi, 2012).

13.4.3 Software as a Service (SaaS)

SaaS employs a model where a service provider hosts applications and offers them to clients over a network, typically the internet. SaaS is increasingly adopted as the underlying technologies that support web services and service-oriented architecture (SOA) mature, and new development approaches gain popularity. SaaS is often associated with a pay-as-you-go subscription licensing model. Furthermore, SaaS applications need to effectively interact with diverse data and other applications across various environments and platforms. SaaS shares close ties with other service delivery models. In most cases, SaaS is implemented to provide commercial software functionality to enterprise customers at a reduced cost, enabling them to enjoy the benefits of commercial software without the complexities of internal installation, administration, support, and licensing (Godse & Mulik, 2009).

13.5 RESULTS

13.5.1 IoT and Cloud Computing

Cloud computing and the IoT, two fundamentally distinct technologies, have seamlessly woven themselves into the fabric of our lives. Their anticipated continued adoption and widespread use are poised to elevate them to significant components of the future internet. Clouds function as an intermediary layer between devices and applications, effectively concealing the intricate complexities and functionalities required for the implementation of these latter. This framework is poised to exert a substantial impact on future application development, ushering in new challenges related to the gathering, processing, and transmission of information.

These challenges will need to be effectively tackled, particularly within the context of a multi-cloud environment (Aguzzi et al., 2013). The complementary nature and integration of the cloud and IoT are demonstrated in Table 13.1.

13.5.2 Trial Healthcare

IoT and multimedia technologies have made their way into the field of health through assisted environmental living and telemedicine. Smart devices, mobile internet, and cloud services contribute to continuous and systematic innovation in

Table 13.1 Cloud and IoT Complementary and Integration

	IoT	*Cloud*
Displacement	Pervasive	Centralized
Reachability	Limited	Ubiquitous
Components	Real World Things	Virtual Resources
Computational Capabilities	Limited	Virtually Unlimited
Storage	Limited or None	Virtually Unlimited
Role of the Internet	Point of Convergence	Means for Delivering Services
Big Data	Source	Means to Manage

healthcare, enabling cost-effective, efficient, timely, and high-quality ubiquitous medical services (Wang et al., 2014).

13.5.3 Smart City

IoT can provide a unified environment for future-oriented smart city services (e.g., 3D representations through RFID sensors and geotagging), exposing information uniformly. Several recently proposed solutions suggest utilizing cloud architectures to facilitate the discovery, connection, and integration of sensors and actuators. This creates platforms capable of providing and supporting ubiquitous connectivity and real-time applications for smart cities (Mitton et al., 2012).

13.5.4 Smart Home and Smart Metering

IoT finds significant application in household environments, where heterogeneous embedded devices enable the automation of common internal activities. In this context, the cloud is well-suited for creating flexible applications with minimal code, simplifying home automation (Kamilaris et al., 2011). To enable various smart homes, where individual family members access reusable services over the internet, the resulting solution must meet three crucial requirements: internal network interconnection (allowing any digital device in the home to interface with others), remote smart control (enabling intelligent management of appliances and services from any device, anywhere), and automation (connecting interconnected household appliances to cloud-oriented smart home services) (He et al., 2014).

13.5.5 Video Surveillance

Intelligent video surveillance has become a vital tool for various safety-related applications. Complex video analytics, as an alternative to internal autonomous management systems, necessitate cloud-based solutions to meet storage requirements (secure central storage of media, fault-tolerant, scalable, and high-speed access) and processing needs (utilizing video processing, computer vision algorithms, and pattern recognition modules for scene analysis). The proposed solutions intelligently store and manage video content from cameras (IP and analog) and efficiently deliver it to multiple user devices via the internet, distributing processing tasks across physical server resources in a load-balanced and fault-tolerant manner (Held et al., 2012).

13.5.6 Smart and Automotive Mobility

As an emerging technology, IoT holds the promise of transforming transportation systems and car services (intelligent transport systems, ITS). Integrating cloud technologies with WSNs, RFID, satellite networks, and other intelligent transportation technologies presents a promising approach to addressing current challenges. It is possible to develop and implement a new generation of vehicle data clouds based on IoT, offering numerous business benefits such as improved traffic security, reduced road congestion, traffic management, and vehicle maintenance or repair recommendations (He et al., 2014).

13.5.7 Cloud-IoT Challenges

The intermediary layer between objects and applications is cloud storage, which conceals intricacies and functionalities. We all understand that the Internet of Things constitutes a network of interconnected artifacts, with various applications engaging these objects. While each application's issues are distinct, they generally fall into similar categories. Addressing these challenges requires heightened focus on security concerns and the evaluation of the implications of modern technologies. Following the integration of the cloud and the Internet of Things, concerns have arisen regarding distrust and comprehension of cloud providers, as well as the physical location of information transmitted to the cloud through different IoT protocols. Questions have arisen about multi-cloud service storage for system tenants. Storing multiple consumer data in a single installation may compromise confidentiality and result in the unauthorized leakage of information. Owing to mistrust in cloud service providers, this vulnerability is considered an insider threat and stands as one of the most unforeseen issues in the IT industry. The key challenges of cloud-IoT are explained next:

a) **Security**: Securing IoT data stored in the cloud for processing and retrieval involves data encryption, ensuring the security of data during transmission or storage in cloud-based repositories, as well as during access and utilization of the cloud. The level of information obscurity in cloud computing leaves data owners unaware of the physical location of their data. Given that data are intricately connected to our surroundings, data safety within the cloud-IoT paradigm takes center stage.
b) **Computational Performance and Storage**: Ensuring high-performance targets for storage and computing plans involving cloud-based IoT devices requires meeting demanding performance requirements. Given that cloud-based IoT devices operate across numerous applications, meeting these specifications in all environments poses challenges.
c) **Reliability**: IoT devices rely on the cloud to support time-critical applications, with any disruptions directly impacting program outcomes. This applies, for example, in domains like surgical instruments for automobiles or safety.
d) **Big Data Storage**: By around 2025, nearly 50 million IoT devices will be operational, creating a substantial hurdle for cloud service providers to access this voluminous data efficiently and securely.
e) **Maintenance**: Building upon the insights gained from preceding sections, highly efficient technologies and strategies are imperative to oversee and

manage security and efficiency in the cloud environment, accommodating the anticipated influx of up to 50 billion IoT devices.

f) **Computing at the Edge**: Latency restrictions, mobility limitations, and geographically distributed IoT implementation necessitate immediate responsiveness from the cloud. Consequently, edge computing serves as a compromise between conventional computing and cloud computing. Although it offers closer proximity to implementation, merging it presents challenges as it requires location awareness.
g) **User-Assisted IoT Devices**: In this type of IoT implementation, users must contribute details and benefits to offset their participation in communication. This poses a formidable challenge as it intertwines with social factors, with users making contributions within their respective contexts.
h) **Interaction with Devices**: Cloud IoT systems demand information from diverse devices for processing and execution. In this scenario, specifications such as cloud-based storage space and computing power can prove challenging to fulfill.

13.6 CONCLUSIONS

In conclusion, it has been determined that the fusion of the IoT and cloud computing, coupled with their ongoing development, has the potential to forge an IT powerhouse capable of generating currently unimaginable technologies and pioneering new research domains that will benefit all of humanity, irrespective of their involvement in the realm of technology or other fields.

Both IoT and cloud computing are poised for continuous growth and evolution in the future. Presently, a significant challenge lies in establishing effective communication channels between them. As demands for requests increase, so does the need for greater bandwidth. The advent of 5G connections is expected to facilitate the expansion and advancement of the technologies discussed in this document.

This analysis provides a clear understanding of the functioning of the Internet of Things and cloud computing. Consequently, future studies could be undertaken to propose the utilization of these two technologies for the betterment of society. While not without challenges, this endeavor can be pursued with dedication and investment, offering the potential for substantial achievements.

The future scope of analyzing the IoT and cloud computing with a focus on multimedia integration is promising and holds numerous opportunities for innovation and advancement. This convergence of IoT and multimedia technologies can lead to transformative applications across various industries. Here are some potential future directions and opportunities:

a) Enhanced User Experiences: Integrating multimedia data (such as images, videos, and audio) with IoT devices can enhance user experiences significantly. For instance, smart home environments could provide real-time video feeds from security cameras, multimedia-based notifications, and interactive visualizations of IoT data.

b) Healthcare and Telemedicine: IoT-enabled medical devices could capture multimedia data like images and videos for remote patient monitoring. Cloud-based analytics could help healthcare professionals make informed decisions by analyzing this data in real time.
c) Industrial IoT (IIoT): Integrating multimedia data in the IIoT context can lead to more comprehensive monitoring and control of industrial processes. Visualizing sensor data through multimedia elements can aid in identifying anomalies and optimizing operations.
d) Smart Cities: Multimedia integration can contribute to smarter urban planning and management. IoT sensors can capture multimedia data related to traffic, waste management, and environmental conditions, which can then be analyzed in the cloud to optimize city services.
e) Agriculture and Precision Farming: IoT devices combined with multimedia capabilities can be used to monitor crops and livestock through images and videos. Cloud-based analytics can process this multimedia data to provide insights into crop health, pest detection, and yield predictions.
f) Retail and Customer Engagement: Retailers can leverage multimedia data from IoT devices to analyze customer behavior and preferences. Cloud-based analytics could help tailor marketing strategies and improve customer engagement.
g) Entertainment and Gaming: Multimedia integration can enhance interactive gaming experiences by incorporating real-world data from IoT devices. Cloud computing can process this data to create immersive and dynamic gameplay.
h) Environmental Monitoring: IoT sensors equipped with multimedia capabilities can capture visual and auditory data from various ecosystems. Cloud-based analysis could aid in monitoring biodiversity, identifying species, and studying environmental changes.
i) Security and Surveillance: Multimedia-enabled IoT devices can provide more comprehensive surveillance solutions. Cloud analytics can analyze video and audio streams for suspicious activities and facilitate rapid response.
j) Data-Driven Insights: Cloud-based analytics can process large volumes of multimedia IoT data to extract valuable insights and patterns. This can lead to data-driven decision-making in various domains.

To leverage these opportunities effectively, several challenges must be addressed, including data privacy and security concerns, bandwidth limitations for multimedia data transmission, standardization of multimedia formats across IoT devices, and the development of advanced cloud-based analytics and machine learning algorithms for processing multimedia IoT data. Collaborations between experts in

IoT, multimedia technologies, cloud computing, and various application domains will play a crucial role in realizing the full potential of multimedia integration in the IoT landscape. As technology continues to evolve, innovative solutions in this space are likely to shape the way we interact with and benefit from IoT devices and cloud computing resources.

REFERENCES

Aazam, M., Khan, I., Alsaffar, A. A., & Huh, E.-N. (2014). Cloud of Things: Integrating Internet of Things and cloud computing and the issues involved. In *Proceedings of 2014 11th International Bhurban Conference on Applied Sciences & Technology (IBCAST)*, Islamabad, Pakistan, 14–18 January 2014, pp. 414–419.

Aguzzi, S., Bradshaw, D., Canning, M., Cansfield, M., Carter, P., Cattaneo, G., Gusmeroli, S., Micheletti, G., Rotondi, D., & Stevens, R. (2013). Definition of a research and innovation policy leveraging cloud computing and IoT combination. Final Report, European Commission, SMART 37.

Alzakholi, O., Shukur, H., Zebari, R., Abas, S., & Sadeeq, M. (2020). Comparison among cloud technologies and cloud performance. *Journal of Applied Science and Technology Trends*, *1*(2), 40–47.

Armbrust, M., Fox, A., Griffith, R., Joseph, A. D., Katz, R., Konwinski, A., Lee, G., Patterson, D., Rabkin, A., & Stoica, I. (2010). A view of cloud computing. *Communications of the ACM*, *53*(4), 50–58.

Ashfaq, R. (2021). Study and analysis of 5G enabling technologies, their feasibility and the development of the Internet of Things. In *Intelligence of Things: AI-IoT Based Critical-Applications and Innovations*, pp. 101–143. Springer.

Atzori, L., Iera, A., & Morabito, G. (2010). The Internet of Things: A survey. *Computer Networks*, *54*(15), 2787–2805.

Doukas, C., & Maglogiannis, I. (2012). Bringing IoT and cloud computing towards pervasive healthcare. In *2012 Sixth International Conference on Innovative Mobile and Internet Services in Ubiquitous Computing*, pp. 922–926.

Emeras, J., Varrette, S., Plugaru, V., & Bouvry, P. (2016). Amazon elastic compute cloud (EC2) versus in-house hpc platform: A cost analysis. *IEEE Transactions on Cloud Computing*, *7*(2), 456–468.

Godse, M., & Mulik, S. (2009). An approach for selecting software-as-a-service (SaaS) product. In *2009 IEEE International Conference on Cloud Computing*, pp. 155–158.

Gubbi, J., Buyya, R., Marusic, S., & Palaniswami, M. (2013). Internet of Things (IoT): A vision, architectural elements, and future directions. *Future Generation Computer Systems*, *29*(7), 1645–1660.

He, W., Yan, G., & Da Xu, L. (2014). Developing vehicular data cloud services in the IoT environment. *IEEE Transactions on Industrial Informatics*, *10*(2), 1587–1595.

Held, C., Krumm, J., Markel, P., & Schenke, R. P. (2012). Intelligent video surveillance. *Computer*, *45*(3), 83–84.

Hwang, K., Dongarra, J., & Fox, G. C. (2013). *Distributed and cloud computing: From parallel processing to the Internet of Things*. Morgan Kaufmann.

Jadeja, Y., & Modi, K. (2012). Cloud computing-concepts, architecture and challenges. In *2012 International Conference on Computing, Electronics and Electrical Technologies (ICCEET)*, pp. 877–880.

Kamilaris, A., Pitsillides, A., & Trifa, V. (2011). The smart home meets the web of things. *International Journal of Ad Hoc and Ubiquitous Computing*, *7*(3), 145–154.

Mitton, N., Papavassiliou, S., Puliafito, A., & Trivedi, K. S. (2012). Combining Cloud and sensors in a smart city environment. *EURASIP Journal on Wireless Communications and Networking*, *2012*(1), 1–10.

Perera, C., Zaslavsky, A., Christen, P., & Georgakopoulos, D. (2013). Context aware computing for the Internet of Things: A survey. *IEEE Communications Surveys & Tutorials*, *16*(1), 414–454.

Rajan, A. P. (2013). Evolution of cloud storage as cloud computing infrastructure service. ArXiv Preprint 1308.1303.

Wang, C., Bi, Z., & Da Xu, L. (2014). IoT and cloud computing in automation of assembly modeling systems. *IEEE Transactions on Industrial Informatics*, *10*(2), 1426–1434.

Zhang, Q., Cheng, L., & Boutaba, R. (2010). Cloud computing: State-of-the-art and research challenges. *Journal of Internet Services and Applications*, *1*, 7–18.

Index

I

K

L

M

For Product Safety Concerns and Information please contact our EU representative GPSR@taylorandfrancis.com
Taylor & Francis Verlag GmbH, Kaufingerstraße 24, 80331 München, Germany

www.ingramcontent.com/pod-product-compliance
Lightning Source LLC
LaVergne TN
LVHW020615110826
845149LV00002B/475

* 9 7 8 1 0 3 2 7 6 8 3 3 5 *